LATIN AMERICAN STUDIES

SOCIAL SCIENCES AND LAW

Edited by
David Mares
University of California,
San Diego

A ROUTLEDGE SERIES

LATIN AMERICAN STUDIES: SOCIAL SCIENCES AND LAW

DAVID MARES, *General Editor*

THE POLITICS OF THE INTERNET IN THIRD WORLD DEVELOPMENT

Challenges in Contrasting Regimes with Case Studies of Costa Rica and Cuba

Bert Hoffmann

ROUTLEDGE
New York & London

Published in 2004 by
Routledge
270 Madison Avenue
New York, NY 10016
www.routledge-ny.com

Published in Great Britain by
Routledge
2 Park Square
Milton Park, Abington
Oxon OX14 4RN
www.routledge.co.uk

Copyright © 2004 by Taylor & Francis Group, a Division of T&F Informa.
Routledge is an imprint of the Taylor & Francis Group.

10   9   8   7   6   5   4   3   2   1

Library of Congress Cataloging-in-Publication Data
Hoffmann, Bert.
    The politics of the Internet in Third World development : challenges in contrasting regimes with case studies of Costa Rica and Cuba / Bert Hoffmann.
        p. cm. — (Latin American studies)
    Includes bibliographical references and index.
    ISBN 0-415-94959-9 (hardback : alk. paper)
    1. Information technology—Political aspects—Developing countries.
2. Information technology—Political aspects—Costa Rica. 3. Information technology—Political aspects—Cuba. 4. Technology and state—Developing countries. 5. Technology and state—Costa Rica. 6. Technology and state—Cuba. 7. Economic development—Political aspects. I. Title. II. Series: Latin American studies (Routledge (Firm)

    HN981.I56.H63 2004
    303.48'33'091724—dc22   2004009247

Contents

Part III: Latin America's 'Socialist Model': Cuba

Chapter Five

Chapter Six

Chapter Seven

List of Tables

List of Figures

List of Abbreviations

2G	Second Generation (mobile phones)
3G	Third Generation (mobile phones)
ABT	Agreement on Basic Telecommunications
AID	Agencia de Información para el Desarrollo (Information for Development Agency, Cuba)
AIN	Agencia de Información Nacional (National Information Agency, Cuba)
ANAP	Asociación Nacional de Agricultores Pequeños (National Small Peasants' Association, Cuba)
APC	Association for Progressive Communication
APIC	Agencia de Prensa Independiente de Cuba (Cuban Independent Press Agency)
ARCOS	Americas Region Caribbean Optical-ring System
ARESEP	Autoridad Reguladora de Servicios Públicos (Agency for the Regulation of Public Services, Costa Rica)
ARETEL	Autoridad Reguladora de Telecomunicaciones (Agency for the Regulation of Telecommunications, Costa Rica)
ARPA	Advanced Research Projects Agency (United States)
AT&T	American Telephone and Telegraph Company (United States)
B2B	Business-to-Business (E-commerce)
B2C	Business-to-Consumer (E-commerce)
BDI	Bundesverband der Deutschen Industrie (Association of German Industry)
Bitnet	'Because it's Time' Network
CAFTA	Central American Free Trade Agreement
CAGR	Compound Annual Growth Rate
CANF	Cuban-American National Foundation (United States)

Caprosoft	Cámara de Productores de Software de Costa Rica (Association of Costa Rican Software Producers)
CATIE	Centro Agronómico Tropical de Investigación y Enseñanza (Center for Research and Teaching on Tropical Agronomics, Costa Rica)
CBI	Caribbean Basin Initiative
CCSS	Caja Costarricense de Seguro Social (Costa Rican Social Security Fund)
CDMA	Code Division Multiple Access ('American'mobile telephony standard)
CDR	Comités para la Defensa de la Revolución (Commitees for the Defense of the Revolution, Cuba)
CD-ROM	Compact Disc / Read-Only-Memory
CEA	Centro de Estudios sobre América (Center for the Study of the Americas, Cuba)
CENIAI	Centro Nacional de Intercambio Automatizado de Información (National Center for Automated Information Exchange, Cuba), changed to: CENIAInternet
CENPO	Centro Nacional de Planificación y Operación de Electricidad (National Center for Electricity Planning and Operation, Costa Rica)
CEPAL	Comisión Económica para América Latina (United Nations Economic Commission on Latin America)
CERN	Conseil Européen de Recherche Nucleaire (European Council for Nuclear Research; renamed to: European Organization for Nuclear Research)
CID	Centro de Investigación Digital (Center for Digital Research, Cuba)
CIDA	Canadian International Development Agency
CIGB	Centro de Ingeniería Genética y Biotecnología (Center for Genetic Engineering and Biotechnology, Cuba)
CINDE	Coalición Costarricense de Iniciativas de Desarrollo (Costa Rican Coalition for Development Initiatives)
CITMA	Ministerio de Ciencia, Tecnología y Medio Ambiente (Ministry of Science, Technology, and the Environment, Cuba)
CITMATEL	abbr. from CITMA & Telecomunicaciones; official name: Empresa de Tecnologías de la Información y Servicios Telemáticos Avanzados (Company for Information Tecnology and Advanced Telematics Services, Cuba)

CNIC	Centro Nacional de Investigación Científica (National Center for Scientific Reasearch, Cuba)
CNP	Consejo Nacional de Producción (National Production Council, Costa Rica)
CoCom	Coordinating Committee for Multilateral Export Controls (United States)
CODESA	Corporación para el Desarrollo, S. A. (Corporation for Development, Costa Rica)
COMECON	Council of Mutual Economic Assistance
COMTELCA	Comisión para las Telecomunicaciones en Centro América (Central American Telecommunications Commission)
CONICIT	Consejo Nacional de Investigaciones Científicas y Tecnológicas (National Council for Scientific and Technological Research, Costa Rica)
CRIS	Communication Rights in the Information Society
CRNet	Red Nacional de Investigación de Costa Rica (Costa Rican Research Network)
CTC	Central de Trabajadores de Cuba (Central Organization of Cuban Trade Unions)
CTC	Companía Telefónica de Chile (Chilean Telephone Company)
Cuba-NIC	Cuban Network Information Center
DNI	Dirección Nacional de Informática (National Direction of Informatics, Cuba)
DNS	Domain Name System
DSL	Digital Subscriber Line
EARTH	Escuela de Agricultura de la Región Tropical Húmeda (Agricultural School for the Humid Tropical Region, Costa Rica)
ENIAC	Electronic Numerical Integrator And Computer
ENTEL	Empresa Nacional de Telecomunicaciones (National Telecommunications Company, Chile)
ETECSA	Empresa de Telecomunicaciones de Cuba, S.A. (Cuban Telecommunications Company)
EZLN	Ejército Zapatista de Liberación Nacional (Zapatista Army for National Liberation, Mexico)
FCC	Federal Communications Commission (United States)
FDI	Foreign Direct Investment
FECON	Federación Costarricense para la Conservación del Ambiente (Costa Rican Federation for Environmental Preservation)

FEEM	Federación de Estudiantes de la Enseñanza Media (Federation of Secondary School Students, Cuba)
FEU	Federación de Estudiantes Universitarios (Federation of University Students, Cuba)
FLACSO	Facultad Latinoamericana de Ciencias Sociales (Latin American Faculty for Social Sciences)
FMC	Federación de Mujeres Cubanas (Federation of Cuban Women)
FOD	Fundación Omar Dengo (Omar Dengo Foundation, Costa Rica)
FOSUTEL	Fondo Universal para los Servicios de Telecomunicaciones (Universal Fund for Telecommunication Services, Costa Rica)
G2C	Government-to-Citizens
G-7	Group of the Seven Leading Industrial Nations
G-8	G-7 plus Russia
GATS	General Agreement on Trade in Services
GATT	General Agreement on Tariffs and Trade
Gbps	Gigabits per second
GDP	Gross Domestic Product
GET	Grupo de la Electrónica del Turismo (Electronics Group for the Tourism Sector, Cuba)
GII	Global Information Infrastructure
GNP	Gross National Product
GSM	Global Systems for Mobile Communications ('European' mobile telephony standard)
HDI	Human Development Index
HPI-1	Human Poverty Index
IANA	Internet Assigned Numbers Authority
IBM	International Business Machines Corp. (United States)
ICANN	Internet Corporation for Assigned Names and Numbers
ICE	Instituto Costarricense de Electricidad (Costa Rican Electricity Institute)
ICRT	Instituto Cubano de Radio y Televisión (Cuban Radio and Television Institute)
ICT	Information and Communication Technologies
IDA	Instituto de Desarrollo Agrario (Institute for Agrarian Development, Costa Rica)
IDB	Inter-American Development Bank

IICA	Instituto Interamericano de Cooperación para la Agricultura (Inter-American Institute for Agricultural Cooperation, Costa Rica)
ILO	International Labour Organization
IMF	International Monetary Fund
INBIO	Instituto Nacional de Biodiversidad (National Institute for Biodiversity, Costa Rica)
INCAE	Instituto Centroamericano de Administración de Empresas (Central American Institute for Business Administration, Costa Rica)
INSAC	Instituto Nacional de Sistemas Automatizados y Computación (National Institute for Automated Systems and Computers, Cuba)
IP	Internet Protocol (often used as abbr. for: TCP/IP)
ISDN	Integrated Services Digital Networks
ISI	Industrialization via Substitution of Imports
ISP	Internet Service Provider
IT	Information Technologies
ITCO	Instituto de Tierras y Colonización (Institute for Land and Colonization, Costa Rica)
ITCR	Instituto Tecnológico de Costa Rica (Technological Institute of Costa Rica)
ITT	International Telephone and Telegraph Company (United States; today: ITT Industries, Inc.)
ITU	International Telecommunication Union
JCCE	Joven Club de Computación y Electrónica (Youth Computer and Electronics Clubs, Cuba)
Kbps	Kilobits per second
LAN	Local Area Network
LANIC	Latin America Network Information Center (at the University of Texas at Austin, United States)
LASA	Latin American Studies Association
LINCOS	Little Intelligent Communities
M-26	Movimiento 26 de Julio (26 July Movement, Cuba)
Mbps	Megabits per second
MCCA	Mercado Común de Centroamérica (Central American Common Market)
MIC	Ministerio de la Informática y las Comunicaciones (Ministry for Informatics and Communication, Cuba)

MICIT	Ministerio de Ciencias y Tecnología (Ministry of Science and Technology, Costa Rica)
MINCOM	Ministerio de las Comunicaciones (Ministry for Communications, Cuba)
MININT	Ministerio del Interior (Ministry of the Interior, Cuba)
MINJUS	Ministerio de Justicia (Ministry of Justice, Cuba)
MINSAP	Ministerio de Salud Pública (Ministry of Public Health, Cuba)
MINVEC	Ministerio para la Inversión Extranjera y la Colaboración Económica (Ministry of Foreign Investment and Economic Cooperation, Cuba)
MISTICA	Metodología e Impacto Social de las Tecnologías de la Información y de la Comunicación en América (Methodology and Social Impact of the Information and Comunication Technologies in America)
MIT	Massachusetts Institute of Technology
NAFTA	North American Free Trade Agreement
NBC	National Broadcasting Company (United States)
NED	National Endowment for Democracy (United States)
NGO	Non-Governmental Organization
NIC	Network Information Center
NICT	New Information and Communication Technologies
NII	National Information Infrastructure
NTIA	National Telecommunications & Information Administration (United States)
NWICO	New World Informations and Communications Order
OAS	Organization of American States
OCB	Office of Cuba Broadcasting (United States)
OECD	Organization for Economic Cooperation and Development
ONC	Oficina Nacional de Café (National Coffee Office, Costa Rica)
PAC	Partido Acción Ciudadana (Citizen Action Party, Costa Rica)
PAHO	Panamerican Health Organization
PC	Personal Computer
PCC	Partido Comunista de Cuba (Cuban Communist Party)
PLN	Partido Liberación Nacional (Party of National Liberation, Costa Rica)
PRI	Partido Revolucionario Institucional (Institutional Revolutionary Party, Mexico)
Procomer	Promotora del Comercio Exterior de Costa Rica

PTT	Posts, Telegraphs and Telephones
PUSC	Partido Unidad Social Cristiana (Party of Social Christian Unity, Costa Rica)
PVP	Partido Vanguardia Popular (Popular Vanguard Pary, Costa Rica)
PYME	Pequeña y Mediana Empresa (Small and Medium-Size Enterprise)
R&D	Research and Development
RACSA	Radiográfica Costarricense S. A. (Costa Rican Radiographic Company)
RAM	Random Access Memory
RCP	Red Científica Peruana (Peruvian Science Net)
RedHUCyT	Red Hemisférica Inter-Universitaria de Información Científica y Tecnológica (Hemispheric Inter-University Network for Scientific and Technological Information)
RIA	Red Internet Avanzada (Advanced Internet Network)
RPN	Red Pública Nacional (National Public Network)
SIME	Ministerio de la Industria Sideromecánica y Electrónica (Ministry of Metallurgy and Electronics, Cuba)
SMS	Short Message Service
TCP/IP	Transmission Control Protocol / Internet Protocol
TDMA	Time Division Multiple Access ('American' mobile telephony standard)
TRIPS	Trade-Related Aspects of Intellectual Property Rights
UBPC	Unidades Básicas de Producción Cooperativa (Basic Units of Cooperative Production, Cuba)
UCR	Universidad de Costa Rica (University of Costa Rica)
UJC	Unión de Jóvenes Comunistas (Union of Young Communists, Cuba)
UMTS	Universal Mobile Telecommunications System (third generation mobile standard)
UNA	Universidad Nacional (National University, Costa Rica)
UNCSTD	United Nations Commission on Science and Technology for Development
UNCTAD	United Nations Conference on Trade and Development
UNDP	United Nations Development Program
UNEAC	Unión Nacional de Escritores y Artistas de Cuba (Cuban National Union of Writers and Artists)
UNED	Universidad Estatal a Distancia (State Distance Learning University, Costa Rica)

UNESCO	United Nations Educational, Scientific and Cultural Organization
UPEC	Unión de Periodistas de Cuba (Cuban Journalists' Union)
URL	Uniform Resource Locator
US-AID	United States Agency for International Development
USIA	United States Information Agency
UUCP	Unix-to-Unix Communication Protocol
VNIIPAS	All Union Scientific Research Institute for Applied Computerized Systems (USSR)
VoIP	Voice-over-IP (Internet telephony)
WAN	Wide Area Network
WAP	Wireless Application Protocol
WHO	World Health Organization
WIPO	World Intellectual Property Organization
WSIS	World Summit on the Information Society
WTO	World Trade Organization
WWW	World Wide Web
X.25	Data transmission protocol, principal standard for computer networks before introduction of TCP/IP

Acknowledgments

Much of the research for this work was carried out in the context of a 2½-year research project on the regulation and use of the NICT in Latin America, which was based at the Institute for Iberoamerican Studies in Hamburg, Germany. I particularly want to thank my colleagues in this project, Roman Herzog, Michael Krennerich, and Markus Schulz, for the fruitful cooperation and stimulating discussions. Furthermore, I want to thank the Institute's Director, Klaus Bodemer, for chairing this project. Heartfelt thanks go to the staff of the Costa Rican counterpart institution, the Fundación Acceso in San José, for their help and hospitality; very special thanks go to my Cuban friends who, over many years now, have turned my visits on the island into such personal experiences. The empirical research during the field stays was only possible because a great number of Cubans and Costa Ricans dedicated some of their time to share their knowledge and their viewpoints with me. To all of them my most sincere thanks. Finally, sincere thanks are also due to the Volkswagen Foundation, whose generous funding made the project possible.

This work was accepted as a dissertation thesis at the Political and Social Science Department of the Free University of Berlin, Germany. I want to thank my colleagues at the Free University's Latin American Institute and its Otto Suhr Institute for Political Science, who have provided me with a highly supportive professional environment. Elmar Altvater and Urs Müller-Plantenberg, the tutors of this thesis, deserve thanks not only for their support and confidence in this project, but most of all for their invaluable stimuli for my intellectual and academic formation over many years. They knew how to be teachers in the best sense of the word. I want also to thank David Mares for his valuable comments on the manuscript, and Alejandro Mares, who took up the task of revising the text for the present book publication.

Very special thanks go to my parents for their support over all these years. Greatest thanks of all to Barbara, for everything. The work is dedicated to Paula and Antonia, who have been such a lively part of the conditions under which it was written.

As to all mentioned, the usual caveats apply: of course, the author is solely responsible for the present work.

Introduction

In December 2003, the United Nations staged the first 'World Summit on the Information Society,' bringing to the forefront of public attention the political, social and development issues associated with the new information and communication technologies (NICT). The rapid diffusion and the specific characteristics of the new, digitally based technologies, with the Internet their flagship, bring about new challenges for economic development, social usage, and political regulation worldwide. Their decentralized and cross-border nature opens up new channels of communication and media use, transforming the structures of the public sphere. How to use the opportunities and how to cope with the challenges presented by the NICT are objects of debates, conflicts and complex decision-making processes involving diverse actors and interest groups. It is these 'politics of the Internet,' that are the subject of this work.

While the development of the NICT has been dominated by the industrialized nations, the technologies have since made their way into all countries, and their effects and implications have a global reach. The distribution of the benefits of the new information and communication technologies, however, presents dramatic disparities between North and South, between urban and rural areas, and between different social groups. The dominant discourse falls short of providing an adequate analytical framework when it conceives of these inequalities as a 'digital divide,' isolating the issue of the NICT from their much broader economic and social context and suggesting 'digital solutions.' In contrast, this work argues that the disparities in NICT access and use must be understood as an expression of the general 'development divide' with its sharp economic and social polarization. The challenge therefore is less 'to bridge the digital divide,' and more to ask how the NICT can contribute to fostering development, not merely in terms of economic growth, but as overcoming social, economic and political structures of exclusion and inequality.

It is the premise of this investigation that, as historian Melvin Kranzberg once put it, "Technology is neither good nor bad, nor is it neutral" (Kranzberg 1985: 50). A computer is a very different kind of tool in a country like Germany than it is in rural Africa, where 70 percent of households do not have electricity. As much in its development as in its application and use, technology is intrinsically shaped by its economic, political and social environment (cf. Williams/Edge 1996). Context matters. Nevertheless, most studies on the impact of the NICT have focused on the industrialized nations or on global trends. The lack of substantial studies in Third World countries, with their very different socio-structural conditions, economic problems and developmental needs, has been pointed out by a number of authors (e.g. Wilson 1997; Press 1996, 1997; Hamelink 1997). This work contributes to redressing these shortcomings, inserting the question of the political implications of the NICT in the broader context of the political economy of Third World development. In doing so, it highlights the specific challenges posed by the NICT in the context of different political regimes and development strategies.

We must begin by comparing the particular characteristics of the NICT to the 'old ICT,' such as traditional mass media or the 'plain old telephone.' These different characteristics include the greater tendency of the NICT to minimize previously existing barriers of time and space, including national boundaries. We will also focus on the possibilities of the computer network-based NICT, as a new type of many-to-many media, to circumvent the filter mechanisms of the established mass media.

Given the cross-border character of the Internet, it has become a common argument that the Internet, as an intrinsically pluralist medium, will inevitably have a 'democratizing' effect on non-pluralist systems. As one author put it, demonstrating the triumphalism often associated with this idea: "The virus of freedom, for which there is no antidote, is spread by electronic networks to the four courners of the earth" (Wriston 1997: 172). However much *en vogue,* such generalizing and technologically determinist approaches lack analytical substance. If the NICT have specific social or political effects, it is so because specific actors use them in specific ways. These interactions need to be explored in empirical studies.

Turning against the 'conventional wisdom' that the Internet is inherently a force for democracy Kalathil/Boas (2003: 2) argue:

> "Proponents [of the 'democratization thesis'] see the Internet as leading to the downfall of authoritarian regimes, but the mechanisms through which this might occur are rarely specified. Instead, popular assumptions often rest on anecdotal evidence, drawing primarily on isolated

examples of Internet-facilitated political protest. Subsequent assertions about the technology's political effects are usually made without consideration of the full national context in which the Internet operates in any given country."

While much of the literature has focused on the 'democratizing effect' of the Internet under authoritarian regimes (e.g. Kedzie 1997, Ferdinand 2000), the questions of democracy associated with the NICT in fact go back much earlier. The NICT are of central importance for pluralist democracies as well, since the core of democracy for the individual is citizenship. Access to information and the right to communicate are vital aspects of meaningful citizenship. To the degree that the NICT increasingly become central channels of information and communication, they are never only 'products' in the economic sphere, but by their very nature they are an eminently political issue. An equitable social distribution of access and use of the NICT is not just one among many desirable material development goals, because it also touches on the very substance of civic and social citizenship and, hence, democracy.

Since information and communication are the essential building blocks of the public sphere, the state and other public and private actors find themselves with important societal responsibilities and regulatory functions regarding the NICT, just as they do with traditional mass and individual media. As a result, particular problems arise from the contradiction between the essentially national character of political organization, in which laws and norms are established and enforced, and the inherently transnational character of the World Wide Web and other NICT applications of global reach.

In this study we will guard against a widespread approach in the literature on the NICT that seems blinded by their 'newness' and which, by emphasizing the 'revolutionary' character of the Internet and related technologies, tends to become (deliberately or by negligence) ahistoric. Instead, we shall pay particular attention to previous experiences of computer technologies, mass media and telecommunications in Third World development, considering such experiences indispensable to understanding the particular challenges posed by the new ICT.

Such an approach recognizes that the NICT build as much on existing technological infrastructure as on administrative and organizational schemes established in the telecommunications sector. A central focus in our analysis, therefore, corresponds to the structures and transformations of what we have termed the telecommunications regime. By this we will understand the specific configuration of economic, social and political actors, stakeholders, and regulatory frameworks—including structures of ownership and market, pricing schemes and patterns of diffusion and consumption, legal rules, administrative norms and technological standards of the telecommunications sector in any given country or region.

Far from the hype and sensationalism often associated with the 'new economy' or the 'Internet age,' the present analysis is based on two in-depth empirical country studies. It may be surprising that we chose the nation-state as the object of analysis, since a withering away of national competencies and the erosion (if not the end) of the nation-state are central themes of the current literature on globalization in general and on the NICT in particular. However, we think that the structures of the nation-state persist as a prime arena of political and social conflict. Wilson (1997: 10) writes:

> "However transnational the issues may be, national politicians and pol-
> icymakers first calculate their consequences at the national level. The
> most interesting questions are institutional and political, posed and an-
> swered at the national level."

To this, we need to add two things: first, it is not only 'politicians and poli-cymakers' but social, political and economic actors of all sorts that deserve our attention; and second, if the focus of analysis is on the actors, structures and decisions within the political realm of the nation-state, these have to be analyzed precisely in their interrelation with global processes and interna-tional forces.

Given the research focus outlined above, for the empirical case studies we chose two Third World countries with strongly contrasting political and economic systems. In fact, the two countries have become perhaps the most celebrated 'models' of their respective ideological sides. The first is Costa Rica, which represents one of the most stable regimes of political democracy and one of the few examples of a relatively inclusive model of capitalist de-velopment in the Third World. The strategy of industrialization via the sub-stitution of imports (ISI) Costa Rica followed since the early 1950s was the dominant paradigm not only in Latin America, but also for many of the de-colonizing nations in Africa and Asia. The second country is Cuba, which after the Revolution of 1959, under the leadership of Fidel Castro, became the most prominent model of state-socialist development in the Third World. Most countries radically breaking from capitalist economic struc-tures adopted some variant of state-socialist development.

In the 1980s and 1990s, the time-span that saw the public breakthrough and mass usage of the core technologies of the NICT, both of these develop-ment paradigms entered into crisis, and, in the context of the Third World's debt crisis and the transformations generally associated with the term 'global-ization,' they largely gave way to the dominance of the neo-liberal paradigm.

The demise of the Soviet Union and the end of the Cold War opened the way for a profound reconfiguration of international politics, with the United States as the world's only 'super-power.' Economically, the now globalized world market became, more than ever, the ultimate benchmark of the standards of production and consumption in any part of the world. For Third World development, liberalization and deregulation became the guiding stars as promoted by the 'Washington Consensus' (Williamson 1990).

It has become a standard wisdom to identify the NICT as a key element of the ongoing process of globalization. At the same time, their diffusion itself has been greatly conditioned by these global transformations. Therefore, this study will have to pay particular attention to the implications for the NICT of these processes of crisis and transformation of the ISI and state-socialist development models. One aspect of central importance in this is the transformation of the telecommunications regime, where the privatization and liberalization of state monopolies became a prominent issue in the shift to the neo-liberal paradigm.

For this analysis of the 'politics of the Internet in Third World development,' the following research-guiding questions for the empirical country studies are considered: Which issues associated with the introduction, diffusion, use, regulation, and production of the NICT have become central topics in the political arena? What actors and which factors can be identified as shaping the political decision-making processes? To what extent do the NICT open up new forms of articulation and participation for social or political actors? What is the role played by international actors and global governance structures?

In which ways do the specific social, political and economic contexts shape the use and regulation of the NICT? How do political actors resolve the tensions stemming from the national character of the political and social institutions in any particular country and from the inherently global character of the NICT? What implications do the transformation of the telecommunications regime and, in particular, the privatization and liberalization of the former state telecommunications monopolies have for NICT use, diffusion and regulation?

What national NICT policies are formulated? Within these, what is the role of the state, and what is that of other public and private actors? What is done to pursue the goal of 'universal access'? What efforts are made to insert the country in the global economy not only as consumer, but also as producer of NICT, and how is this addressed in national development strategies?

Of course, not all of these questions apply to all countries to the same degree. An issue that may be the center of heated debate in one nation may pass almost unnoticed in another. If in Cuba, it had been a central and long disputed political issue whether the country should or should not connect to the Internet, in the Costa Rican case there is simply no analogous discussion. Given the greatly differing cases, the two analyses will not follow an identical form, but will emphasize the most relevant issues for the particular country.

As noted in the acknowledgements, much of the work leading to this study was carried out during a 2½-year research project at the Institute for Iberoamerican Studies in Hamburg, Germany (cf. Herzog/Hoffmann/Schulz 2002). The empirical country studies are based on field stays in Costa Rica (April/May 2001), Cuba (May 2000 and April 2001), and the United States (March 2000), as well as on a comprehensive survey of the available primary and secondary literature. The field stays facilitated numerous formal and informal interviews with a broad range of social and political actors, as well as with NICT users.

Following the introduction, this work is divided into three parts. Part I (chapters one and two) provides an extensive analysis from a political perspective of questions raised by the NICT and Third World development, which will take up many of the threads briefly outlined above. Having used the term 'new information and communication technologies' so frequently already, we will start out with a more precise definition and the exploration of the essential characteristics of these technologies. In the ensuing step, we will assess the present disparities in global NICT access and use and, in critique of the 'digital divide' concept, propose an alternative conceptualization that better inserts the issue of the new information and communication technologies into the context of Third World development. Following this, the contrasting development experiences of the ISI model and the state-socialist development model are compared, with particular attention to the role of communications, technology and telecommunications. Part I closes with an analysis of the transformations of the telecommunications regime in the 1980s and 1990s and an outline of the essential political issues regarding the NICT in Third World Development.

On this background, the work proceeds to the empirical country studies, first of Costa Rica (Part II, including chapters three and four) and then of Cuba (Part III, including chapters five and six). Chapter three discusses the structures and transformation of Costa Rica's development model path and its political regime. In this context, the failed liberalization of the country's telecommunications regime can be analyzed as a paradigmatic conflict, which provoked major social tensions and led Costa Rica to enter the 21st century, very much against the international trend, with a strengthened state monopoly that includes Internet services and mobile telephony. Chapter four examines the state-led efforts for modernization and socially inclusive diffusion of the NICT, and it explores the structural turn in Costa Rica's development conception associated with the attraction of a large-scale Intel computer chip plant at the end of the 1990s.

The following country study on Cuba also opens with an outline of the structures and transformations of the country's state-socialist development, as the profound crisis since the late 1980s made a major re-articulation of the socialist system and the partial opening to the world market inevitable. We will then analyze how this process included a far-reaching transformation of the telecommunications regime, in which the state monopoly company was turned into a joint venture with foreign capital and a significant modernization and expansion of the long-neglected telecommunications system was undertaken (chapter five). Chapter six analyzes the thorny political decision-making process leading Cuba to join the Internet in a situation in which the Cuban government fiercely defended the state monopoly over mass media while the U.S. administration hailed increased communication as a prime means to promote regime change on the island. The chapter goes on to analyze the difficulties and ambivalences arising from the approach adopted by the Cuban government, which seeks to actively promote the 'Informatization of Society' while emphasizing control and restricting individual Internet use.

Finally, chapter seven sums up the findings of the country studies in comparative perspective and proposes conclusions to be drawn from these contrasting cases for the analysis of 'the politics of the Internet in Third World development.'

Part I

Third World Development and NICT in Political Perspective

Chapter One

The New Information and Communication Technologies (NICT): Comparative Experiences and Present Disparities

NICT—OR WHY WE INSIST ON SUCH A LENGTHY TERM

If the title of this work refers to 'the politics of the Internet,' this is a somewhat metaphorical formulation. In fact, 'the Internet' is just the most prominent aspect of what more correctly is called the 'new information and communication technologies,' or abbreviated: NICT. Other authors omit the 'new' and just speak of the 'information and communication technologies,' or ICT, others of 'information technologies,' or IT.

Although we have no major interest in semantic disputes, we want to insist on the importance of including the aspect of 'communication,' and not only 'information.' In the recent international discussion, this topic has risen strongly: the widely publicized term of the 'information society' has been explicitly criticized and confronted with the call for 'communication rights,' which emphasizes the human rights dimension of the issue and implies shifting the discourse from 'users' or 'consumers' to 'citizens.' Cees Hamelink, Professor of International Communication at the University of Amsterdam and for many years a leading voice in the field, noted in his keynote address to the first preparatory meeting for the 2003 World Summit on the Information Society (WSIS):

> "It is disconcerting that in the context of the 'Information Society' the notion 'communication' has disappeared. Yet, the real core question is how we should shape future 'communication societies.' Oddly enough, the UN World Conference on Human Rights (Vienna, 1993) did not

refer in its Final Declaration to communication. There was only mention of information and news. Such an essential omission should not be repeated in the Final Declaration of the 2003 WSIS. (. . .) The Summit would make a real difference if a human right to communicate would be formally recognized. (. . .) The key challenge is to ensure that citizens - their rights, freedoms and responsibilities- should guide the outcome of the Summit."[1]

The focus on 'communications rights' has become a rallying call of civil society organizations working on these issues. A major political campaign launched by a number of articulate non-governmental organizations is in fact named 'Communication Rights in the Information Society,' CRIS. Importantly, too, in common usage 'communication' tends to imply much more a bilateral process than 'information' which more often than not is perceived as a one-way process of provision and diffusion of information.

Finally, when speaking of the new information and communication technologies, the use of the word 'new' deserves a note of explanation. We are aware of the relativity of this adjective, not only because it gives a vague temporal idea which inevitably is devaluated with the passing of time, but also because today's 'new' information and communication technologies are the fruit of a rather long evolutionary process. In addition, the 'new' ICT are inseparably bound up with the 'old' ICT: to provide Internet access, the 'new' computer-modem uses the 'old' copper wires of the telephone system etc. In particular, any analysis of the political issues concerning the NICT will have to include the telephone system, which not only provides essential infrastructure for most of the NICT, but whose structures and regulations also determine much of the conditions under which the NICT are introduced and used.

Being aware of how blurred the dividing line is between 'old' and 'new' ICT, we nevertheless feel it necessary to use the term 'new' in order to make it clear from the beginning that the central interest of this study is not on printed press, radio, television or other 'classic' ICT but rather on the political implications of those new information and communication technologies that found their flagship in the Internet and whose definition and characteristics we shall discuss in the following section.

THE CHARACTERISTICS OF THE NICT—DIGITIZATION AND ITS IMPLICATIONS

What are we speaking of when we talk of the 'new information and communication technologies'? To put it in one sentence: The term NICT refers to the converging set of technologies in microelectronics, computing, telecommunications, and broadcasting, whose common feature is digitization.

Digitization means the process by which information, whether in form of text, data, sound or image, is broken down into the digital, binary language of computers, into '0'or '1,' 'yes' or 'no,' 'on' or 'off.' This conversion of information from very different sources into the binary digits universally readable by computer opened up the possibility for the integration of products, industries and media that hitherto had been developed and used separately. Thus, digitization brings together the formerly separate spheres of telecommunication companies, broadcasting, computer manufacturers and data processing industries, entertainment business, office functions etc. It is precisely this ability to create an interface between the most diverse fields through a common digital language that has been crucial for the enormous pervasiveness and ubiquity of digital technologies, which can be found in nearly all spheres of life, from kitchen appliances to space-based weapon systems. Due to the convergence and integration of these digital technologies, the impact of any specific innovation is not limited to specific technological and social areas but potentially reaches an immensely wide field of technology and social organization. Given the centrality of this aspect for society as a whole, authors speak of the "digital era" (Hewitt de Alcántara 2001), the "digital world" (Negroponte 1995), "digital capitalism" (Schiller 1999) etc.

A widely accepted definition defines "information and communication technologies" in general as

> "all those technologies that enable the handling of information and facilitate different forms of communication among human actors, between human beings and electronic systems, and among electronic systems" (Hamelink 1997: 8).

We can adapt this definition by adding the aspect of digitization to arrive at a formal definition of what we will understand as NICT in this work: All those technologies on digital base that enable the handling of information and facilitate different forms of communication among human actors, between human beings and electronic systems, and among electronic systems.[2]

If the convergence and integration of different technologies is a central feature of the NICT, this is accompanied by a second aspect of fundamental importance: the dramatic increase in capacity and speed of information transmission through global computer networks (with the corollary of sharply decreasing costs), greatly reducing hitherto existing barriers of time and space. If traditional media and communication structures largely developed on a national or regional level, the transnational nature of digital networks creates new challenges for national regulation, from legislation—what is illicit in the sending country can be perfectly acceptable in the receiving country—to price setting or security issues. If there have always been transnational flows of communication of some kind, their dimension is dwarfed by the immensely greater velocity, flexibility and quantity of cross-border communication and information transmission potentially facilitated by the NICT.

Another important characteristic of the NICT has been the fast pace of technological change, the most cutting-edge technological innovations passing into obsolescence within a number of years, rapidly devaluing costly hardware and acquired user knowledge. As to the manufacturing side, the brief marketability of NICT products makes the orientation on as large a market as possible more imperative than ever in order to obtain sufficient economies of scale.

Of course, many more aspects of the NICT are derived from the above-sketched process of digitization and its consequences.[3] One rather little noted aspect should be pointed out: digital technologies also dilute earlier dividing-lines between mass and individual media, between one-to-many communication (radio, television etc.) in the public sphere and one-to-one communication (telephone, letters etc.) in the private sphere. E-mail, for instance, can be both, a one-to-one medium as much as a message broadcast to a million receivers at the same time. (Again, this dividing-line was blurred before, too, but the NICT have further undermined this distinction.)

A final clarification should be made. The excitement surrounding new information and communication technologies has led many to excessively extend the meaning of these terms. Therefore it should be noted that in this study we will not include as NICT genetic engineering, as suggested by Castells (2000: 29, 54–59) with the unconvincing argument that "it is focused on decoding, manipulating, and reprogramming of the information codes of living matter" (ibid: 29); nor do we include progress in agricultural or medical science, as does the United Nations Development Program's Report on "Making new technologies work for human development" (UNDP 2001); nor do we find it useful to combine the issue of the NICT with advances in science as such or with a diffuse category of 'knowledge,' as does the World Bank's recent "Knowledge for Development" report (World Bank 1998: 1), where this concept includes everything from birth control to software engineering to information problems about the creditworthiness of a firm.

THE NICT EVOLUTION

In regard to the NICT, the term 'revolution' appears frequently, underscoring a notion of historical discontinuity. A typical example is Manuel Castells, whose trilogy on "The Information Age" (Castells 1997, 1998,

2000) has become a standard reference and probably the most widely cited social science work on the topic. The opening chapter announces "The Information Technology Revolution," and it starts precisely by making a case against historical gradualism and localizing the "end of the twentieth century" as "one of these rare intervals in history" adequately termed "revolution" (Castells 2000: 28f).

We, however, prefer to keep some distance from the term 'revolution' in considering the NICT. This is by no means to minimize the great dynamics of change associated with these technologies—that upheaval, in fact, is a central motivation of this work. But the emphasis on the historical discontinuity risks underestimating the continuities involved in the process—as much in the technological innovations themselves (as we will see in the following pages) as in their adaptation in the social, political and economic context (a subject to be examined in detail in the empirical country studies). Terms like the 'information technology revolution' etc. tend to imply that the technological discontinuity goes hand in hand with a 'revolutionary' discontinuity in social and political relations. Such an implication often blurs the dividing line between the analysis of empirical evidence and—especially where the empirical evidence does not live up to the proclaimed epochal dimensions of change—postulates of what *should* change, which opens a wide field for ideological preferences of all sorts.

If we focus in this work on the time from the late 1980s / early 1990s until today as the time-span which saw the public breakthrough and mass usage of the core technologies of the NICT, this is not to ignore that this process was based on a long list of technological innovations and evolutions that took place earlier. Effectively, if the NICT are characterized by a complex process of technological convergence there is no single historical moment to be named as the starting-point.[4]

The early breakthroughs in computer technology can be traced back to the first mechanical, programmable binary computing machine constructed by Konrad Zuse in the living room of his parents' apartment in Berlin in 1938, and the IBM-supported development of an electro-mechanical calculator in the United States in 1939. The first general purpose computer, the 30-ton ENIAC, was revealed in 1946, but in fact it had been developed during World War II under the auspices of the U.S. Ministry of Defense. Since then, the evolution of computer technology has been marked by a process of seemingly endless miniaturization of the hardware components with at the same time exponentially increasing computing power and massively decreasing costs. Crucial steps were the invention of the transistor in 1947; the introduction of silicon as a path-breaking new material in 1954; the first integrated circuit in

1957; the production of the first microprocessor by Intel in 1971; the first commercially successful microcomputer from Apple in 1976; and IBM's Personal Computer (PC) introduced in 1981, soon cloned in South East Asian countries on a large scale to become the prime international standard for microcomputer hardware. These hardware developments were accompanied by the creation of a breakthrough operating system by the founders of Microsoft, laying the foundations of today's dominating software giant.

Essential innovations also were made in the telecommunications sector. Since the late 1960s and 1970s, digital switches and digital transmission facilities were developed. The era of orbital satellites was inaugurated when the Soviet Union successfully launched the Sputnik in 1957. Shocked by this technological leap of their Cold war adversary, large-scale investments by the United States and Western countries in space technology followed, and notably in the 1970s and 1980s, satellites became increasingly important in the transmission of international telephone calls, television programs, and other data traffic. In the 1980s, besides the inauguration of mobile telephony, crucial advances for NICT development were also made in optoelectronics, namely fiber optics and laser transmission, and in digital packet transmission technology.

In computer networking, the origins of the Internet in the United States are directly linked to the military logic of the Cold War confrontation with the USSR. The 'Sputnik shock' prompted the U.S. Department of Defense to create the Advanced Research Projects Agency (ARPA) in the late 1950s; and it was the Cuban Missile Crisis of 1962, when the two super-powers came closer than ever before to a nuclear conflict, that led the U.S. Air Force to commission a study by the RAND corporation, the U.S. military's most prominent think-tank, on how to maintain a workable communication structure in the aftermath of a nuclear exchange. This study, titled "On Distributed Communication Networks" (Baran 1962), laid the theoretical foundations for what was to become the Internet, and ARPA was the institution to follow up on this work in practice. Finally, in 1969, the year of the first manned landing on the moon, the milestone in computer networking was set by the inauguration of the ARPANET, the 'mother of the Internet,' explicitly designed to provide the country's military with a decentralized communications system invulnerable to nuclear attack.

Crucial for the further success of the NICT, however, were also breakthroughs far beyond military considerations, especially in the routing architecture and in the network communication protocols. Between 1978 and 1980 the Transmission Control Protocol / Internet Protocol (TCP/IP) was developed that still as of today is at the base of the world's Internet traffic.

Another decisive element for the rapid diffusion of computer networking was an invention by two young Chicago students in 1978: the modem, which modulates digital computer signals into analogue telephone signals and vice versa, thus providing for a low-cost solution to connect personal computers to the global networks via ordinary telephone lines. In terms of applications of computer networking, a crucial early push came from the spread of e-mail communication, first introduced in 1971.

It has been argued that the 1970s mark the "technological divide" (Castells 2000: 53) in that in this decade a set of crucial technological innovations came together. However, it was not until the late 1980s and 1990s that the core technologies of the NICT fully emerged and saw their public breakthrough and diffusion. A highly symbolic moment for this was in 1990, when scientists from the European Organization for Nuclear Research (CERN) presented the concept of the World Wide Web with its ingenious 'hypertext' architecture (they also set up a universal standard address format for the web, the uniform resource locator, URL). Together with the invention of a graphical web browser (Mosaic first, then Netscape) in 1993/94, these steps were decisive in facilitating the transformation of computer networking from an expert issue into an everyday application for a wide public.

Since then, the active use of the NICT increased rapidly in virtually all sectors of society. Major applications such as e-commerce began to take shape in the mid-1990s. In 1993, the U.S. administration was the first government to establish its own website (www.whitehouse.gov), with institutions like the United Nations and the World Bank following in the same year and a myriad of public and private institutions all over the world in the years to come. In 1994, commercial users for the first time outnumbered academics on the Internet; search engines were developed that added tremendously to the practical usefulness of the Internet; multimedia applications began to become popular and broadband communication started to spread beyond institutions and business to residential users. The World Wide Web increasingly became world wide indeed: By 1997, almost all countries of the world had established some form of Internet connection and had acquired their national domain name.

In telecommunications, the 1990s saw the rapid increase in the diffusion of mobile telephony, first as an elite tool in the industrialized countries, but soon expanding to wider use and to Third World countries. In 1997, third generation cellular phones were presented which facilitated mobile Internet access. Internet telephony—also known as 'Voice-over-IP' (VoIP)—became technologically possible and caught the attention of U.S. telecommunication companies which in 1996 asked the U.S. Congress to ban the competing technology.

In the history of NICT development, much of the public imagery is oc-
cupied by the sagas of individual pioneers, from that mythical Californian
garage, where two high school drop-outs hand-crafted the first micro-computer to become a leading global company within only six years, to the personal success story of Bill Gates, and others. However, as much as individual
creativity in counter-cultural milieus has contributed to the NICT development, the fundamental role played by the U.S. state and military and by big
business and 'big science' cannot be overlooked. This goes well beyond the
development of the ARPANET. In fact, military contracts and the space program were essential markets and funding sources for the U.S. computer and
electronics industry as well as for networking innovations. Even the Reagan
administration, however anti-statist in its neo-liberal discourse, in practice
was decisively state interventionist in the technological field, with the 'Star
Wars' missile defense program creating enormous flows of government
money for research and development by public institutions and private companies. "The Internet's emergence"—concludes Schiller (2000: 8)—"had
nothing to do with free-market forces and everything to do with the Cold
War military-industrial complex." Also Castells, in a somewhat surprising
turn after his strong emphasis on the crucial importance of networking
structures and countercultural milieus in the innovation process, resumes:
"The state, not the innovative entrepreneur in his garage, both in America
and throughout the world, was the initiator of the Information Technology
Revolution" (Castells 2000: 69).

As for today, there still are celebrated niches in the NICT industry
where small actors may outsmart much bigger ones. However, this should
not disguise the fact that, as a consequence of the globalized large-scale production and the high research and development costs, in most segments of
the NICT only a very limited number of transnational companies dominate
90 percent or more of the world market. The anti-monopoly suits against
Microsoft in U.S. courts—though effectively cut short by the Bush administration in 2002–brought the dangers of such a concentration of economic
and technological power to broad public attention.

GLOBAL TECHNOLOGIES AND GLOBAL DISPARITIES: A FIRST ASSESSMENT OF THE UNEQUAL DISTRIBUTION OF THE NICT

The dynamics of the worldwide NICT diffusion have been indeed impressive. If in August 1981 there was a worldwide total of 213 Internet host
computers—that is, a computer or server directly linked to the global
Internet network, usually serving a variety of other client computers—and a

few thousand users, these numbers since have multiplied at a breathtaking speed. For the end of 2001, the International Telecommunication Union (ITU) counted no less than 157 million host computers worldwide (ITU 2003), and the estimates for global Internet users passed 600 million (ibid).[5]

However, the distribution of growth was far from equal. The nations with highest host density are the industrialized or high-income OECD countries—and 73 percent of all Internet host computers are concentrated in only one country, the United States (ITU 2003; see also UNDP 2002, Norris 2001). In contrast, the seven most populous countries of the Third World, China, India, Indonesia, Brazil, Pakistan, Nigeria and Bangladesh, home to more than half the world's population, share only 1.6 percent of worldwide Internet hosts (ibid). Figure 1 shows the worldwide distribution of Internet hosts by world regions.

We get a similar, though not quite as drastic picture when looking at Internet user numbers, which show a strong positive correlation with GNP per capita (ITU 2003). At the end of 2002, high-income OECD countries, though making up only 14 percent of the world's population, accounted for more than two thirds of worldwide Internet users (ibid) whereas the 14 percent of the world's population living in Nigeria, Bangladesh, Indonesia, the Philippines, Brazil, Ethiopia and Pakistan account for less than 0.5 percent. The distribution by world regions shows that 28.1 percent of the world's 600 million Internet users are living in the United States or Canada, 27.7 percent in Europe, 35.0 percent in Asia, 1.7 percent in Australia and Oceania, 6.0 percent in Latin America, and a mere 1.5 percent in all of Africa (see figure 2).

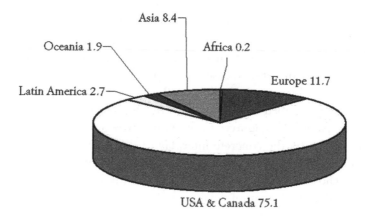

Figure 1 Internet Hosts by World Regions, December 2002

Source: ITU 2003

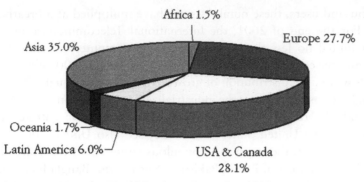

Figure 2 Internet Users by World Regions, December 2002

Source: ITU 2003

Behind these aggregate regional figures, however, lie great differences within the continents. Asia includes nations with high internet access rates like the OECD countries Japan and Israel and the South East Asian "tiger states," as well as many of the poorest and least connected countries of the world. In Europe, there is a strong imbalance between the Western countries and the former state-socialist countries of Central and Eastern Europe. In Africa, only two countries, South Africa and Egypt, account for more than half of the continent's Internet users (and for 83 percent of its host computers). In Latin America, the intra-regional contrasts are not quite as sharp as in Africa, but still strong. (For a country-specific breakdown with detailed data, see table 4 on page 123).

If we have emphasized that the global disparities in NICT use and diffusion tend to follow the existing fault lines of social and economic inequality, we have to stress that this is not to say that NICT development takes place in a uniform way across all Third World countries or that it is a mere function of GDP or similar indicators. This certainly is not so. In fact, we insist on the necessity of national case studies precisely because of the fact that form and pace of diffusion and use as well as the political, social, and economic consequences of the NICT are greatly shaped by the specific conditions of any particular country: by its political regime and development model, as well as by the concrete interplay of different actors, norms and regulatory arrangements and decisions made at the national level. In the two case studies chosen in this work, we will see this variation very clearly.

As revealing as the national data are, they do not reflect the strong disparities in NICT diffusion and use within each country. Breaking down national categories further demonstrates a strong concentration in the major urban centers, and a generally low diffusion beyond these: Brazil's biggest city, São Paulo, alone uses more international bandwidth than the entire African continent (UNDP 2001: 4). Especially in many African and Asian countries, full Internet access may be limited to the capital cities, whereas in provincial centers access is limited to e-mail services; in many rural areas, hardly any telecommunications are available. Overlapping with these regional discrepancies, the available data show a strong social stratification of NICT diffusion and usage with a heavy inclination towards the affluent, educated and urban sectors of society. In addition, in the use of the NICT there is a clear age bias against older people, and also a considerable gender bias (ILO 2001: 58f, CPSR 2002). A study on Internet users in Latin America concludes that these are "primarily urban, male, white, middle-aged, upper-class and somewhat English proficient" (Gómez 2000: 73).[6]

For a third indicator of NICT use and penetration we may look to international bandwidth—that is, the data exchange capacity of existing network connections. The resulting diagram most clearly underscores which areas of the world are at the core of international data traffic, and which are not (figure 3).

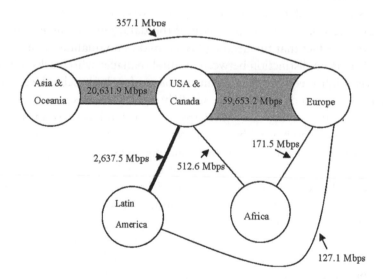

Figure 3 International Bandwidth between World Regions, 2000 (in Megabits per second)

Note: In the category of 'Asia & Oceania,' the eight 'OECD plus tiger' states make up the bulk of bandwidth use, while the rest of the continent would show a bandwidth volume slightly higher than that of Africa.

Source: Press (2001: 10)

It is also important to note that international connections of Third World countries are fundamentally reliant on links to First World countries. Even the data traffic between neighboring countries often utilizes networks in the United States, Europe or the developed nations in Asia and Oceania. Thus, the map of international IP connections of the Third World largely reproduces the 'classical' patterns of center-periphery dependency (Press 2001: 9).

WHAT THE DATA TELL US—AND WHAT THEY DO NOT

The data on host computers and on Internet users have become the two most frequently used indicators for worldwide NICT development. However, they have to be read with caution. The data for hosts "can only be considered an approximation," as the International Telecommunication Union readily admits (ITU 2001: 105). Still more problematic, however, are the user data. First of all, it is a rather rough estimate; this is already reflected in the ITU's technical notes to its statistical data: "[The category] *Users* is based on reported estimates, derivations based on reported Internet Access Provider subscriber counts, or calculated by multiplying the number of hosts by an estimated multiplier" (ibid). More frankly still, the widely cited Nua surveys on Internet user numbers call their data an "estimated guess" (www.nua.ie/surveys).

Beyond this general imprecision of the data, more complex problems arise from the fact that the heading 'users' makes no qualitative differentiation. There is no distinction between an Intel manager who works the entire day over high-speed Internet connections and a high school student in Nairobi who spends part of his small budget on occasional e-mail and Internet use in a cyber-café. (Perhaps even a rural activist in a remote Andean village who has neither computer nor phone line, but whose friends in a nearby NGO send and receive e-mail massages, which they pass to him in printed form, is also a user.) Instead of merely looking at binary classifications of ,"How many are online?"—the title of the Nua surveys—the critical gaze must extend to the differences in the quality and in the form and function of the NICT use.

These qualitative differences are markedly greater in the case of the NICT than in traditional telephone communication. In the case of the latter, the common 'universal service' approach more appropriately treated 'access' as a clear-cut issue: the installation of a telephone main line, which for all users brought essentially the same analogue connection to the network and the same relatively simple phone set. Special features were available, such as fax or answering machines or multiple phone sets connected to one main line, but they were minor additions to the base functionality of the network (Kahin 1995: 8). Also, many aspects of the poor quality of the telephone net in many

Third World countries tended to affect all users alike (though it must be noted that rural areas often showed higher rates of incomplete or interrupted phone calls). The essential qualitative difference in regard to access was defined by price, that is, the volume of phone call minutes a household could afford.

In the case of Internet-based communication, the picture is quite different; here, 'access' or 'connectivity' becomes a much more complex issue. A connection through a dial-up line and modem, with which navigation in the World Wide Web easily turns into 'world wide waiting,' provides a very different functionality than a 24-hour broadband leased line.[7] A user who can only use 'store and forward' e-mail communication has a very different instrument from one who is an active producer of web content through his own homepage or domain. All this finds no expression at all in the statistical data given under the title 'users.'

GLOBAL DISPARITIES: IS THE GAP NARROWING?

In the overall data on Internet users and host computers, the last few years have shown a gradual improvement towards a somewhat less polarized distribution. If the United States and Canada at the end of 1997 concentrated more than half of the world's Internet users, this fell to 28.1 percent at the end of 2002 (ITU 2003); similarly the African, Asian and Latin American countries have increased their share to today's low figures from still lower figures some years ago. Optimistic perspectives see this as evidence that, though slowly, the 'digital divide' is narrowing, and that the Third World is "catching up" (UNDP 2001: 40).[8]

This, however, seems an all too static view that overlooks the strong differences in the quality of access—a problem the study by the United Nations Commission on Science and Technology for Development (UNC-STD) authored by Mansell/Wehn (1998) has termed the "emergent access dualisms" (ibid: 161). The ITU has come to acknowledge this essential aspect in its most recent World Telecommunication Development Report 2002, stating that "the nature of the digital divide is shifting, from quantity to quality" (ITU 2002b: 12).

However, the gap does resist narrowing quantitatively as well, if we take bandwidth as an indicator. Bandwidth has indeed shown high growth rates in the Third World in recent years: in Latin America, for instance, at the end of 1996 only Mexico had an international Internet connection exceeding one megabit per second (Mbps). Since then international bandwidth has increased greatly, reaching 48.3 Mbps for the entire continent in 1999, and 71.0 Mbps in 2000–a growth of 47 percent in a single year (Press 2001: 5). However, these high growth rates are dwarfed by the increases in North

America and Europe, which in the same time period were no less than 178 percent and 102 percent, respectively (ibid).

In 2002, the ITU writes somewhat triumphantly: "At last some over-used clichés can now be put to rest. It was often said, for example, that 'Tokyo has more telephones than the whole of the African continent.' While this may have been true some 20 years ago (. . .) today there are more than twice as many main telephone lines in Africa as in Tokyo" (ITU 2002b: 6). If this result still is far from announcing equitable development, the look at bandwidth gives an impressive update on the not only continuing, but increasing disparities. Here the ITU has to present sobering results: "International Internet bandwidth (or IP connectivity) is a good measure of users' experience with the Internet. The greater the bandwidth, the quicker the response times. The 400,000 citizens of Luxembourg between them share more international Internet bandwidth than Africa's 760 million citizens" (ibid). The bandwidth per capita of a country like Belgium is 8,000 times higher than that of Bolivia, and 80,000 times higher than of Bangladesh, (ITU 2002a: 29), a type of comparison that can be repeated *ad nauseam*.

Using a complex set of indicators, a recent study by the World Bank's InfoDev Program concludes that "1) All developing countries, even the poorest, are improving their access to and use of modern ICTs (. . .) 2) However, the gap between the rich OECD countries and the poor developing countries is growing (. . .) the global equity problem is getting worse" (Rodríguez/Wilson 2000: 3f). This should not be too astonishing. Technological modernity is by definition a positional good. Since the central impetus of technological innovation in a capitalist economy lies in the achievement of competitive advantages over others (other companies or other nations), the reproduction of a technological divide—though at ever-changing levels—is an inherent part of the system's normal functioning.

Given the far higher investments in research and development activities in the industrialized and newly industrialized countries, there is little doubt that the process of technological innovation proceeds at a much faster rate in the industrialized countries than in the Third World. The discourse proclaiming the 'digital inclusion' of the Third World is part of a global economic model that does hold possibilities of ascension for some, but that hardly allows for an equal inclusion of all.

THE 'DIGITAL DIVIDE': SECOND THOUGHTS ON A PROMINENT CONCEPT

In light of the sharp inequalities in the access to the digital technologies, as shown by the data in the preceding chapter, the term 'digital divide' has become a leitmotif in the international debate. The need to address this problem has achieved consensus in international rhetoric, illustrated, for instance, in the 'Okinawa Charter on the Global Information Society,' signed by the heads of state from the so called 'G-8' group of the leading industrialized countries (G-7 plus Russia). The charter promises to give high priority "to bridge the international information and knowledge divide" and lead a "continued drive toward universal and affordable access" (G-8 2000; see also OECD 2001). More recently, the rhetoric of 'bridging the digital divide' has been prominently displayed at the United Nations' "World Summit on the Information Society"(WSIS), whose first phase was celebrated in Geneva in December 2003 (cf. www.itu.int/wsis).

Certainly, the discussion about the 'digital divide' has been helpful in focusing attention on the problem of global inequality regarding the NICT. However, there are good reasons to argue that the concept of a 'digital divide' provides an insufficient framework for either analysis or policymaking regarding this issue.

The concept of the 'digital divide' tends to isolate the issue of the new information and communication technologies from the much broader question of uneven economic and social development (between North and South and within each country) of which it is a part and a symptom. Utilizing the same logic as that of the 'digital divide,' one could speak of the 'access to drinking water divide.' That the 'digital divide' builds upon a much larger context of structural inequalities also becomes evident when we look at some of the basic conditions for Internet use (at least in the dominant individual access model of the North; alternative approaches that try to circumvent these access barriers are discussed later in this work): one needs a computer plus modem; stable electricity supply to put these into operation; access to the telephone net; the knowledge to operate the computer, with literacy the most elementary necessary, though far from sufficient, condition; and the money to pay for the current costs of telephone line use, Internet service providers, and maintenance of the equipment. In most countries of the Third World, a large part of the population is lacking one or even all of these conditions.

Some data to illustrate these issues: 38.4 percent of the world's personal computers are located in the United States and Canada, but less than 2 percent are in Africa (ITU 2002a: A-67). In the cross-country comparison, the data for PC density—that is, personal computers per 100 inhabitants—largely parallel the general indicators for economic development (ibid; for Latin America see table 4 on page 123). In Third World countries, ownership of a personal computer tends to be limited to institutions, business and higher-income sectors of society: many of the least developed countries have less than one computer per 100 inhabitants (ibid). The adult literacy rate for the least developed countries is barely above the 50 percent level (UNDP 2002: 185). In Africa, 70 percent of the population in rural areas live without access to electricity, while on the Indian subcontinent, more than half of homes are not connected to the electricity net (Afemann 2001: 27).

Secondly, as with the term 'Internet user,' the notion of the 'digital divide' implies a binary division into those who have access and those who don't, failing to consider the qualitative dimension of what this 'access' might entail. Hargittai (2002), in her analysis of the differences in people's online activities in the United States, has tried to adapt the concept by speaking of these qualitative differences as a "second-level digital divide."[9]

These efforts to stretch the 'digital divide' concept react to the fact that meaningful access to NICT encompasses far more than providing computers and connections. Warschauer (2002) points to the limits of such a stretching of the term, arguing, that

> "this original sense of the digital divide term—which attached overriding importance to the physical availability of computers and connectivity, rather than to issues of content, language, education, literacy, or community and social resources—is difficult to overcome in people's minds."

Framing the problem as 'the digital divide' tends to connote 'digital solutions'—i.e., the provision of more or better computers and telecommunications, "without engaging the important set of complementary resources and complex interventions to support social inclusion, of which informational technology applications may be enabling elements, but are certainly insufficient when simply added to the status quo mix of resources and relationships" (ibid). Seen from this perspective, the developmental goal is less to overcome 'the digital divide,' but rather to further a process of economic, social and political development by using the NICT.

CONCEPTUALIZING THE ISSUE: NICT FOR DEVELOPMENT

This critique of the 'digital divide' concept in fact echoes calls from civil society groups from Third World countries. In a recent document synthesizing discussions of Latin American NGOs working on the issue, we read:

> "We are tired of the so-called 'digital divide' problem. The real problem is how we are going to use the strategic opportunities that ICTs offer for closing social divides (. . .) This is not a semantic problem, but a vision that embraces all our objectives, methods and efforts for using ICTs to the benefit of sustainable human development" (Gómez/Casadiego2002).[10]

The authors—and in fact many of the engaged social groups in the field—strongly emphasize that connectivity and a specific ranking in NICT user data are not ends in themselves, but that they can be, at best, tools to reach development goals. The essential change in perspective is to frame the issue not from the side of technology ('digital divide') but from that of development, "understood here to mean promoting democracy with social justice, economic prosperity with equity, and realization of the full human potential" (ibid). The questions then are how and under what circumstances the NICT can become instrumental to these ends. In this work, we share this conceptual approach, focusing on the NICT from a development point of view that includes emancipatory economic, social and political goals.

In fact, the current discussion on Internet access in many ways recalls the debate on telephone access two decades ago when the 'Maitland Report' stated the goal of telephone access for all by the year 2000 (Maitland 1984: 4). However, more than half the world's population still has never made a telephone call. The entire African continent has less than 2 percent of the world's telephone main lines, India has fewer than Great Britain, and the 145 million people in Pakistan share fewer telephones than the 8 million inhabitants of Austria. If in Germany, the telephone density—that is, main telephone lines per 100 inhabitants—is 65.1, in Cambodia it is 0.26 and so on (ITU 2003). With some exceptions and in a less extreme form, the picture is very similar to that painted by the NICT data. Also reminiscent is the fact that in Third World countries, telephone main lines are largely concentrated in the major urban centers, while the rural areas often show an extremely low telephone density.

As the telephone is certainly not very new technology, dating back to the 19th century and achieving massive diffusion in the Western industrialized countries by the 1970s, the look at the 'telephone divide' is particularly helpful to show that the 'digital divide' is not a surprising new phenomenon, but rather that it continues established patterns of structural inequality. Rather than revealing a 'digital divide,' the stark inequalities in NICT use and diffusion should be seen as expressions of "the development divide in the digital era," as Hewitt de Alcántara (2001) suggests. If so, the challenges posed by the NICT for Third World countries must be inserted in a much wider debate about development, technology and politics.

Therefore, before discussing the political issues associated today with the new information and technologies, we will first turn to a critical analysis of the development experience of the past. The two countries chosen for the empirical studies have followed different economic development strategies which effectively represented the two dominant development paradigms for Third World countries in most of the Cold War era: Costa Rica, along with almost all Latin American countries and many others in Africa and Asia, adopted the so-called ISI strategy (Industrialization via Substition of Imports), involving an active role of the state within an essentially capitalist economic context. Cuba, in turn, pursued a variant of the state-socialist model with state ownership of the means of production and central planning of the economy. In both countries, the economic, social and political struc- tures resulting from these development paths greatly shape the way in which the issues raised by the new information and communication technologies are being dealt with today.

In the analysis of both the ISI strategy and the state-socialist develop- ment model, we will pay particular attention to the telecommunications regime and the role played by the 'old ICTs,' from the telephone to traditional mass media to the production of high technology products. This look back is all the more necessary since much of the literature on the NICT tends to be blinded by their 'newness.' By emphasizing the 'revolutionary' character and the singularity of the new information and communication technologies, many ignore the earlier experiences with development, communications tech- nology and politics in the Third World on which the current debate on NICT for development necessarily has to build.

DEVELOPMENT, TECHNOLOGY AND COMMUNICATION: THE ISI EXPERIENCE

It was the world economic crisis of 1929 which became the point of depar- ture for industrialization in the Third World. With the breakdown of world trade Third World countries no longer found sufficient markets for their tra- ditional exports, primarily natural resources and agricultural products. As a consequence, they could not continue to import industrial products as be- fore. In this situation, more forced by circumstances than led by strategic de- cision, a number of Latin American countries initiated a process of substituting industrial imports through domestic production (cf. Thorp 1984, Bulmer-Thomas 1998).

It was not until after the end of World War II that, out of this process, there evolved an explicit development strategy of industrialization via import- substitution, or ISI, elaborated by the United Nations' Economic Commission on Latin America, CEPAL, and its leading economist, Raúl Prebisch.[11] From the early 1950s onward the ISI (also called *cepalismo* after the institution that had promoted it) became the dominant development paradigm in Latin America and, in the 1960s and 70s, also for many of the newly decolonized nations of Africa and Asia.[12] Following Prebisch's pioneering work, which denounced the chronic decline of the terms of trade for primary commodities in regard to industrial products, the ISI strategy challenged the liberal economic wisdom built on the classic Ricardean argument, according to which trading Portuguese wine for manufactured British cloth would be to the equal benefit of both. Instead, the *cepalistas* argued, development would only be possible if Third World countries could transform their productive structure through a process of industrialization and to turn from outward to inward-oriented growth (from *crecimiento hacia afuera* to *crecimiento hacia adentro*).

Although conceived within the parameters of a capitalist economy, the ISI strategy called for a relatively high degree of state intervention. The sharp inequalities and widespread poverty in the Third World made it imperative that the 'invisible hand of the market' be corrected by the state, presumed to be the only institution able to act on behalf of society as a whole. This was all the more necessary, according to the argument, since colonial or post-colonial experiences left Third World countries with deformed economic and social structures, with underdeveloped markets and institutions, and with weak entrepreneurial classes. Therefore, the state had to step in and act as engine and coordinator of national development; hence the term '*desarrollista* state' (developmental state)[13] was coined.

The precise elements of the policy arsenal of the *desarrollista* state varied over time and between countries, but they typically included: a tariff policy to protect nascent national industry from foreign competition; an active credit policy to foster specific economic sectors; macro-economic and/or sector planning structures; an overvaluation of the nation's currency and/or multiple exchange-rates; the emergence of a state-bureaucracy to execute these policies and the creation of public institutions in a broad spectrum of economic and social activities; the procurement of material infrastructure, from roads to electricity, and of public services in areas like education and health; the partial regulation of prices, especially in the form of subsidies for certain products; and the direct engagement of the state in the production of goods and services via publicly owned companies. Transnational companies often were banned from sectors perceived as 'strategic,' and in other fields operational controls on their activities were established; at the same time foreign investments were regarded as crucial to the acceleration of the industrialization process and for the knowledge and technology transfer needed. In particular, developmental hopes rested on the dynamics of forward and backward production linkages (Hirschman 1958; 1971; and 1981) that were to emerge from these investments.

The ISI strategy often went together with a Pan-Latin American, Pan-African or Third World rhetoric, as well as by a number of regional trade integration schemes. Nevertheless, the fundamental framework in which it was conceived was always, economically as well as politically, the nation-state. In fact, the industrialization process tended to be regarded as an essential expression of national political sovereignty. To stress the importance of the national character of the ISI strategy, Altvater/Mahnkopf (1996: 405) explicitly speak of the "national developmental state."[14]

ISI IMPLEMENTED: PERIPHERAL FORDISM AND ITS POLITICAL REGIME

The general 'role models' for ISI development were the Fordist production structures and consumption patterns of the industrialized capitalist countries. Basically, the term 'Fordism' refers to the rationalization of production methods as first introduced by Henry Ford in car production; it allowed for the mass production of complex goods, turning former luxury items into products of mass consumption.

Beyond this, however, the mode of capitalist regulation theoretically described as 'Fordism' (Aglietta 1979; Lipietz 1985a; Boyer 1990; Hurtienne 1984: 265–315; Altvater 1987: 24–36) implies a much broader arrangement which emerged in the industrialized countries, essentially since the 1920s and especially after World War II. In this fuller view, the Fordist paradigm extends to the political and social regime and, eventually, touches upon practically all spheres of life, from consumption patterns to gender relations. At its core, within an essentially Keynesian macro-economic approach and within the borders of the nation-state, mass demand for mass consumer products was made possible through the provision of relatively high salaries for a relatively broad part of the population. This went hand in hand with new forms of state intervention, expanded social security systems and state-sanctioned mechanisms to institutionalize class conflicts and distributional struggles, which contributed to the relatively high degree of social cohesion characteristic of the Fordist mode of regulation.

Though this vision of Fordism may have been the goal, it certainly was not what resulted from ISI policies in the Third World. What turned out instead can be termed 'peripheral Fordism' (cf. Lipietz 1985b, Hurtienne 1986): a regime which copied parts of the Fordist model without being able to assume its broader organizational scheme, particularly failing to provide high incomes for a majority of the population, the regulatory framework of the social welfare state, and an economically sustainable scheme of inclusive social integration. Formal Fordist labor relations never came to cover the population at large; instead of becoming integrating elements of society, labor relations were based on the persistence of a broad sector of informal employment, thus deepening rather than overcoming the structural hetero-geneity typical of Third World capitalism (Córdova 1973) .[15]

The deformed character of peripheral Fordism also becomes evident in the fact that, despite its goals of social progress and its 'national project' rhetoric, the ISI strategy did not lead to a progressive redistribution of in-come but, to the contrary, produced a further concentration of wealth. The paradigmatic example is Brazil, which followed a longer and more profound ISI process than almost any other country, and which at the same time reached one of the most highly polarized income distributions in the world (Fritz 2002: 60–84).

Finally, the social and political context of peripheral Fordism also dif-fered greatly from that of the First World role model. With the changes in the social structure brought about by the industrialization drive, the politi-cal regimes implementing the ISI were faced with the crucial task of forming '*desarrollista* alliances' (Cardoso/Faletto 1988: 107) between the industrial entrepreneurs, the enlarged state bureaucracy, organized workers and large urban 'popular sectors' which were not integrated into formal work rela-tions. The Fordist mechanisms of institutionalized class compromise in the North found its Third World counterpart in populist regimes that integrated workers and entrepreneurs in essentially corporatist arrangements, imple-menting distributionist measures in favor of workers' interests while main-taining authoritarian control over their organizations (and generally, too, over many other aspects of society). As a result, in the 1950s and 60s in Latin America the ISI strategy came to be closely associated with '*desarrol-lista* populism' (ibid: 106) and the emergence of charismatic leaders, who were able to mobilize society around a 'national cause.'

However, if the deprivation of power of the old landed oligarchy had been an essential pre-condition for the development of the Fordist model in the First World, in most countries of the Third World the ISI strategy re-mained economically dependent on a continued flow of external revenues from the traditional agrarian exports. Due to this dependency, an agrarian reform which would have attacked the entrenched interests of the agrarian elites was deemed politically inviable. The failure to take such a step under-mined the power base of the *desarrollista* alliances, and it cemented the eco-nomic and social exclusion of large parts of the rural population.

In addition, the state generally did not maintain sufficient autonomy vis-à-vis the emergent class of national industrial producers. By using their weight in the political alliance to negotiate continued and favorable tariffs and price policies, the national entrepreneurial class acquired an increas-ingly rentist character, to the detriment of innovations in the production process or the final product (cf. Krueger 1974, Müller-Plantenberg 1972). As a result, the state apparatus itself became 'colonized' by the different social forces and their particular interests (cf. O'Donnell 1979a: 23).[16] Given the weight of these conflicting interests, Cardoso/Faletto in their classic work explain the failure of the ISI strategy within a wider critique of the political and economic structures of Third World capitalism: "All intents to push forward industrialization must fail, if they are not accompanied by profound changes in the political structures" (Cardoso/Faletto 1988: 140).

Since peripheral Fordism fundamentally lacked the inclusive character that had characterized the Fordist formation in the First World, the distributional struggles could not reliably be contained within the established corporatist arrangements. When the elites felt that the popular sectors 'got out of control' and infringed too much on elite interests, they resorted to the military as their 'armed branch' (for which they could count on the more or less open support of the United States). With the military coups of the 1960s and 70s, the political regime in the majority of Latin American countries changed from '*desarrollista* populism' to dictatorships which O'Donnell (1973, 1979b) has described as 'bureaucratic-authoritarian states.' These reacted as much to the perceived threats from the popular sectors as to the need for capitalist 'deepening' and an internationalization of the national economy, which would go beyond what the *desarrollista* alliance would support. In Latin America, it was the Pinochet regime in Chile that emphatically spearheaded the neo-liberal transformation of economy and society.

INDEBTED INDUSTRIALIZATION: THE WORLD MARKET AS ULTIMATE BENCHMARK OF 'INWARD-ORIENTED DEVELOPMENT'

If the ISI strategy signaled a certain dissociation from the world market, it must be stressed that this separation was a very partial one. The crucial paradox of the ISI strategy is precisely that, though its goal was to substitute and thus to reduce imports, it resulted in a surge of imports, particularly production inputs, equipment, machinery and other capital goods. The deepening of the ISI scheme proved no solution. Though some larger Third World countries went on to establish heavy industries producing capital goods, the growing complexity and technological exigencies of these industries nevertheless increased the need for imports.

These imports theoretically should have been financed through an increasing export capacity in the newly created industries themselves, according to the classic infant-industry argument as formulated by Friedrich List in 19th-century Germany (List 1982). In this view, protecting, through high external tariffs, newly established industries from world market competition during an 'infancy' period would allow them to build up and consolidate modern production structures and technological competence. Following this initial period, international competitiveness should be achieved, leading to rising industrial exports and the gradual reduction of protective tariffs. However, in the Latin American ISI experience, the temporary character of the tariff system proved unenforceable, leading the mentor of the ISI strategy, Raúl Prebisch, to show profound disappointment as early as 1963:

> "An industrial structure virtually isolated from the outside world thus grew up in our countries (. . .) tariffs have been carried to such a pitch that they are undoubtedly—on an average—the highest in the world. It is not uncommon to find tariff duties of over 500 percent. As is well known, the proliferation of industries of every kind in a closed market has deprived the Latin American countries of the advantages of specialization and economies of scale, and owing to the protection afforded by excessive tariff duties and restrictions, a healthy form of internal competition has failed to develop, to the detriment of efficient production" (Prebisch 1963: 71).

Under these conditions, the Latin American industries remained far short of world market standards for price, quality and technology. In effect, the ISI strategy was a remarkable success story in that it indeed led to the emergence of large industrial sectors in many countries, and that for a considerable time it generated high growth rates, employment and income. However, it largely failed in making these industries internationally competitive and in generating industrial exports.

Direct investments from transnational companies aggravated rather than mitigated the problem, since they did not invest in order to produce for the world market, but rather to benefit from the protected national or regional markets. Therefore, the industrialization project, with its high import needs, continued to rely on the revenues from traditional exports, ironically strengthening precisely those ties of dependency that the ISI had aimed to overcome. In addition, the ISI governments resorted to foreign credits on a very large scale, accumulating enormous foreign debts. Between 1972 and 1982, total Third World debt increased eight-fold, from US$100 billion to US$800 billion (Altvater et al. 1988: 21), leading critics to describe the ISI process as 'indebted industrialization' (Frieden 1981), 'crippled industrialization' (Fajnzylber 1983) or even as 'underdevelopment through industrialization' (Müller-Plantenberg 1972: 349).

In consequence, although the proponents of cepalista industrialization proclaimed the 'inward orientation' of the strategy, the world market remained its ultimate benchmark. The ISI's partial dissociation from world trade went hand in hand with a dramatic increase in dependence on global capital and credit markets, with productive investments unable to generate sufficient exports to service and repay these credits. The resulting outbreak of the debt crisis in the 1980s led to the so-called structural adjustment programs under the auspices of the IMF and World Bank. These programs not only had high social costs for the affected societies, they also dismantled much of the recently created industrial structures[17] and swept away the ISI strategy as the dominant development paradigm.

THIRD WORLD COUNTRIES AND FIRST WORLD TECHNOLOGIES

The ISI strategy, as promoted by the CEPAL, was inspired by the historical optimism of modernization theory, which saw the task of Third World development essentially as one of 'catching up' to the industrialized countries.[18] Industrialization under these terms implied taking over the technologies used and developed in the First World, with the aim of reproducing them domestically. In this context, the creation of national computer industries and the quest for 'technological autonomy' became an ambitious part of *desarrollista* policies in a number of countries with large domestic markets, namely in India (Grieco 1984), Mexico (Borja 1995) and Brazil (Meyer-Stamer 1996, Bastos 1992, Porto/Pochmann 2001). These projects can serve as paradigmatic examples of the difficulties encountered by the ISI strategy.

In Mexico, the *Programa de Cómputo*, started in 1981, established a de-facto market reserve through a system of tariffs, quotas and import licences. Foreign companies such as IBM, Hewlett Packard, and Apple started operations in Mexico, clearly orienting production towards the domestic market. Although the sector showed high growth rates, its downsides were severe. The high import needs weighed heavily on the balance of payments. The production companies never fulfilled their obligations for research and development investments. The linkages to the rest of the economy remained low and the local content percentage reached only one fifth of what had been planned. By international standards, the nascent Mexican computer industry remained on a small scale and, in consequence, prices were estimated to be 30 to 50 percent above international prices; technologically, Mexican production came to lag one or two years behind world market competitors (Borja 1995: 95). Finally, under the impact of the 1982 debt crisis, domestic demand collapsed, state funds dried up, and a gradual displacement of the 'nationalist approach' began. The transnational companies turned to produce for the U.S. market, and in 1990, a Presidential

Decree sealed the fate of the *Programa de Cómputo*, adopting instead an unmistakably neo-liberal policy.

Though far more radical than the Mexican approach, Brazil's nationalist 'informatics policy'—which had fostered a large and diversified national computer industry—eventually gave way to neo-liberal policies in the 1990s (cf. Porto/Pochmann 2001; and Meyer-Stamer 1996). As a result, most of the local firms either disappeared or were bought out by foreign companies; a number of those remaining shifted into specialized niches. Computer manufacturing today persists as an economic activity of some significance, but the loss of domestic backward linkages to internal sources of components and technology was severe (Tigre/Botelho 2001). Moreover, the work force employed in the 50 largest manufacturers of information technology products was almost halved between 1989 and 1999, falling from 74,000 to about 38,500 (ILO 2001: 173).

For production aimed at the world market, economies of scale have great weight, creating high incentives for the substitution of current costs (labor) through fixed capital (machinery). In addition, due to the relatively high level of salaries and the relatively easy availability of capital in the First World, the technological innovations of the production process inherently tend to favor capital intensity over labor intensity. For Third World countries, however, where salaries tend to be low, un-skilled labor abundant, and capital scarce, priorities are different. Particularly where industrial development is oriented towards the domestic market, which permits only limited economies of scale, the reduction of the costs of technological equipment is much more important than the increase in output per unit.

With this background, calls for 'appropriate technology' gained ground in the 1970s. Although the quest for alternative technologies has narrow limitations as an overall strategy, it still today holds important possibilities, if not as a general substitute for First World technology, then at least as an important complement. The new information and communication technologies certainly provide examples of successful uses of appropriate technology. For instance, initiatives promoting free or open source software[19] like Linux fight a brave uphill battle against the software dominance of transnational companies, with potentially significant benefits for the Third World (Stallman 2002).

Another reason appropriate technologies may be necessary is that NICT devices developed in the First World generally presuppose literacy. For Third World countries with high rates of illiteracy, the keyboard is a socially exclusionary technology. A prominent example of an appropriate technology approach are the recent efforts of Indian scientists to develop a

'Simputer' (*Simple* Com*puter*), which is not only low-cost but also not key-board-based, thus enabling illiterate users (cf. www.simputer.org). Again, however important these efforts are, alternative technologies will not often fully replace conventional ones. It will also be particularly important to search for socially and economically appropriate ways to use dominant technologies, as in the case of collectively used Internet access centers in countries where the Northern pattern of individual residential use is unfeasible for much of the population.

The history of industrialization efforts has no shortage of failed initiatives to introduce First World technologies to developing countries. One evocative term is the 'rusty tractor syndrome' (Panos Institute 1998: 11): shipments of tractors or trucks, backed by hopes of transforming the prospects of developing countries, end up rusting and useless because of their inadequacy in the local conditions. This issue is no less crucial for the NICT than for tractors.

MASS MEDIA FOR NATIONAL DEVELOPMENT

All through the 1950s and 60s, almost all Third World countries gave much higher priority to the build-up of mass media infrastructure, particularly TV and radio broadcasting, than to telecommunications. The widespread view was that an extension of telephones to residential users had little positive developmental effects and was instead a luxury for the upper and middle classes. In contrast, mass media such as TV and radio promised to reach a nation-wide audience of all social backgrounds, which was regarded as crucial to the integration of post-colonial societies and the process of 'nation-building' (cf. Schramm 1964, Pye 1963). Supported by the modernization paradigm, these efforts were primarily concerned with a) the role of media in the creation of national identities; b) bringing remote communities into the national space; and c) sharing information domestically in a vertical way, which is informing the people of the decisions of their governments, spreading new technical knowledge to peasants, etc. The proliferation of this approach was greatly helped by the fact that, for many governments, such an emphasis on vertical one-to-many-communication (in which the 'one' typically was the state, a public institution or a state-linked company) was politically much more attractive than the horizontal many-to-many-communication of the telephone system.

Thus it was mass media, not telecommunications, that was the focus of Third World countries raising the call for a 'New World Information and Communications Order' (NWICO) in the 1970s. Echoing the demand for a 'New World Economic Order,' the NWICO essentially aimed at counteracting the unequal power structures dominating the international news and media business, claiming 'information sovereignty' for the countries of the South. The United Nations Educational, Scientific and Cultural Organization took up these calls, in 1984 publishing the programmatic MacBride Report (UNESCO 1984).

This report criticized the extent to which Western commercial media dominated news and media worldwide and it suggested a number of regulatory policies to redress this situation. This triggered an intense political reaction, particularly from the Reagan administration in the United States and the Thatcher government in Great Britain, which condemned the proposed policies as an infringement on the freedom of press. The issue of the NWICO became caught up in a heated ideological confrontation and the controversy led to an institutional crisis within the UN body: in 1985, the United States ceased its payments to UNESCO, in 1986 Great Britain followed. The dispute was effectively settled in November 1989, when UNESCO's General Conference officially abandoned the central propositions of the MacBride Report. In their place, it adopted a new communication policy, explicitly stating

> "that all efforts should be made to ensure the free flow of information at the international as well as national level, and a wider and better balanced dissemination of information, without any obstacle to the freedom of expression (. . .)" (UNESCO 1989).[20]

Ignoring the ideological bias of the debate, the NWICO conflict revealed the limits of an exclusively governmental approach. With such an approach, authoritarian governments, in the name of 'information sovereignty,' might well repress independent and critical media domestically, not so much attacking Northern media domination as its own society and its citizens' rights.

However, the outcome of this experience was not the expansion of the regulatory approach to include civil society groups to ensure media pluralism. What happened instead was that 'freedom of the media' was essentially equated with freedom of the market forces in the media business. In the name of defending the 'free flow of information' the developmental goals of overcoming the disparities between North and South and of increasing South-South and South-North communication were simply abandoned or reduced to symbolic measures.

THE TELECOMMUNICATIONS REGIME OF PERIPHERAL FORDISM

Before independence, the telecommunications companies in most African and Asian countries were monopolies that were either part of the colonial

administrative system or in the hands of companies from the colonial powers. At independence most countries quickly nationalized telecommunications.[21] In Latin America, in the course of the *desarrollista* project of the
post-WWII period, national governments took over foreign-owned telecommunications companies. Telecommunications were generally regarded as an
integral part of the nation's public infrastructure and national ownership
was seen, as it was in the industrialized countries during the 19th and most
of the 20th century, as key to national sovereignty—though not to economic
development.[22]

Although the state-owned monopolies did not see the telephone system as a priority of national development, they did, however, perceive it as
a means to expand employment in the public sector, and as a major source
of revenues for the state. In most countries, the telecommunications monopoly proved to be a profitable cash-cow, chronically transferring as much
as 30 to 40 percent of its revenues to the state treasury, money which could
be used to finance other spending or to cover the losses incurred in other
fields (Saunders et al. 1994: 34). As a consequence, whereas investments in
physical infrastructure in areas such as transportation or energy generation
received great public attention and were often pushed forward on the basis
of foreign credits, telecommunications companies suffered a chronic capital drain that greatly reduced their capacity to invest in the expansion and
innovation of the system.[23] Accordingly, the diffusion and quality of telephone networks in the Third World generally remained very low. The
telecommunications regime of peripheral Fordism also included a general
practice of cross-subsidization within the telephone system, with high costs
for long-distance calls subsidizing lower tariffs for local calls and for main
line installation.

It was not until the mid-1970s that telecommunications won greater
appreciation, with focus on the argument that improved telecommunications could lower transportation needs and costs. Parallel to UNESCO's
Mac Bride commission, the International Telecommunication Union in
1982 established the Maitland Commission to address the issue of telecommunications in the Third World. In its report entitled 'The Missing Link'
(ITU 1984) the Maitland Commission concluded that "in most developing
countries, the telecommunication system is not adequate even to sustain essential services." At the same time it underscored the critical importance of
information and communication technologies for development—not only
for the modern and urbanized sectors of society, but for the development of
rural areas as well. The report noted that the developmental impact of an
improvement in telecommunications is greatest for communities that are

isolated or geographically distant from urban centers. In its recommendations, the Maitland Report stressed the need for a much higher priority for investments in the telecommunications sector and for more transfer of technology. At the same time, the inefficiencies of most state telecommunication monopolies received criticism, with calls for administrative and organizational improvements (ibid).

In the industrialized countries, the Fordist telecommunications regime had been based essentially on public monopolies and a high degree of state regulation, with gradual expansion towards the goal of 'universal service:' the provision of residential telephone main lines for every household, with affordable rates for local calls. By and large, this was not a cost-based system, but one based on different patterns of cross-subsidization as pricing strategy. In peripheral Fordism, parts of the First World model—public telecom monopolies, cross-subsidization schemes—were copied. From the mid-1970s onward, telecommunications began to expand throughout the Third World (Saunders et al. 1994: 5), to a great extent financed with foreign credits, thus contributing to the growing debt problems.[24] Nevertheless, the social and distributional effects were very different, showing a strong bias geographically, in favor of the major urban areas to the detriment of rural areas, and socially, in favor of the upper and middle classes (e.g. Clippinger 1976).

Even with the basic growth of Third World telecommunications networks in the 1970s and '80s, the goal of 'universal service' (providing a telephone to each home) was so unrealistic that it was not even seriously considered. Instead, where efforts for the social diffusion of telephony were undertaken, the objective was 'universal access,' generally meaning the provision of public telephone service within walking distance of each person (Panos 1997). Beginning in the 1970s, one form of access, aimed at the lower income population, was the collective use of telephones via public coin box telephones or public call offices (PCOs), though this was generally not pursued on a large scale. These public call offices—at times installed in public facilities or post offices, at times set up on concessionaire basis—can well be considered the direct precursors of Internet cafés and telecenters two decades later.[25]

By the end of the 1980s, telecommunications had grown to become a strategic factor of development. Increased efforts for modernization and expansion had led to a notable expansion of telecommunications in the Third World, raising its share of the world's telephone lines from about 7 percent in 1969 to 12 percent in 1988 (Saunders et al. 1994: 5). However, these figures come with various caveats. First, with 75 percent of the world's population, the Third World's 12 percent share of the world's telephone main lines still reflects a deep disparity. Secondly, these overall figures hide—as do the current data on NICT diffusion—strong disparities between different countries and regions within the Third World. In 1988, the telephone density (that is the number of main lines per 100 inhabitants) was 0.2 in a country like Tanzania or Nigeria, compared to 25.0 in Taiwan or 20.7 in South Korea; in Latin America, telephone density for the whole continent was 5.9, but it ranged from 0.5 in Haiti to 10.6 in Uruguay. For the African continent excluding South Africa, it was still as low as 0.7, and for Asia excluding the OECD countries and the four South East Asian 'tigers' it was a mere 0.9 (ibid: 6–7).

Thirdly, the quality of the telephone systems remained generally poor. Whereas in the 1980s the industrialized countries were racing to install fiber optic cables and integrated services digital networks (ISDN), which greatly increased capabilities for NICT use, traditional technologies remained dominant in most Third World countries and telephone users had to fight with frequent incomplete calls, interruptions, line-crossings and other deficiencies. In addition, often year-long waiting lists for main line telephone installation became the landmark illustration of the state-run monopolies' inability to meet demand. Such failure facilitated calls from political and economic forces taking their cue from the overall demise of the ISI paradigm to liberalize and privatize telecommunications.

GLOBALIZATION, THE NEO-LIBERAL PARADIGM, AND THE LIMITS OF THIRD WORLD DEMOCRATIZATION

The end of the ISI model as the leading paradigm for Third World development came with the debt crisis of the 1980s, which ended access to fresh credits and made it imperative to turn to export promotion in order to generate hard currency earnings. Moreover, afflicted with acute insolvency, the countries were left with little room to maneuver in their negotiations with the creditors and the International Monetary Fund (IMF). The ensuing 'structural adjustment programs' imposed—in at times more and at times less radical form—a new economic model styled on the neo-liberal paradigm.

At a different level of analysis, the demise of the ISI paradigm can be seen as part of a larger process generally called 'globalization.' When speaking of globalization and the role of the NICT in this process, it is necessary to note our distance from a wide-spread tendency to describe this process as an objective, seemingly actorless process driven by the anonymous forces of technological progress.[26] In contrast, in this work we conceive of globalization as a process driven and shaped by economic and social actors, interests and power structures. This process runs into increasing contradictions between the essentially national character of the political sphere and the global character of the economic sphere, eroding the economic underpinnings of the 'national developmental state' in Third World countries, as well as those of the Keynesian state, including much of the social regulation framework of the Fordist regime in the industrialized countries. The development and spread of the new information and communication technologies play an important role in this process by providing crucial infrastructure for the expansion and acceleration of world trade, international financial markets and global production structures. Where national or regional production units had once found 'natural protection' in the costs and complications of physical distances, these 'tariffs of distance' have shrunk greatly.

However, globalization as we know it today would not have been thinkable without major transformations in international politics. In fact, only since the demise of the Soviet Union and the state-socialist countries of Eastern Europe has the world market truly become 'global.' In addition, three interrelated developments marked the profound transformation of world economics in the last quarter of the 20th century: the liberalization of international trade, investments and financial markets; the opening and deregulation of national domestic economies; and the privatization of state-owned or publicly controlled enterprises. In consequence, more than ever before the world market acts as the universal benchmark for national industries, with the power to define prices, production structures and rates of return for invested capital. In this sense, at the core of the globalization process is the market-mediated push for the worldwide homogenization of economic standards (Altvater/Mahnkopf 1996: 98f).

In this process, international organizations such as the International Monetary Fund, the World Bank and the World Trade Organization have been the most powerful intermediaries in framing and structuring market transformations, in ways largely coinciding with the interests and ideological preferences of the dominant economic and social forces. For Third World development, the neo-liberal paradigm was summed up in the "Washington Consensus" (Williamson 1990), in reference to the U.S. government and the Washington based IMF and World Bank. Its main components are well known: opening of national economies, reduction of tariff barriers, promotion of export production, privatization of state-owned or public companies, down-sizing of the state's role in the economy, attraction of foreign capital, fiscal austerity policies, etc.

Politically, the 1980s were the 'decade of transition' in Latin America, when almost all nations returned to civilian rule and pluri-party systems, after the military dictatorships of the 1970s. Numerous studies, greatly influenced by the actors-centered work of O'Donnell/Schmitter/Whitehead (1986), have analyzed these transitions. Nevertheless, this literature paid little attention to the role of the media or communications.[27]

Generally, broad enthusiasm accompanied the end of authoritarian rule, yet disenchantment typically set in when it became evident that these political transitions failed to significantly expand social inclusion.[28] While much of the mainstream literature initially blamed the defects of the new democracies on its authoritarian legacies, the persistent lack or even loss of democratic substance over time in formally democratic regimes led to a shift in the focus of democratization studies. Beyond institutions and procedural arrangements such as elections, the individual, everyday experience of democracy depends on the concept of citizenship, which includes the social contextualization and factual effectiveness of the rights formally granted by the political regime (e.g. O'Donnell 1998, 1999a, 1999b, 2000). It is because of the need for such a substantial citizenship that the quality of access to information and of communication rights, whether referring to traditional media or the NICT, becomes an essential aspect of democracy.

It is noteworthy that the transitions to democracy became possible precisely when the political threats from the radical left had vanished or fallen victim to repression, and a sufficient part of the dominant sectors of society felt assured that their economic interests would be safeguarded even without military rule. This elite security was due to the 'externalization' of much of the authority over economic policy that accompanied the liberalization of Third World economies. However, limiting the factual reach of the national policymaking process tended to withdraw fundamental economic decisions from the realm of the national demos. In their influential work, Linz/Stepan (1996) explicitly see "a sovereign state as a prerequisite to democracy" (ibid: 17); yet while they address the greatly varying quality of 'democracy,' it does not occur to the authors to ask what the varying quality of 'sovereignty' might mean for democracy.

Habermas (1998) dealt with precisely these implications, warning of the dangers for democracy in what he calls the 'post-national constellation,' fearing that national politics will be reduced to a more or less intelligent management of an adjustment to the financial and economic imperatives dictated by the forces of the world market (ibid: 95). While Habermas formulates this as a prospect for developed countries like Germany, his description certainly recalls the experience of many Third World countries in the wake of the 1980s debt crisis—and it sheds light on the 'disenchantment with democracy' so widely diagnosed since the 1990s.

DEVELOPMENT, TECHNOLOGY AND COMMUNICATION: THE STATE-SOCIALIST EXPERIENCE

The Austrian Marxist Otto Bauer once noted that after the October Revolution of 1917, the pressing task was "less the construction of a socialist society but rather the industrialization of Russia" (cited in Altvater 1991: 356). Indeed, Soviet state-socialism[29] can be understood as a project of modernization and late industrialization of a society marked by social and economic characteristics that, some decades later, would have been called 'underdeveloped.' The famous equation 'socialism = Soviet power + electrification' was coined and in the 1930s Stalin pushed forward with a massive industrialization drive, tripling the country's industrial output within a decade (Hobsbawm 1995: 96). The terribly high human costs of this operation are all too well known, and it led to social and political deformations which would mark state-socialism until its demise. In more immediate economic terms, however, it succeeded in changing the world map of industrialization profoundly: the Soviet Union's share of the world's manufactured output increased from 5 percent in 1929 to 18 percent in 1938.

Of course, Stalin and Raúl Prebisch are worlds apart ideologically. The cepalista strategy remained within the context of a capitalist economy and—in principle, though we have noted the deviations in practice—a liberal political order; the state-socialist model developed in the Soviet Union was based on a centrally planned economy and an unquestionably vertical political structure under the binominal rule of party and state. Nevertheless, both were based on a partial dissociation from the world market—with neither able to escape it altogether; both also based their development project on the Fordist industrial structures and technological schemes of the developed countries of the West—with neither able to copy Western Fordism in any but a deformed way.

Whereas Marx had imagined socialism would succeed in the most highly developed industrialized countries of his era, it triumphed in one of the most backward countries of Europe, the Russian Empire. Here, socialism was not a project for passing from developed capitalism to a superior stage of social organization: rather, it was a vehicle for late development. It is this historical experience that has made state-socialism particularly attractive as a development model for Third World countries of Asia, Africa and Latin America.[30]

In the Soviet Union, the 'statization' of the productive apparatus, the planned economy and the state monopoly on foreign trade were at the core of a development model that sharply reduced commercial relations with the capitalist world economy. With the world revolution not in sight and Stalin formulating his 'socialism-in-one-country' thesis, Soviet late industrialization was one of import-substitution within national boundaries. In contrast to the cepalista strategy, this model did not start out with the substitution of relatively simple consumer goods, with production of capital goods coming in a later stage: it followed the opposite sequencing. Stalin put absolute priority on producing the means of production—heavy industry, the extraction of natural resources, and capital goods—at the expense of the production of consumer goods. The repressive political context allowed imposing severe material sacrifices on the population, with the promise that this order of priorities was necessary only temporarily: eventually, communism would lead to a luminous future with an abundance of consumer goods.

At the time, as Hobsbawm (1995: 96) points out, the early, breath-taking achievements of Soviet industrialization gave this promise credibility. After little more than a decade since the proclamation of the 1929 five-year-plan, the Soviet industry was able to supply the military hardware capable of defeating the invading armies of Nazi-Germany. For its proponents, this 'success story' proved the superiority of the Marxist-Leninist model; for its opponents, the impression was no less strong, as it instilled fears that eventually led to the major restructuring of Western capitalism along Keynesian lines.

In 1949, mainland China joined the ranks of socialist nations. Many more Third World countries would follow in the coming decades. For nations which had suffered colonial domination or post-colonial dependency, the adoption of some form of state-socialist model was attractive for a number of reasons. First and foremost, the failure of capitalism to provide for socially acceptable forms of development in the periphery made the quest for alternatives urgent. State-socialism not only offered a path to late development, it specifically provided mechanisms to overcome foreign-dominated economic structures and to consolidate the power of the new regimes by expropriating the property of transnational companies and local elites. This form of 'original socialist accumulation,' fueled by the partial dissociation from the capitalist world market and the elimination of market mechanisms, allowed the state to accumulate enormous resources as much for redistributive measures as for large-scale economic investments in ambitious development projects and social infrastructure. (Other, less noble purposes often included military build-up and the enrichment of new elites.) In addition, the authoritarian organizational model of the Leninist one-party system provided a formidable instrument for securing internal political power and for legitimating the leaders and liberation movements that had come to the command posts of government as a result of the anti-colonial or anti-capitalist struggles. Moreover, cooperation with the Soviet Union could in varying degrees lead to non-market-based international economic relations, diplomatic support, as well as military aid and protection.[31]

In 1949 the Soviet Union and its Eastern European allies founded the Council of Mutual Economic Assistance (Comecon), which, similar to the regional integration schemes typical of the ISI strategy, was to be essentially an extension of socialism's 'domestic economy.' With the foreign trade monopoly banning outside competition, import-substituting industrialization now was organized at a supra-national level. Industrial exports beyond this wider 'domestic market' to the hard currency sectors of the capitalist world economy remained very limited.

As the ISI strategy did in the South, the state-socialist model of industrialization for a long time provided enviable growth rates,[32] leading to Khrushchev's famous proclamation that the Soviet Union would "catch up and overtake the United States" in industrial production. This quote also best illustrates how important the benchmark production and consumption patterns of the capitalist industrialized countries were for Soviet leaders: the Fordist model, established and pushed forward by the dynamics of the capitalist world market, provided global standards for development. The Soviet Union tried to catch up to these in isolation from the world market in a sort of 'halved Fordism,' which emphatically implemented the rationalization of industrial work, yet which lacked the crucial counterparts of mass production of consumer goods and dynamic mass consumption (cf. Altvater 1991: 18–48).

To sustain its growth rates, the extensive economic model of state-socialism needed a continuous absorption of additional resources. As this became less and less possible, growth rates slowed significantly in the 1970s and beyond. Though this crisis required a substantial reformulation of the system, passing from an extensive to an intensive use of resources, political motivations blocked this resolution, providing political stability in the short term but irreversibly undermining the economic foundations of the Soviet state-socialist project.

The variations in Third World state-socialist models, between countries and over time, are too complex to discuss in depth. At times, the dividing line blurred between 'classic' state-socialist systems and mixed models, such as Sandinista Nicaragua, and more traditional forms of authoritarian rule, as in a number of African countries. Additionally, relations with the Soviet Union and Eastern European countries ranged from open conflict, as in the Chinese case, to very close cooperation, as in the case of the three Third World countries that were formally incorporated into Comecon:

Mongolia in 1962, Cuba in 1972, and Vietnam in 1978. In all three of these cases, integration rested more on foreign policy considerations than the economic interests of the Soviet Union. In fact, in all three cases—most notably Cuba—the Soviet Union incurred in high economic costs in its bilateral relations. The Soviet Union's motivation was not the colonial-style extraction of economic resources from Third World countries, but their symbolic and geo-strategic value, including allegiance in international politics.

Although the Comecon's division of labor largely gave Vietnam and Cuba the role of agricultural exporters, industrialization did remain the central development goal. The substantial flow of effective subsidies—whether highly preferential prices or super-soft credits—allowed for ambitious industrialization projects, the continuation of the extensive growth model, and investments in social infrastructure (high health and education standards were landmark achievements). In this way Cuba, led by the charismatic personality of Fidel Castro, became the most celebrated example of the state-socialist model in the Third World. We will look at this case in more detail in the second country study of this book.

THE TECHNOLOGY QUESTION:
A FAREWELL TO AUTONOMY

In the wake of the Second World War, the Soviet Union managed to develop its own nuclear weapons program, securing its position as the world's second super power. In the 1950s, when the Soviet Union successfully launched the first ever orbital satellite, Sputnik, it signaled—to the shock of the Western countries—superiority in a crucial field of high technology. Scientific and technical progress remained high on the agenda of state-socialist development: investments in science and R&D activities were considerable and the system of higher education produced a higher percentage of scientists and engineers per capita than in the Western countries. However, over time all this failed to translate into technological innovation on par with that in the West. Due to the logic of central planning and extensive growth, there was little incentive for technological innovation in the production process. In addition, the sharp distinction between the military-industrial complex and the rest of the economy greatly limited the possibility for technological linkages and cross-fertilization.

The case of the Soviet Union's computer industry illustrates these problems.[33] During World War II and in its aftermath, the Soviet Union pushed for domestic development of information technology. In the 1950s and up to the mid-1960s, Soviet computer technology was not far behind the Western equivalents: a proper line of mainframe computers was designed and, in software, Soviet computers were working with the endogenously developed ALGOL language. However, in contrast to the Brazilian case, where the armed forces pressed for an autonomous technological path in the name of national security, in the USSR it was precisely the military that came to press for a departure from the reliance on domestic development; the military feared missing decisive technological steps that would permit U.S. superiority in a strategic sector.

As a consequence, after 1965 the Soviet Union began to import Western computers, a practice which increasingly became standard. Since the computer industry focused its efforts on acquiring the latest U.S. models to copy these by reverse engineering, the result was a chronic time and quality lag. Moreover, these technology imports did not dynamize the technological innovation process in the USSR itself. If Soviet computer technology could not catch up with the innovations of the military-industrial complex of the United States, this problem was exacerbated where technological innovation resulted from a complex interplay of different actors. In the United States, crucial steps in the development of the first micro-computer, the dominant operating software, the invention of the modem, and much of the impetus for the diffusion of the Internet came from outside of the academic-industrial establishments—but in the USSR, there simply were no spaces for technological innovation outside of the official institutions.

MASS MEDIA IN A MODEL OF VERTICAL COMMUNICATION

Third World governments generally pushed forward vertical one-to-many communication in order to reach out to a nationwide audience—and this applies all the more to the socialist state in the Leninist tradition. Mass media received high political priority as ideological 'transmission belts' between the leadership and the population at large. To guarantee this function, all TV and radio stations, as well as the printed press, were property of the state, the party or affiliated mass organizations, and political controls restricted the content of their programs or publications. Media pluralism was rejected as an instrument for the bourgeoisie: only the workers' control of the media—with the state and party seen as the embodiment of workers' interests—would guarantee media at the service of 'the people.' The state media monopoly also legitimated itself by emphasizing the 'national unity' needed in the ongoing confrontation with foreign aggressors and 'traitors' within. A very limited, legal exception to the state's media monopoly was publications allowed for internal circulation in the churches.

For the state's media monopoly to be effective, ownership and control of domestic media had to be accompanied by effective state control over the national media space. During the Cold War, the external opponents indeed did invest considerable sums in media efforts to 'win the hearts and minds' of the people in the socialist countries, with radio programs such as 'Voice of America' or 'Radio Free Europe' trying to reach over the 'iron curtain.' The socialist state could limit citizens' exposure to outside messages by physically preventing access or reception, through controls at border posts or by jamming the TV or radio transmissions; on a second level, the state threatened the sanctioning of those citizens caught reading / listening to / watching foreign media. Contrary to the classic postulates of totalitarianism, a third alternative could accompany these mechanisms: toleration. The most prominent case of toleration probably is the GDR's grudging acceptance of West German television: although disliked by GDR authorities, it became a feature of ordinary people's daily lives in the 1970s and 1980s.

While we have emphasized the role of state monopoly and vertical structure for mass media in state-socialism, this does not mean that the functions or content of mass media was uniform across countries, time, and forms of media. Nevertheless, while the tolerance for debate and deviancies varied, openly oppositional positions did have had to resort to informal or illegal publications (*samizdat*), which only very rarely reached a mass audience, or to seek a public outside of the nation's boundaries.

THE TELECOMMUNICATIONS REGIME OF STATE-SOCIALISM: A MATTER OF LOW PRIORITY

In contrast to mass media, the telephone system was an issue of low priority in all state-socialist countries. The institutional core of the state-socialist telecommunications regime was a national carrier, part of the administrative system under the authority of a ministry dedicated to communications or postal and telephone services. In public institutions and state enterprises, some slow expansion and modernization took place, but telecommunications were not seen as a vital factor of economic development. Residential telephones for the population at large were deemed a luxury with no significant economic benefits; additionally, in systems dominated by a vertical communication model, the potential political and social implications of a horizontal communication technology like the telephone were regarded suspiciously. In fact, it was an open secret that state control measures could infringe upon the privacy of the postal and telephone services, contributing to self-censorship in the use of these communications technologies.[34]

As a result, overall growth rates of the telephone system in the European socialist countries were about 2–3 percent annually, far lower than in the West (Müller 1991: 54). At the end of the 1980s, Eastern European countries had a telephone density of 11 telephone main lines per 100 inhabitants (Saunders et al 1994: 316),[35] about four times lower than in Western Europe. Not only were residential telephones scarce, but the diffusion of public call offices or coin-box telephones also remained low. The tariff structure did not reflect the perception of the telephone as a luxury item, however, as prices for domestic calls were generally very low and not cost-related. Another trend hidden in the national average data is the fact that telecommunications in state-socialism generally showed a very strong concentration in major urban centers whereas access in rural areas remained extremely poor.[36]

Given the low priority attached to telecommunications, revenues were redistributed to other sectors, resulting in a low rate of investment in modernizing the system and an ever widening technological gap to the West. In the 1980s much of the technological equipment needed for digitization simply was not available in the domestic economy and the hard currency squeeze of the Soviet Union and the Eastern European countries at this time sharply limited the possibilities of importing such equipment. Moreover, the CoCom list restricted Western technology exports, including digital telephone equipment, to socialist countries.[37] In consequence, at the end of the 1980s there were almost no digital lines and hardly any potent fiber optic networks in place; similarly, telefax facilities and mobile telephones were extremely scarce. Waiting lists for private telephone lines often exceeded 10 years.

THE COLLAPSE OF STATE-SOCIALISM—AND THE TRANSFORMATION OF ITS SURVIVORS

When growth rates in the socialist states slowed down during the 1970s, external factors put off the threat of 'crisis or reform': sharply rising oil prices increased the Soviet Union's foreign currency revenues; in addition the state-socialist countries turned to large-scale external financing in the form of hard currency credits from the West. These developments only delayed crisis, however. Foreign currency credits are as much a form of integration into the world market as trade relations and they inevitably require increased exports to the hard currency sector of the world economy for the fulfilment of debt service.

A particularly illustrative case is Poland, whose US$28.9 billion debt was the highest of all state-socialist countries (cf. Altvater et al. 1988: 152). Poland experienced an open debt crisis at the beginning of the 1980s—and it was not by accident that this coincided with the imposition of martial law in 1981. This authoritarian response to the workers' demands, raised by the Solidarnosc movement, provided the political background on which the Jaruzelski government could fulfil its debt service obligations: it achieved a positive balance of payments through a rigid reduction of imports, significantly reducing the supply of consumer goods for the population (Wachs 1987: 247–249). As we know, the resulting loss of political legitimacy was severe.

In the Soviet Union, the exports of natural resources prevented the debt load from becoming excessive. Indeed in the 1970s and 1980s the Soviet Union expanded the export of oil and gas, raw materials and precious metals to the point that they made up no less than 90 percent of Soviet exports to the capitalist world. The Soviet Union's integration into the world market was therefore typical of an underdeveloped country. Given this dependency on external revenues through primary exports, the fall in world oil prices in 1986 hit the Soviet economy severely—just as the government had begun a process of economic restructuring.

At the same time, the state-socialist 'survivor countries'—China, Vietnam and Cuba—have shown that there is no automatic connection between economic globalization and the collapse of state-socialism. In different ways these three countries managed to survive the downfall of the Soviet Union and to adapt to the new global constellation while maintaining a one-party political system under Communist Party rule. In Cuba this process of adjusting to global integration led to a painful economic and social crisis, and was accomplished by a rather ambiguous strategy of limited economic reform (see chapter six). In contrast, China in the late 1970s initiated an economic reform process that has since been maintained and gradually extended, providing extraordinarily high growth rates and culminating in China's admittance to the World Trade Organization in 2001. This process so transformed state-socialism that the very term seems to have become inadequate: the Chinese government began to speak of 'market-socialism.' Yet, as much in China as in Vietnam,[38] the state still plays a very active role in the economy, ranging from continued direct ownership of a large segment of production and services to the overall guidance and regulation of economic activities (Herr 2000).

An essential aspect of China's economic strategy since the late 1970s was its turn towards export-led development. On this basis, complemented by an active exchange-rate policy and protectionist measures, continuous high balance of payments deficits could be avoided and the accumulation of foreign debt could be kept at a relatively very low level. If this was one essential pre-condition for the relative autonomy in which the Chinese strategy developed, another was the sheer size of the country: having almost one fourth of the world's population has been an enormous asset in negotiations with foreign companies.

Charting China's new course in 1977, Deng Xiaoping argued that "the key to achieving modernization is the development of science and technology " (cited in Hobsbawm 1995: 461). This change in attitude embraced telecommunications, which came to be regarded as a strategic priority sector for national development, leading to a new, expansive state-socialist telecommunications regime. Starting with one of the lowest teledensity rates in the world, China's annual growth rates passed 10 percent in 1984–far exceeding those of the state-socialist countries of Eastern Europe and eventually reaching stunning rates of up to 50 percent expansion annually in the 1990s (ITU 2002a; Mueller/Tan 1997: 26f.). Vietnam, though with less international attention, has followed a similar course. In terms of teledensity growth, China and Vietnam rank among the top countries worldwide, with China increasing telephone subscribers from 0.6 percent of the population in 1990 to 32.8 percent in 2002, and Vietnam increasing from 0.1 percent in 1990 to 6.9 percent in 2001 (ITU 2003).

Chapter Two

NICT in Third World Development: Political Issues in a Transformed Telecommunications Regime

In September 1993 the U.S. government launched its National Information Infrastructure (NII) initiative. Though this was a broad and rather unspecific 'agenda for action,' it gave worldwide publicity to the issue of developing the NICT infrastructure, invoking the powerful metaphor of the 'Information Super-Highway.'[1] In the following year, the U.S. government extended the NII concept to an international level, calling for a Global Information Infrastructure (GII) initiative (Gore 1994). This spurred most other countries to develop their own version of a national NICT development strategy: by now, almost all Third World governments can point to some program or plan in this field.

There is no shortage of great promises for the impact of the NICT for development and democracy, and the U.S. sponsored Global Information Infrastructure is a prominent example. It forecasts not only "robust and sustainable economic progress, strong democracies, better solutions to global and local environmental challenges, improved health care," but nothing less than a "new Athenean Age of democracy forged in the fora the GII will create" (Gore 1994). Beyond such rhetorical excesses, the questions such discourse raises are valid: do the NICT foster development? Do they foster democracy? Do they advance social participation? However, put in this form of a yes-or-no-alternative, the question in a way ends up asking if the new technologies are 'good' or 'bad'—such a question will hardly lead to useful answers. A retooled question is certainly more meaningful: under what conditions and through which forms of use and regulation can the NICT contribute to what goals of development and democratization?

To outline the contours of the political debates and conflicts over the NICT in Third World development, it is essential to consider the broader context of economic and political transformations these countries have undergone. In particular, the paradigmatic shift to a liberalized telecommunications regime and the implementation of such a regime in the Third World will be critically analyzed since these processes are a central element of the conditions under which the introduction, use, diffusion and regulation of the NICT occur in any given country.

THE CRISIS OF THE FORDIST TELECOMMUNICATIONS REGIME AND THE NEW PARADIGM OF LIBERALIZED TELECOMMUNICATIONS

At the core of the Fordist telecommunications regime in the capitalist industrialized countries had been the conception of telecommunications as a 'natural monopoly': an industry that had such extraordinarily high sunk capital investments, with only long-term amortization, and such a high degree of economies of scale and scope, that it was considered too costly to have competing providers investing in duplicate infrastructure. Classic liberal economic theory long accepted that, given the specifics of the network-based service offered, any single monopoly provider could offer services more cheaply than competing providers. In consequence, public monopolies took charge of telecommunications, usually under the Ministries of Posts, Telegraphs and Telephones (PTTs). Following the state's commitment to achieve universal service, the pricing scheme generally depended on cross-subsidies from long-distance to local calls, from business to residential users, and from urban to rural areas.

The United States was one of the few countries where telecom services were provided by a private firm, the American Telephone and Telegraph Company (AT&T). It operated essentially on a monopoly basis, but since the passing of the Communications Act of 1934, within a sophisticated regulatory framework overseen by the governmental Federal Communications Commission (FCC), the world's first stand-alone telecommunications regulator. The demise of the Fordist telecommunications regime came with the overall ideological shift towards a more aggressive brand of capitalist regulation under the Presidency of Ronald Reagan: on January 1, 1984, the cornerstone of a new, liberalized telecommunications regime was established when the monopoly of AT&T was broken up, with operations initially divided among 22 local companies and with official sanction for private sector competition (Schiller 2000: 7).

With Great Britain, Japan and Canada soon following suit, by the beginning of 1989 more than half of the world's main lines were operated by private or privately dominated companies (Saunders et al. 1994: 19). Most Western European countries were much more resistant to the paradigmatic shift, but in 1994 the European Union began to accede to the model of liberalized telecommunications when it agreed to open basic telephony, the core service offered by the public national operators, to competition by the year 1998.

The mainstream explanation argues that there were 'objective' reasons for the liberalization of telecommunications, due to the technological innovation process (e.g. Saunders et al. 1994: 19). Particularly in the case of long-distance phone calls, technological innovations were indeed used to circumvent the established national telecom systems. The introduction of microwave technology as an alternative to copper wire connections made competition with the established monopoly attractive, since the latter, due to its scheme of cross-subsidization, maintained high long distance prices, while the cost of providing these services diminished (Raventos 1998: 5). Still later, voice-over-IP technology opened up even more possibilities for competition. However, as important as these developments were, the 'driving force' was not the technological innovation itself, but the interests of companies and individuals in minimizing costs and maximizing profits, and their ability to push for change in the regulatory system.

An example of the interaction of private company interests, government action and technological innovation in the internationalization of the new telecommunications regime is the call-back mechanisms for international phone calls. In the 1990s, when the liberalization drive in the United States had already sufficiently eroded the cross-subsidization scheme and tariffs for long-distance calls were lowered, a number of companies began to offer call-back mechanisms that enabled clients in other countries to use the U.S. phone system for long-distance calls, thus circumventing the national telecom company. Technically, this became possible through the digitization of the telephone network. However, politically decisive was the FCC's authorization of these operations as it refused, against the protest of numerous other countries, to rein in the call back companies (Schiller 2000: 50). In consequence, affected countries began to reduce their long distance rates in order to minimize use of call-back mechanisms.

The liberalization drive doubtlessly channeled an enormous amount of capital into the telecommunications sector. In the 1990s, transnational companies, supported by banks, capital funds and the stock exchange market, poured enormous investments into the booming telecom sector, permitting an accelerated pace in the implementation of the infrastructure needed for the new information and communication technologies. However, this went hand in hand with excessive investments by competing companies, which have created largely redundant networks that for years to come will be so under utilized that some observers have spoken of "the greatest destruction of capital in world history" (Fischbach 2002).

After the euphoria of the 1990s, the boom ended abruptly in the first years of the 21st century. Telecom values on the stock markets plummeted and in 2001 alone, no fewer than 470,000 jobs disappeared in the telecom-sector (ITU 2002a: 59). WorldCom, the owner of the world's largest data network infrastructure, which moved over half of the global Internet traffic, declared insolvency in the summer of 2002. Another illustrative case of high profile insolvency involves another major provider of international data networks, Global Crossing. This company had invested US$14 billion since 1997 in the build-up of international data networks; due to worldwide over-capacities their value decreased sharply and, when the company's assets were sold in the insolvency process, the entire network went for merely US$0.25 billion (Financial Times Deutschland, August 12, 2002). In Europe, enormous investments and the undeniably slow start of the new, third generation mobile communication standard, UMTS, are now seen as symptomatic of the bursting telecom bubble.

Parallel to the privatization and liberalization of the existing structures was the creation of new regulatory bodies. These had the task of guaranteeing the establishment of what came to be regarded as fair market conditions in a situation where—given the structures of the 'natural monopoly' noted above—the incumbent company was effectively the sole owner of much infrastructure that it now had to share with its competitors.

Despite the paradigmatic shift to a new telecommunications regime, in which telecommunications services became commercialized services first and foremost, practically all governments have maintained a principal commitment to universal service or universal access to the population at large. Even in the United States, the Telecommunications Act of 1996, while establishing the fundamentals of the liberalized telecom regime, reaffirmed the goal of "preservation and advancement of Universal Service" (FCC 2002). While governments undid the cross-subsidization of the past with relish, they propagated new forms of subsidies. The 'Universal Service Fund' set up by the 1996 Act was established under the direction of the Federal Communications Commission and encompassed three different types of subsidized programs: 1) telephone service discounts to people of low income, either through reduced rates on a limited number of calls or through the reduction in initial connection charges; 2) cost reductions for schools, libraries, and rural health care; and 3) aid to telecommunications companies to compensate for high-cost provision of services in rural and insular areas, where installation costs are disproportionately high (FCC 2002).[2]

In the United States, the newly established 'Universal Service Fund' depends on contributions, based on a percentage of revenues, from telecommunications providers. Many critics have claimed that these compensatory mechanisms, in the United States as in other countries, are far too minimal to prevent a widening of the social and regional disparities. Moreover, these new subsidization schemes illustrate a fundamental change: in the Fordist telecommunications regime, the state's social commitment was interwoven with the economic and administrative logic of the telecommunications operators. In contrast, the new universal service mechanisms are a sort of social welfare program financed through a minor 'taxation' measure, while the overall development of telecommunications is left to market forces—and these are, obviously, favoring those social sectors with the highest effective demand: business, high-income population, urban centers.

THE GLOBAL GOVERNANCE STRUCTURES OF THE NEW TELECOMMUNICATIONS REGIME

A major reorientation of the international institutions and global governance structures concerning telecommunications accompanied the shift from the Fordist telecommunications regime to the neo-liberal paradigm. Amongst these institutions, the International Telecommunication Union (ITU) stands out as the oldest intergovernmental organization in the world, dating to 1865–long before international regimes for world trade, environment or political issues became topics on the public agenda. It is the only specialized agency of the United Nations system devoted exclusively to the telecommunications sector. With 189 member countries, it embraces virtually all states in the world. Its essential tasks stem from the transnational character inherent in telecommunications: setting common technological and operational standards and guaranteeing interconnectivity, facilitating agreements on the sharing of tariffs for international communications, allocating the use of public goods such as radio frequency spectrum or orbital slots for satellites, etc. (Ó Siochrú/Girard 2002: 35–50; see also www.itu.org).

The ITU also maintains an extensive research department and, since all member states report to the ITU, it is the prime source for official telecommunications data. As organizer of the World Summit on the Information Society in 2003, it also has received a new level of public attention. While being an UN affiliated institution obliges the ITU to maintain a certain level of respect towards the different approaches and positions amongst its member states, since the late 1980s the ITU has become an influential advocate for the liberalization of telecommunications (cf. ITU 1989 and 1991).

The World Trade Organization (WTO)—in form of its precursor, the General Agreement on Tariffs and Trade (GATT)—began to include the issue of telecommunications liberalization on its agenda with the start of the 'Uruguay round' in 1986. Since then, the WTO—which has been kept outside of the UN system—has steadily expanded its role and is an extremely powerful actor not only in the drive for general trade liberalization, but also in the implementation of the liberalized telecommunications regime.

By 1994, GATT negotiations concluded in the signing of a General Agreement on Trade in Services (GATS), including an elaborate clause dedicated to telecommunications, and an agreement on Trade-Related Aspects of Intellectual Property Rights (TRIPS). In February 1997, the GATS Agreement on Basic Telecommunications (ABT), also known as the World Telecommunication Agreement, secured the commitment of 69 signatory countries to open their markets to domestic and foreign competition in all areas of telecommunications.[3] Amongst the signatories were almost all countries of continental Latin America; Third World countries from Asia and Africa have been much more reluctant.

Though the ABT contains a number of loopholes and limitations—notably, most states refused to include a binding time schedule for liberalization—this protocol requires much greater liberalization than any previous agreement (Ó Siochrú/Girard 2002: 58). It is also highly ambiguous about the pursuit of social goals through universal service / universal access policies; while it acknowledges that "any member state has the right to define the kind of universal service obligation it wishes to maintain," it does put conditions forward:

> "Such obligations will not be regarded as anti-competitive per se, provided they are administered in a transparent, non-discriminatory and competitively neutral manner and are not more burdensome than necessary (. . .)" (WTO 1996).

The term "not more burdensome than necessary" allows ample space for interpretation and intervention against national policy making deemed 'excessive.' For instance, private companies might seek a case against cross-subsidization schemes that go against the practice of purely cost-based tariffs, considering it "more burdensome than necessary" and therefore "anti-competitive." Under the 1997 Agreement this could be deemed a violation of international trade law and the domestic universal service policy might become a case for WTO arbitration, as Hamelink (1999: 5) argues. The author concludes:

> "The document is far more concerned with ensuring that foreign suppliers have access to national markets for telecommunication services than with guaranteeing access by national citizens to those same services."

In recent years, the WTO has gained influence on a number of areas formerly under the competency of the ITU. Similarly, in intellectual property rights, the long-standing World Intellectual Property Organization (WIPO) has lost political weight to the WTO, under which the TRIPS agreement was signed (and which, *grosso modo,* tips the balance in favor of the rights of the owners and of those engaged in trade over those of the creators and the public).[4]

Other actors with great weight in the issue of telecommunications liberalization in Third World countries are the World Bank and the International Monetary Fund. Both have been crucial in the implementation of privatization policies within the context of 'structural adjustment programs.' Moreover, both have employed political and academic discussion, through high level conferences and publications, to push for telecom liberalization.[5]

There are, of course, a wide number of other institutions involved in the international governance of the telecommunications and NICT sector. Here, however, we shall note only one more, central institution: the Internet Corporation for Assigned Names and Numbers, ICANN. The organizational core of the Internet is the Domain Name System (DNS), which assigns specific number sequences and names to all e-mail and website addresses, giving them a unique, unmistakable identification that is indispensable for communication through worldwide computer networks. The Internet Assigned Numbers Authority (IANA), as the predecessor of ICANN was called, was set up to care for the allocation of these Internet Protocol addresses and to be the overseeing body for the resolution of conflicts concerning these.

Under the present system, attractive Internet names tend to be a scarce resource which may be valued highly. While there are generic top level domains such as '.com' or '.org,' which are open to companies or residents from anywhere and which a private company (VeriSign, under oversight of ICANN) administers, ICANN itself has assigned to each country a top level domain name consisting of a two-letter national code, such as '.mx' for Mexico, or '.de' for Germany. This becomes a political issue precisely in the case of national liberation or separatist movements and countries without full international recognition, for example when the '.pl' domain was officially transferred, though the corresponding country—Palestine—still does not exist as a sovereign state. ICANN also designates in each country one authority for the administration of the domain name system under the respective country code.

In contrast to other international governance structures, ICANN is not an intergovernmental organization, but has been since its inception a private, though non-profit, corporation under U.S. corporate law and linked, via a 'memorandum of understanding,' to the U.S. Department of Commerce. The board of directors consists of functional representatives from the sector. Recent efforts to have at least a minority of its board of directors elected by Internet users worldwide have so far been an unconvincing step towards meaningful institutional democratization of ICANN (cf. Ruggiero 2002, Ó Siochrú/Girard 2002: 99–116, Hofmann 2000, Kleinwächter 2001).

It has become standard wisdom to diagnose a shift of competencies from the national level to supra-national governance structures. However, there is also a less publicized yet considerable shift of competencies from some international organizations to others: a number of traditional organizations in the sector and many of those in the United Nations system that have been relatively open to discussion of the socio-cultural dimensions of information and communication technologies have lost influence. Those that are more closely associated with commercial and economic interests, above all the World Trade Organization, with its stated objective of freeing market forces, have gained international leverage.

The processes of the erosion of the nation-state's competencies as described above should not be misunderstood to imply 'the end of the nation-state,' as the catchy titles of some books suggest (Ohmae 1995, Guehenno 1995). Rather than disappearing, the character of the nation-state and its form of international articulation are transforming. In this process, international organizations and global governance structures are gaining importance, but the nation-state nevertheless remains the fundamental unit of political organization. In fact, most of the international organizations that are reducing the sovereignty of the nation-state are themselves 'inter-nation-state arrangements' and would be unthinkable without a foundation in the nation-state system.

PRIVATIZATION AND LIBERALIZATION OF TELECOMMUNICATIONS IN THE THIRD WORLD

The issue of telecommunications reform in the Third World largely came as part of a much broader process of restructuring state and economy in the context of the 'structural adjustment programs' negotiated with the IMF and the World Bank (Petrazzini 1995:5; ITU 1991). Therefore, the politics

of telecommunications restructuring stemmed from the overall political constellation produced by these programs. Telecom privatization was only one
issue among many, and it certainly was not always the most important one.
On the national level, the decisive actors in the drive for privatization were
the governments, obliged to fulfil commitments to international financial institutions in light of desperate balance-of-payments needs.

In Latin America, it was the military government of General Pinochet
in Chile that became the first to implement the new liberalized telecommunications regime, with the sale of the twin national telecommunications
companies CTC and Entel in 1987 and 1988. Yet although the new paradigm was promoted worldwide, by the count of the ITU (2002a: 4f), in 2002
no fewer than 88 countries had neither fully nor partially privatized their incumbent telecommunications operator. This was due primarily to the resistance of African and Asian countries whereas, as the ITU (2000:1)
emphatically stated, "no region of the world has embraced the privatization
of telecommunications as enthusiastically as Latin America."[6]

This "enthusiastic" support for privatization in Latin America, however, has a simple explanation: Latin America was the continent with the
heaviest and most pressing foreign debt load. In most cases, privatization of
the state-owned telecommunications company was not the clear result of a
policy choice favoring a particular model of telecom organization, but was
rather due to the fact that "a decade of economic crises during the 1980s left
many Latin governments strapped for cash and unable to resist the demands
of financial institutions." (ITU 2000: 2). The ITU makes this argument in as-
tonishingly clear terms: " . . . [these crises] led to further international loans
being dependent upon privatization" (ibid). If this "turned out to be most
lucrative, raising more than US$40 billion for governments of the region"
(ibid), it must be added that a large share of these revenues immediately
went into servicing the foreign debt. Argentina, which in 1989 became the
second country on the continent to privatize its telecom company, presents
an illustrative case: the company was sold solely for debt certificates
(Schvarzer 2000: 13). Privatization of telecoms was therefore not only lucrative for governments, but also for the creditors of the highly indebted countries—who could recover capital otherwise hopelessly lost.

At this point, we must distinguish between the process of liberalization
and that of privatization. Privatization refers to the transfer of state-owned
companies or other productive assets of the state to total or partial private
ownership; liberalization refers to the admission or fostering of competition
between different enterprises in a specific market in which previously competition had been banned or limited. These processes are neither identical nor

always concomitant. To the contrary, as the ITU had to note in looking back on the 1990s, in most Latin American countries "privatization came hand in hand with a lack of competition in basic services" (ITU 2000: 3). The need for maximizing foreign currency revenues overruled other considerations:

> "In most of the countries of the region, the sale of state-owned companies was tied to a period of exclusivity. (. . .) What resulted, however, was merely the replacement of a public monopoly by a private one" (ITU 2000: 2f.)

This outcome has been identified as the "typical Latin American model of privatization" (Raventos 1998: 1). Though public discourse presented the two processes as Siamese twins, in practice in indebted Third World countries, privatization was the priority, not liberalization.[7]

This makes it important to consider a third dimension of telecom restructuring, besides privatization and liberalization: the corporatization of the established telecommunications enterprise. In this scheme, public ownership persists, but the telecom operator is transformed from an administrative structure into a public corporation organized under company law. In this capacity it may have more flexibility in its business contracts, a more commercially oriented approach in the provision of its services, and more autonomy in the reinvestment of its earnings. This option certainly has not been the preferred model of the neo-liberal agenda and it is rather neglected in the international debate, but it is the one that most Asian and African countries have implemented (cf. ITU 2002d). Some see this process as merely an intermediate solution, a stepping-stone in a gradual process leading towards further privatization and liberalization;[8] others, however, support it as a viable long-term alternative to the dominant neo-liberal paradigm, injecting the sector with more efficiency, but maintaining the fundamentally public character of telecommunications.

The public discourse of the liberalization drive has highlighted the lack of efficiency and the poor performance of the state-owned telecommunications monopolies. Even though these aspects were not, as we have stressed, the principal motor for telecom privatization, the shortcomings of the existing telecommunications regime did greatly assist the political process of dismantling the system and implementing the new liberalized model. It was the often year-long waiting lists for main line telephone installation that became the landmark illustration of the state-companies' inability to meet the demand. It is precisely because of the failed 'halved version' of the Fordist telecommunications regime in the Third World, which did not come close to fulfilling the promise of universal service (or at least good service to a relatively large part of the population), that privatization and liberalization could be presented as being in the popular interest. Politically, this argument was crucial to reducing public opposition to telecommunications liberalization.

For quite some time, the experience of the state-monopolies' deficiencies overshadowed the neo-liberal model's shortcomings, stemming from its focus on consumer demand (endowed with purchasing power) rather than citizens' needs. Indeed, problems such as waiting lists for telephone installation largely have disappeared. However, the crucial social issue now is represented, as is typical of markets, in the form of prices. And here, the elimination of cross-subsidies and the 'rebalancing' of tariffs has led to a considerable rise in the rates for domestic calls. Again, the conclusion drawn by the ITU (2000: 3) is sobering:

> "The glitter of privatization has faded. (. . .) Latin America still faces the hard fact that not much more than one-third of the region's households have a fixed telephone. Even though a large proportion of the population is still without a phone, according to conventional statistics, there is hardly any unmet demand for telecommunications services in the region. The main reason for this is that local access prices (monthly subscription) have generally risen, excluding a great part of the population from the market."

In the course of the privatization of telecommunications, the governments established new regulatory bodies (cf. ITU 2000, ch. 5; ITU 2002d). However, just as the Fordist regime translated poorly into a 'halved version' in the Third World, the universal service funds of the neo-liberal model, already inadequate in the industrialized countries, tend to be poor copies of their First World models. The discrepancy between the limited funds available and the size of the task of universal service is striking; in addition, parallel to the declining confidence in the political elite and the widespread experience of corruption, public trust in the administration of these funds is often low.

THE ROLE OF THE NICT IN TELECOMMUNICATIONS LIBERALIZATION

In the First World, in the mid-1980s the principal challenge in telecommunications became the integration of the new information and communication technologies. This task required high investments in the digitization of the network, along with other necessary infrastructure. In Third World countries, in contrast, the telephone network itself still had a very limited reach when the NICT entered the agenda.

When the quest for telecom privatization reached the Third World, even most in the upper and middle classes focused mainly on getting a telephone line and reliable telephone communication. E-mail and the Internet were still widely unknown. Public attention in the privatization process therefore settled primarily on the consequences for traditional main line telephony, the effects on employment and on national sovereignty, etc., and not the potential implications for NICT development. Even as this changed over time, with mobile telephony and Internet services entering public debate in the second half of the 1990s, they were not the most important topics. Nevertheless, the outcome of the restructuring of the telecommunications sector had major consequences for NICT development in these countries, since it determined the basic framework in which the technologies would be used and developed.

In most countries, state telecommunications monopolies seemed to be ill prepared for the new technologies. Computer networking and e-mail use had been pioneered by academic circles, not the telecom operating companies, which were often slow at taking up these new activities. This in part was due to the inefficiencies and the low innovation potential of the often highly bureaucratized state companies, but also occurred because of chronic undercapitalization. Given the inability of the state-owned carriers to meet the demand for traditional main line telephony, many felt it unlikely that the state companies could successfully shoulder the additional investments needed for introducing the new technologies. These investments would require that governments stop siphoning off a share of telecom revenues to cover other state expenditures—and governments were reluctant to take this step given acute hard currency and fiscal shortages. This situation meant that, politically as well as economically, the easiest short-term solution was to sell the state-owned telephone carrier and auction off licenses for mobile telephony and Internet services, in order to both generate instant state revenues and shift the burden of investment to new owners.

Whereas in main line telephony the tendency for privatization to turn state monopolies into private monopolies was strong, this was very much less so in the case of the new technologies: mobile telephony as well as Internet services show a much greater degree of effective competition (ITU 2002a: 4). [9] Even in many of those Third World countries where main line telephony was not liberalized, the new services still were opened up to private companies, typically through the auctioning of licenses (ibid).

MOBILE TELEPHONY: THE MISSING LINK?

ITU's 2002 World Telecommunication Development Report celebrates the benefits of mobile telephony for Third World countries. In reference to the Maitland Commission's classic 'Missing Link' report from 1984, which underscored the importance of providing the people in Third World countries with telephone access, the 2002 report claims in its chapter two title: "We found the missing link: It's mobile communications." We will now take a closer look at this bold claim.

As occurred in the main line telephone services, the process of privatization in mobile telephony implied internationalization: the ITU (2000: 14) concludes that ". . . [p]ractically all new cellular market entrants in Latin America are backed by strategic foreign investors." Given the booming expectations for telecommunications in the 1990s, these transnational companies tended to have little problem obtaining significant volumes of capital for investment in Third World telecom markets.

The example of wireless cell phones enabling telephone access for remote communities that were hitherto unconnected to the main line network is a constant in recent development literature outlining the benefits of the NICT for the Third World. Mobile telephony certainly does hold considerable development potential and it particularly promises to extend access to geographic locations, specific professions (e.g. truck drivers) and ways of life (e.g. nomads) that are not readily compatible with fixed-line phones; Chéneau-Luquay (2002: 11) offers an account of the advantages of mobile telephony to the less house-centered way of life of many people in Africa and the Third World in general. Nevertheless, these positive images are somewhat misleading and the process is far more ambiguous.

On the one hand, growth in mobile telephony has been enormous: Up from just 100,000 subscribers in 1990, and 3.5 million in 1995, the number of mobile cellular subscribers in Latin America soared to 102 millions at the end of 2002 (ITU 2003). According to official data (which is less reliable in the case of mobile telephony than with fixed telephony) in almost 100 countries of the world, the number of mobile subscribers has surpassed main line telephones (ITU 2002a: 13f.). In Latin America, these countries include Paraguay, Mexico, Venezuela, Bolivia and Chile (ibid). However, there are downsides to this trend. A crucial negative is that tariffs are much higher for mobile telephony than for fixed lines. In 2000 the average monthly subscription fee for main line telephones in low and lower middle income countries was US$3.4 and US$3.9 respectively (ITU 2002a: A-28, A-29), compared to US$11.1 and US$20.1 respectively for cellular mobile phones (ibid: A-36, A-37). The difference is still more striking when comparing per-unit costs for phone calls. For the same low and lower middle income countries, the price for a three-minute local call via main line telephony averaged US$0.07 and 0.06 respectively; the same call using mobile telephony was 7 times or more higher, at US$0.33 and 0.49 respectively (that is for off-peak hours, for peak-time calls it is US$0.44 and US$0.68 respectively).

In light of these data, it is clear that the enthusiasm for mobile telephony has its price. And when the ITU hails it as a milestone in telecommunications that "mobile [are] overtaking fixed-line networks" (ITU 2002a: 18), it refers to subscriber numbers but not to the volume of telephone usage; when judged by minutes of communication, fixed-line telephony is still clearly dominant. Nevertheless, a significant percentage of low income sectors of society have a mobile and no fixed line telephone in spite of its higher per-unit costs. Because many in the poorer sectors of society cannot meet the financial criteria for subscription services, the introduction of prepaid cards has been especially important to them. Though per-unit costs for prepaid card users generally are even higher than for subscription users, this form of payment proved more suitable for people without bank accounts or with irregular income (cf. ITU 2002a: 14f). As a result, for users with low income, the costs for telephone use tend to constitute an extraordinarily high burden on their personal budget; if they are to avoid this, they can make only much reduced or only passive use of their mobile phones. Moreover, though these cases of 'mobile phones in the favelas' and other examples have been prominently publicized, in most countries the typical mobile phone user is rather well-off economically, lives in the capital or another urban center and uses the mobile phone not instead but in addition to his main line access.

As in basic telephony, private companies targeted investments in the new services towards those sectors that promised maximum revenues, meaning they concentrated socially on business users and upper and middle class sectors of society, and geographically on major metropolitan areas. When Brazil opened up its mobile telephony monopoly and auctioned ten regional licenses, the São Paulo license alone raised a staggering US$2.5 billion—whereas the Amazon region, where mobile telephone service was most needed simply for basic access (due to the extremely poor main line network), did not meet even the minimum bid (ITU 2000: 10). In order to achieve developmental goals of social and geographical diffusion of telephony that run contrary to private companies' profit expectations, a number of countries have required that mobile telephony operators contribute to compensation funds.

Given these concerns about the extent of services, it is remarkable, how easily the ITU's 2002 World Telecommunication Development Report throws out prior social goals:

> "The Maitland Report ended with a plea that all of humanity be brought into the reach of a telephone by the end of the century. That rather abstract target (. . .) has become a bit outdated, now that we have mobile phones and the Internet."

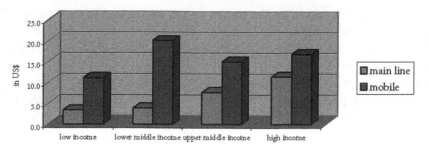

Figure 4 Comparing Tariffs for Main Line and Mobile Telephony: Monthly Subscription Rates, World Countries by Income Groups, 2000

Note: Main line tariff is for residential users

Source: ITU (2002a)

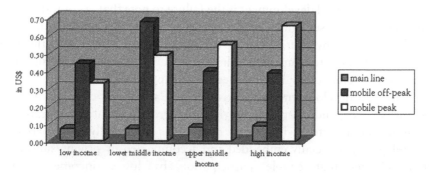

Figure 5 Comparing Tariffs for Main Line and Mobile Telephony: Local 3-Minute Call, World Countries by Income Groups, 2000

Note: For main line calls no distinction between peak and off-peak hours is provided.

Source: ITU 2002a

The report continues by reformulating the coverage target in such a peculiar ways that the notion of 'universal access' loses all its meaning:

> "Mobile is the largest telecommunication network in many countries, particularly lower income nations. It seems appropriate that it be included in universal access determination. Furthermore, mobile cellular has the added feature that accessibility to the network is easy to measure. It would be defined as the percentage of the population within the reach of a terrestrial mobile cellular signal, regardless of whether they are subscribers."

This definition, however, is simply absurd. Someone who sees cars pass by on the highway does not have access to automobile use. Someone without a television set does not share in the use of this technology only because TV transmission signals pass over his home. The same is obviously true for mobile telephony. Yet the ITU's new formula for mobile telephony 'access' is no minor footnote or mistaken wording: it is one of the central messages of the 2002 report. It sets as a high-priority development goal to be achieved by 2006 "mobile population coverage > 90 percent" in lower-middle and low income economies, reiterating that "coverage refers to the ability to receive terrestrial mobile cellular signals" (ibid: 20). By this redefinition, 'universal access' is indeed upon us. However, it will be an insubstantial statistical triumph, with little relation to the Maitland Report's concern that "all of humanity be brought into reach of a telephone by the year 2000." Though this objective was not fulfilled in time, it is certainly not "outdated."

Another misleading aspect of the current euphoria over mobile telephony for Third World development is the failure to note that, as of today, mobile telephones essentially offer one-to-one voice communication. It is fixed-line telephony that opens up the possibility of accessing, through modem and computer connection, the core of the new information and communication technologies: the Internet, e-mail applications etc.

From a development perspective, mobile telephony certainly has the potential to expand telephone service to currently isolated geographical areas and social sectors with specific needs. This is especially true where mobile telephony is socially embedded and accompanied by pricing/subsidization schemes that enable participation by lower income groups. Nevertheless, it is far from being the promised panacea for Third World communications. High costs prevent mobile telephony from effectively reaching out to large sectors of the population, and instead of being a "substitute to fixed-line telephones" (ITU 2002a: 20), mobile telephony is in reality much more a complementary technology for those who already have main line access. The fundamental prerequisite for more meaningful and large-scale NICT diffusion is still fixed-line telephony.

INTERNET SERVICE PROVIDERS: STRUCTURE AND TRANSFORMATION

Most Third World countries, including many that did not privatize their telecommunications companies, have opened Internet service providers (ISP) for private companies. Most of the early pioneers of networking activities in the Third World came from academic institutions or NGOs. These groups built up networks without commercial interest, in a situation where neither the telephone company nor the national law-makers showed much interest. As a result of this early start, and since Internet services require relatively little capital investment and generally have low market entry barriers, initial competition to the incumbent telecom operators in this sector came from many small, local start-up companies often attached to academic institutions or non-governmental organizations (ITU 1999: 13).

In Peru, the *Red Científica Peruana* (Peruvian Science Net, RCP) remained the country's leading ISP throughout the 1990s, in open defiance of the privatized monopoly telecommunications company *Telefónica del Perú* (TdP), owned by the Spanish *Telefónica de España,* which began to compete for the Peruvian ISP market in 1996.[10] In Mexico, more than 30 percent of users accessed the Internet through schools or other educational institutions (IDC 1999) in 1999. However, with the diffusion of Internet usage, in most cases the academic networks left the general users' market to commercial providers, returning to their original function of providing network access to the academic sector.

Since the mid-1990s, a wide variety of commercial ISPs have sprung up, leading to estimates of 1,000 ISPs in 2000 in Latin America (ITU 2000: 21). However, behind this large number of ISPs hides a highly concentrated market. The privatized telecommunications incumbent generally holds a large percentage of Internet subscribers: in Chile, for instance, the privatized twin carriers, CTC and ENTEL, hold 95 percent of the Internet access market (ibid). In addition, transnational companies have entered Third World markets, taking over local ISPs and gaining increasing market shares. Thus, predictions for Latin America are that the great number of ISPs "will consolidate to some dozen or so major players via mergers, acquisitions and strategic partnerships" (ibid), and most of the smaller, nationally owned ISPs will disappear, be taken over or survive in small market niches.

A specific sort of competition for ISPs came from the provision of free Internet access by business from other sectors in order to court customers. The Brazilian *Banco Bradesco* started this trend by offering free Internet access to its 8.4 million bank account holders in 1999, but many of these initiatives have been short-lived since the results did not cover the costs. Still, even before these initiatives lost steam, their implications for the social diffusion of Internet use had been quite limited since they did nothing to change the high costs of hardware and telephone line use that mostly restricted residential access to rather well-off middle-class sectors of society. (Web-based 'free-mail' programs, such as those offered by Hotmail or Yahoo, are not competitors for national providers since they only provide a specific application, the e-mail program, but do not provide access to the Internet as such.)

In general, regulations for ISPs tend to be far less clearly spelled out than in the case of fixed-line or mobile telephony. Regulation does have a crucial responsibility, though, at the national level in the system of domain name registration—that is, the decision about structure, registration rights and fees for domain names under the two-letter national code assigned to each country by ICANN. This code signals the presence of the nation-state within 'cyberspace,' and its regulation is entirely within the competencies of nation-state sovereignty. A number of small countries have actually privatized their country's top-level domain, turning it into a prime source of financial revenue. For instance, the small Pacific Island nation of Tuvalu is marketing its country code—the commercially attractive ending '.tv'—through an agreement which gives a private company ('.tv Corporation,' belonging to VeriSign) the exclusive registry rights for all names in the .tv domain. Addresses like www.gay.tv or www.net.tv recently sold for US $100,000 each, according to the company (www.tv). Similarly, the Pacific Island of Niue is commercializing its country domain (".nu," which can be read as "new"), advertising the 'product' with the argument that

> "the .nu Domain offers a robust but low-cost alternative to the .com domain name in an international model which targets end users and the small office/home office market, as well as Internet users in countries with restrictions on the local country code domain name registration process" (www.nunames.nu).

A particularly thorny regulation issue everywhere is Internet telephony (the so-called 'voice-over-IP'), which enables a form of telephone calls over the Internet—with voice input split up in rapidly exchanged data packages. While the quality of this kind of call is inferior, the cost advantage is clear, as any international communication may run through a cheap local connection to the nearest Internet service provider.

While the practical dividing line between telephone service and data transfer is blurred, it remains a key legal question whether voice-over-IP is considered a basic telephone service or merely data transfer over computer networks (ITU 2001). This seemingly semantic distinction actually touches on the fundamental conditions under which privatization of telecommunications were organized. Incumbent monopolies were granted exclusivity periods for telephone services and practically everywhere new entrants had to pay sizeable license fees; in contrast, ISPs offering Internet telephony neither pay licenses for this service nor are they subject to contribution requirements to universal service funds or other regulation schemes dealing with telephone service. In consequence, many countries have banned voice-over-IP or have only legalized it for licensed telephone operators, but not for ISPs. These restrictions, however, are generally difficult to enforce.

In the push for the liberalized telecommunications regime, IP telephony was a prime example in the argument that technological innovations are 'objectively' eroding the model of state-monopoly telecommunications— and thus privatization was imperative. However, IP telephony undermines the privatized telecommunications model just as much, leading governments to the same administrative restrictions regardless of the adopted telecommunications regime.

BEYOND 'TRADITIONAL ACCESS'? CRITICAL NOTES ON TECHNOCENTRIC PROMISES

Insisting on main-line telephony as the basis for extending NICT use may sound conservative. There is certainly no lack of authors in the development debate that emphasize the promises of other technological approaches. For instance, a recent CEPAL study on 'Latin America's path into the digital age' emphatically concludes:

> "Bottom-line is that the access to the network of networks is by far not restricted to a telephone line and a computer. Actually, Latin America's existing e-frastructure [sic] and its economic characteristics disfavor this traditional access alternative, setting the focus on technological enovations [sic], like television set-up devices, digital television or 3G [i.e. third generation mobile phones]. (. . .) Also powerline could prove as a useful alternative to gain some ground" (Hilbert 2001: 26).

However, such a techno-centered optimism is poorly grounded if the goal is to extend access to the NICT to larger parts of the Third World population in the near or mid-term future. Advocates of mobile telephony put a lot of confidence in coming technological innovations that will bring together Internet access and mobile telephones, but this is a distinctly distant perspective. A large share of cell phones in the Third World are still analog (the so-called 'first generation'); digital second generation (or '2G') mobile phones with different standards and protocols[11] are increasingly entering the markets, and SMS (Short Message Service) has become a popular application. WAP (Wireless Application Protocol) services, which would allow these second generation mobile phones to receive at least some Internet content, promised more, but Internet access was so rudimentary that WAP flopped in the industrialized countries—and it is practically irrelevant in the Third World.

More advanced third generation ('3G') mobile phones have not yet made a significant entry in the First World, and the expectations of their usability and impact have dropped considerably, as the debate on the slow introduction of UMTS in Europe shows. Doubtless they are many years from offering a usable response to the challenge of the social diffusion of Internet access in the Third World. Even further away are powerline solutions, which promise to offer Internet access over the electricity net.

A realistic option already in use is Internet access through cable TV, allowing high speed access via cable modem. Here the problem is not technological, but—with all of Latin America totalling 13 million cable TV subscribers (ITU 2000: 23)—social, as the ITU admits:

> "It is unlikely that cable Internet will dramatically extend access. Most cable television subscribers are in the higher-income bracket and are already likely to be subscribing to dial-up Internet" (ibid).

Internet via cable TV, thus, is more likely to improve the quality of access— higher speed and reliability and an 'always on' connection—to a minority that already has access than to expand Internet use to a broader public.

Finally, satellite Internet connection can be highly useful in reaching regions which are isolated from or poorly connected to the telephone network. But although the costs for satellite-based communications have fallen in the last years, they still remain prohibitively high for private and communal users in peripheral areas of Third World countries.

Nevertheless, each of these different technological approaches can be helpful under specific circumstances, particularly if they are used to complement more traditional forms of access. In marginalized rural areas with dispersed settlements, a combination of individual mobile telephones with a community-based, easy-access telecenter connected via satellite may be an adequate solution if building telephone mainline connections is deemed too costly. Alternatively, in a country with extraordinarily high cable TV penetration like Argentina (with figures as high as 60 percent of households— Hilbert 2001: 26), cable-modems could indeed do a lot to extend Internet access. But jumping from these possibilities to an overall emphasis on technological innovation as the principal solution for a wider social diffusion of the NICT is ill-advised. The widely championed wireless access technologies, be they mobile phones or satellite connections, are far better at overcoming geographical barriers than the social and economic obstacles to Internet access in the Third World.

In Latin America, no less than 93 percent of users were accessing the Internet through main line telephony in 1999 (Hilbert 2001: 21). With similar figures in most parts of the Third World, the argument that main line telephony is 'not so important' to the development debate about NICT is simply not convincing. An important reason why the argument has nevertheless become fashionable in the mainstream development discourse is probably that it distracts from the fact that the liberalized telecommunications regime is unlikely to achieve a socially inclusive extension of access via telephone main lines.

PUBLIC ACCESS, MEANINGFUL USE: THE ROLE OF TELECENTERS

If digital technologies are to overcome the economic and social divide, this will not be through pure technological innovation, but rather innovative social and organizational responses. As we noted earlier, the model of the industrialized countries, where private Internet access is conceived predominantly as residential use, is highly exclusionary in the poorer countries of the South, effectively limiting access to the well-off sectors of society. For the social diffusion of NICT use, public access centers are crucial.

These public access centers can have many appearances. The two principal types are: 'Internet cafés,' commercial businesses offering Internet services to the public at rates that allow profitable operation; and so-called 'telecenters,' which normally are non-profit centers for collective use, organized by NGOs or communal structures, and often located in community houses, cultural centers, schools, etc.[12] Both types normally offer the basic use of computers and related hardware such as printers (for word processing or other applications) as much as Internet-based communication services (e-mail, web navigation, chat, etc.). In many cases, services include photocopying, fax, and telephone. Telecenters also generally provide some sort of training and user support, which Internet cafés normally only do to a much lesser extent. From a social development point of view, Internet cafés are often regarded as of limited importance, since they are profit-oriented and tend to seek a clientele with sufficient purchasing power. Nevertheless, in practice, the distinction between both types is often less clear. One of the most successful experiences, the *cabinas públicas* established by the Peruvian Science Net (RCP) in the mid-1990s, is an illustrative example of how an NGO-based telecenter initiative, which was clearly designed from a social development point of view, evolved into a franchising scheme with commercial success—and still had a great social and developmental impact.

Another model that is found in some countries involves telecom or postal companies integrating Internet access into the services available in their offices. In some cases public institutions, typically in local administrative buildings or libraries, offer Internet services for free or at low cost to the public. A different approach is presented by a number of institutions or organizations that offer Internet services only to their clientele or affiliates; typical examples are professional associations and NGOs, which may offer these services to their members, but also computer rooms in public schools or universities which provide access for the institution's students, but not the general public. While the public character of these access centers therefore is limited to a specific target group, they still can play an important role in the social diffusion of NICT use.

All these approaches, however, have to consider that 'access' has a qualitative dimension, as discussed earlier. Connectivity in itself is only a necessary but not sufficient condition for NICT use in development. Much of the social impact public access centers can make will depend on both their social embeddedness and how readily the new technologies can be integrated into already existing social practices. In their criticism of a superficial NICT for development approach, Latin American activists and scholars have designed an alternative three-step scheme for Internet use that goes beyond 'connectivity' to ensure a) equitable access, b) meaningful use, and c) social appropriation of ICT resources (Camacho 2001, Gómez/Martínez 2001).

In this scheme, equitable access means:

> "the ability to connect at a reasonable price and the ready availability of basic training in the use of the tools so that an increasing number of people can use these resources, regardless of their sex, class, religion, language, or race" (Gómez/Martínez 2001).

The concept of 'meaningful use' goes further to describe:

> "the ability to effectively use ICT resources and combine them with other appropriate forms of communication. Meaningful use also includes the possibility of people producing their own content and having access to other useful content in their own language. People make meaningful use of ICTs when they know how to combine Internet resources with community radio, face-to-face meetings, printed materials, and video, among others" (ibid).

Finally, the concept of 'social appropriation' links the use of Internet resources to goals for development:

> "Social appropriation occurs when Internet resources help transform daily life by contributing to the solution of concrete problems. Evidence of appropriation is not found in the use of ICTs, but rather in the changes that they have brought about in the real world. Only when Internet resources become useful tools for transforming everyday life do

ICTs reach their full development potential. The social appropriation of ICTs for development can be demonstrated in a number of ways, such as: by offering better medical information to patients; improving the quality of education through the use of innovative teaching resources; introducing varied, relevant programming into community radio broadcasting; increasing sales of local products in the marketplace; disseminating the results of local research; and coordinating action among diverse groups with common goals" (ibid).

One aspect often overlooked in the discussion of telecenters and public access centers in general is the issue of privacy, which should be considered an essential element of communication rights. Though for reasons of time and space we have not discussed the manifold and serious problems regarding privacy rights in the NICT in due detail in this work,[13] it is important to note that public access centers are specifically sensitive to this issue because of their non-individual character. Depending on the overall circumstances, the practical situation in the center of access, and the purposes of the user, a lack (or perceived lack) of privacy might represent serious limitations on the practical usability of the NICT. With the possibility of others looking over their shoulders, users may feel inhibited in writing an e-mail letter, calling up homepages of oppositional political groups, or searching the web for practical advice on pregnancy or AIDS.

In general terms we could conclude that a) in public access centers minimal standards for privacy in Internet usage should be safeguarded; b) availability of multiple public access centers can limit the individual user's dependency on a single institution providing the service; and c) the more restrictive the local and national environment is, the more problematic the infringement on privacy rights in NICT use in telecenters tends to be.

E-COMMERCE AND EXPORT CLUSTERS: ADAPTING AND PRODUCING NICT IN THIRD WORLD ECONOMIES

In a curious form of dualism, the mainstream development discourse emphatically proclaims the beneficial impact of the NICT while at the same time issuing frank threats that those countries that fail to integrate the new technologies will be utterly left behind internationally. The successful adaptation of NICT thus is framed as an issue of safeguarding the international competitiveness of the country's economy. The following will consider two central aspects of the economic implications of the NICT: the adaptation of the NICT for e-commerce activities, and the possibilities for production of NICT related products through export clusters.

As to e-commerce, expectations have far exceeded any real impact, and the gap between the industrialized countries and the Third World in e-commerce continues to dwarf the gap in Internet user data (UNCTAD 2002: 6). Latin America, for instance, though accounting for more than 6 percent of Internet users, makes up merely 0.5 to 1.0 percent of world e-commerce (UNCTAD 2002: 7, 11; see also UNCTAD 2001). Here, as almost everywhere else in the Third World, the development of e-commerce has lagged due to, in addition to relatively low income levels, the low rate of credit card diffusion and generally poor logistics and fulfilment services. Consumers of all income levels prefer traditional face-to-face shopping over on-line buying where the infrastructure for the distribution of goods is weak and delivery is not considered trustworthy.

As a result, business-to-business (B2B) transactions account for the bulk of Third World e-commerce activities. For Latin America, these reached a volume of US$2.85 billion in 2000, whereas business-to-consumer (B2C) e-commerce remained much lower at an estimated US$724 million (ibid: 11f.)—with Brazil accounting for about 50 percent. B2B transactions are not the norm, however, and the limited demand is not due to a simple lack of entrepreneurial dynamism. In the formal sector of the economy, an estimated 50 to 70 percent of Latin American businesses have Internet access and yet, though e-mail and web services are widely used for business contacts and information gathering, only a minority engage in online business (UNCTAD 2002: 11). It is noteworthy that neither B2B nor B2C was spearheaded by dotcom start-ups in the tradition of Amazon or E-bay; they were instead a function of the traditional big players seeking to optimize their supply chains. The dominant sectors were the transnational companies of the car manufacturing, electronics, chemical and pharmaceutical industries, where business procedures are largely standardized (Altenburg et al. 2002: 64); among the domestic companies, leaders were banks and the large retail-store chains.

Beyond these sectors, e-commerce-applications for Third World countries have also advanced in those sectors exporting or directing their services to First World countries. Tourism is one industry that increasingly employs NICT in advertising and booking, and it is also an example where active involvement by the state and other actors is needed in order to give small and medium enterprises better chances to participate. The national tourism board, for instance, might establish a central web portal through which all businesses of the sector can channel their presentations and services at low cost and with full technical assistance. To avoid the online activities of larger competitors taking market share from smaller domestic companies, similar initiatives should be considered in other sectors of the economy as well.

As is true in case of telecenters, hardware and connectivity alone are a far from sufficient condition for adequate NICT use by small and medium enterprises. At least as important as credits for necessary equipment are context-sensitive training and affordable, accessible technical support programs. And though NICT statistics usually imply that more is better, this is not always so. Given the high pace of technological innovation (and obsolescence), the integration of NICT not only requires an initial investment but also a continued expenditure to update equipment and know-how. While e-mail communication will substitute expensive phone calls, fax or postal transmissions, other direct cost reductions will be the exception rather than the rule with NICT investments. Though they certainly can bring about efficiency gains or even new business opportunities, integrating new technology will not always translate in a direct way to increased earnings that recover the invested capital. If financed through credit, NICT investments will often signify quite an additional debt burden for businesses. As a result, the sheer amount of money invested in NICT may not be decisive—at times, less can be more. What is crucially important is an adequate balance of resources allocated to NICT versus other needs, plus the meaningful use and adequate embedding of the new technologies in existing organizational structures and practices.

One group that has created a strong incentive for better integration of the NICT is the émigré population, typically living in more developed countries. In a number of Third World economies, family remittances have become one of the principal forms of hard currency earnings and Internet-based communication eases personal communication over distance and time, helping to sustain transnationalized family structures. The NICT also provide improved money transfer facilities, and they have enabled some enterprises to find particular market niches in the emigrant community.

A different aspect to the economics of the NICT in Third World countries is the question of what conditions assist these countries in producing NICT related products. As the liberalization paradigm became predominant in the 1990s, discussion deliberately turned away from the conception of 'technological autonomy' as formulated within the ISI strategy. Instead, export-orientation, market liberalization, foreign direct investments and the focus on promoting industrial 'clusters' became cornerstones of the new model's approach to high-tech production.

The dominant role model was the famous Silicon Valley or, in more sober terms, Santa Clara County, 30 miles south of San Francisco. Here, the interaction of successful start-up computer and networking companies with a number of high-level public and private research institutions created a high technology and innovation center, making Silicon Valley the world's premier address for micro-electronics development.[14]

If Silicon Valley became the paradigmatic example,[15] the works of Michael Porter (1990 and 2000), who emphasized that crucial competitive advantages can be created through the formation of industrial clusters at the local level, became the standard theoretical reference in the worldwide diffusion of the strategy of cluster production.[16] Building on and expanding the linkage concept developed by Hirschman (1958), the concept of industrial clusters sees in the spatial and sectoral concentration of firms and other economic or public agents a vital instrument for maximizing forward and backward linkages. Unlike the linkage approach in the ISI model, however, this model emphasized that these industries should be outward-oriented and also rejected infant industry protectionism, advocating world market competition from the very start.

The cluster approach gives particular emphasis to the development of inter-firm and inter-institutional networking structures; the generation of external economies in the creation of specialized technical, organizational and financial services; and the processes of learning and innovation through implicit and explicit coordination. However, in the popularization of the cluster and networking concepts, inter-firm cooperation at times appears in rather idealized colors. It is therefore important to remember that, in a liberalized capitalist economy, the local agglomeration of industries in similar production areas is designed not only to favor the exchange of information, but also, as Porter (1990: 83) unmistakably argues, to intensify competition between these firms, since competition is regarded as the quintessential motor of innovation.

Although within the neo-liberal framework these clusters should be led by private business oriented towards the world-market, the clusters concept underscores the importance of university-industry-government cooperation, meaning an active role for the state and public institutions. This is the background linking the cluster approach to systemic competitiveness approaches, which emphasize the importance of the 'meso-level,' that is, the whole set of public and private institutions and organizations—from business associations through civil society groups to the institutions of the educational system—between the micro-level of individual enterprises and the macro-level of the state and the national economy (Messner 1997). Clusters in this sense serve as mechanisms for transferring essential parts of the question of systemic competitiveness (Eßer et al. 1996) from the national to a regional or sub-regional level.

In a number of Third World countries, regions have emerged with concentrations of NICT producers or related industries: the Indian city of

Bangalore is the most famous example. Still, with the popularization of the cluster concept, the term has become widely used and its precise contours diluted.[17] More important than semantic disputes, however, are the discussion of the developmental and social impact of clusters, and the question of positive linkages and spill-over effects for the region or the nation as a whole. In their comparative study on innovation clusters in Latin America, Bortagaray/Tiffin (2000) highlight equally Costa Rica's *Valle Central* and the Havana area in Cuba, two cases which we will analyze in more detail in the course of the country studies presented in this work.

E-GOVERNMENT: MODERNIZING THE ADMINISTRATIVE SYSTEM

Recent development literature has put quite some emphasis on what has been termed 'e-government,' that is the incorporation of NICT in government and administrative structures. E-government initiatives have been hailed as important factors in the quest for international competitiveness, since the globalization process entails increased pressure for more efficient public administration, especially in relation to the business sector (e.g. Haldenwang 2002: 2, UNDPEPA/ASPA 2002, or the World Bank's e-government website: http://www1.worldbank.org/publicsector/egov). In fact, now most ministries and major public institutions in Third World countries maintain their own websites, administrative processes are becoming increasingly NICT-based, and numerous documents and forms are available in digitized and downloadable form through the Internet.

Although all these initiatives certainly can be helpful, it would be false to assume that e-government initiatives inherently imply an improvement of governmental services. They can certainly reproduce redundancies and inefficiencies, competing competencies or intransparent procedures just as much as traditional media communication can—without sufficient political will to implement a transparent and citizen-centric approach, no new technologies will ever bring about such change. In fact, a major international research program recently concluded that "the approach in most of South America is moving toward prioritizing service delivery to businesses potentially at the expense of individual services to citizens" (UNDPEPA/ASPA 2002: 40).

The same ambiguous relationship exists between the NICT and elections. Brazil has become a worldwide leader in computerizing election processes using network-connected voting machines, which proved highly efficient in the 2002 presidential elections, won by the Workers' Party candidate Luis Inácio 'Lula' da Silva (cf. Rohter 2002; see also German 1999: 54–57). However, Latin America has also witnessed how the NICT can be used to manipulate the electoral process. The suspicious computer black-out in the national vote counting center during the Mexican elections of 1988 was an early case in point; more recent (and technologically more sophisticated) were the Fujimori government's manipulations of the computerized data transmission in the first round of Peru's elections in 2000 (Herzog 2002: 307–309).

How far current e-government initiatives pay off in economic, political or social terms; how far they augment public sector productivity, enhance transparency and increase citizens' satisfaction with public bureaucracies; and how helpful and important they eventually are in advancing overall development goals-all are open questions. On the negative side, e-government initiatives imply considerable capital investments and continuous costs which will have to be diverted from other areas of government expenditures, and a substantial part of these costs will be in hard currency. Where these investments do not translate in increased external revenues, they signify an additional burden on the country's balance of payments situation. Besides financial costs, e-government may entail social costs. As long as inequities in Internet access keep much of the public from accessing public on-line services, government-to-citizens (G2C) initiatives will reach a minority group, mostly well-off and well-educated. To the degree that these NICT-based services are better, faster or more reliable than traditional systems, they may deepen social discriminations rather than help overcome them. Again, the social adequacy of the initiatives is far more crucial for success than the number of administrative procedures digitally available.

CIRCUMVENTING ESTABLISHED MEDIA FILTERS: NICT AS DECENTRALIZED AND TRANSNATIONAL MEDIA

The new information and communication technologies have become important tools of political communication for diverse social and political forces, from guerrilla groups to governments, from human rights activists to lobbying efforts by entrepreneurial associations. The decentralized and transnational characteristics of the NICT have been particularly important to such efforts.

An important element in the use of the NICT as political media is their potential to partially counterbalance established economic and media power structures. Compared to traditional radio, television or print media, only limited material resources are necessary for reaching a worldwide audience. Transnational companies are all over the Internet and many have considerably extended their presence in hitherto 'far away' markets (a good example is *CNN en Español*, whose website has become a central reference for news on Latin America). However, a striking political impact of e-mail, Internet & co. has been the strengthening of voices of those groups that had only limited access to established mass media, and who found in the NICT a platform from which they could evade some of the prevailing 'filters' in traditional mass media.

Although there have been efforts of transnational companies to tilt NICT use towards more vertical forms of communication (such as 'streaming' or 'content-pushing'), the potential for horizontal communication in many of the NICT media applications is very evident. This has not only enabled individuals and civil society organizations to establish and maintain frequent, easy communication over large distances, which has been a boon to practical organization, it has also created considerable potential to circumvent established media and power structures at the national and international level.

A particularly crucial 'filter' on traditional media has been the set of regulations defined by the nation-state. With the NICT, a fundamental contradiction emerges between the inherently cross-border nature of the Internet and the national regulatory framework in place for the media. This issue is not limited to authoritarian regimes, as is often suggested (e.g. Drake et al. 2000): For instance, in Germany the distribution of Nazi propaganda is prohibited by law. Whereas it is quite difficult to obtain a print version of Hitler's 'Mein Kampf,' it only takes a simple *Google* search on the Internet to find the book as a downloadable zip-file, located on servers in one of many countries where distribution of such material does not violate any law.

If the cross-border characteristics of the Internet pose challenges for any country, these are particularly thorny where national sovereignty encompasses an emphatic notion of 'media sovereignty,' which restricts the entry of foreign media according to more or less narrow limitations on acceptable political content. The state monopoly over mass media typical of state-socialist systems represents a particularly 'thorough' version of the concept of 'media sovereignty.'

This is not to say that traditional media had been completely kept out by national borders: books were smuggled across the border, leaflets were dropped from the air, and radio and TV broadcasts reached out to foreign territories. However, the new media based on global computer networks doubtlessly represent a qualitative difference, since they are transnational by their very structure. Here, no extra effort is needed to reach across national borders: to the contrary, once connected to the global network, it takes a major effort to prevent any Internet-based media from entering. In the case studies in this work, the political challenge presented by the transnational and decentralized character of the NICT based media is a much less important issue in the pluralist system of Costa Rica than in state-socialist Cuba, which maintains a state monopoly over all mass media.

Probably the most spectacular and instructive example of political NICT use in the Third World comes from the Zapatista guerrillas and their unexpected uprising in Mexico's Southern state of Chiapas on January 1, 1994. E-mail and Internet enabled the Zapatistas, entrenched in the remote Lacandonian rainforests of Chiapas, to communicate their declarations and other texts to national and worldwide audiences.[18] The result was that they were able to link up to international media and find support from a loose, world-wide 'alliance' of initiatives, united by their protest against globalized neo-liberalism. While the Zapatistas also reached out to Mexico's civil society, the extraordinary international echo was crucial to invalidating the 'traditional' response of the Mexican government—ordering army and police forces to suppress the insurrection while downplaying its size and causes (see Schulz 1998, Huffschmid 2001, and Cleaver 1995).

However, while some have called the Zapatistas "the world's first Internet guerrilla,"[19] it is important to remember that though the Zapatistas used new technologies, these were neither at the core of their project nor do they explain their success. No one would have cared to read any text by a 'Sub-Comandante Marcos' posted on the web, if it weren't for the uprising. Not as many would have kept on reading, if the Zapatistas had not so spectacularly departed from the traditional conception of guerrillas as a revolutionary vanguard determined to take power militarily, instead conceiving themselves as a self-limited movement with a high communicative capacity. Moreover, if they aimed to "seize the word, not the power" (Huffschmid 2001), they used it to call attention to and increase bargaining power for a radical, but ultimately 'reformist' social and political agenda. It was crucial that heterogeneous sectors in Mexico and in the international public opinion could share in an agenda targeting the social plight of the indigenous groups in Mexico; the anti-democratic and repressive practices of Mexico's political system; and the negative social impact of the neo-liberal economic program—the uprising was timed to begin on the day that Mexico's integration into the North American Free Trade Agreement (NAFTA) took effect. Moreover, the broader political context of the uprising was of central importance, since the Zapatistas found a unique 'window of opportunity,' due to the declining social support and international reputation of Mexico's corporatist regime led by the 'Party of the Institutionalized Revolution' (PRI). The use of the NICT, as important as it was for the Zapatistas, was a tool, but it was not the cause of their pubic and political impact. Many other guerrilla groups and social movements would quickly realize that the inclusion of a website and mailing lists in their public relations work alone will not make them any more prominent or successful.

Part II

Latin America's 'Mixed Model': Costa Rica

Chapter Three
The Costa Rican Development Model and Its Telecommunications Regime

Since this study analyzes the implications of the NICT in the broader context of political, economic and social transformations, it is imperative to start by analyzing the principal structures of Costa Rica's development model which provided for an atypically inclusive form of development within the Third World.

For half a century, the political and ideological importance of the 'Costa Rican model' has far outweighed the limited economic significance of this small Central American country, which has a population of barely 3 ½ million people. With its stable political democracy and its relatively inclusive social system, Costa Rica has been strongly highlighted as Latin America's 'mixed model,' and as a 'third way' between the political extremes of right-wing military dictatorships and left-wing revolutionary and socialist projects (e.g. Mesa-Lago 2000). This characterization corresponds with a widespread perception in the country itself: as modern Costa Rica's 'founding father' José 'Pepe' Figueres once put it: "Costa Rica is not a country. (. . .) It is a project. It is an experiment" (cited in: Ernst/Schmidt 1986: 58).

In order to outline the structures and dynamics of the Costa Rican development model, we will divide it into two long periods. The first one includes the foundational phase, the short rule of the first governing *junta* under Figueres from 1948–49, and the ensuing process of consolidation and expansion of the model, ranging from 1950 to the outbreak of the economic and debt crisis in 1982. The period since has seen the crisis and transformation of the model. In this second period, a gradual process of economic liberalization is eroding the structures of the 'old model,' though it falls far short of full-scale implementation of the neo-liberal paradigm so dominant in Latin America. This analysis will include an examination of the social structures resulting from this development path.

Armed with an outline of the political, economic and social contours of the Costa Rican development model, we will then turn to the specific structures of the country's exceptional telecommunications regime and the political struggles over its transformation.

FOUNDATION, CONSOLIDATION AND EVOLUTION OF THE COSTA RICAN MODEL

The foundations of the Costa Rican model were laid in the so-called 'Revolution of 1948,' when a governing *junta* under the leadership of José 'Pepe' Figueres ended up victorious in a short civil war (cf. Bell 1971). It is part of the peculiar story of the Costa Rican case that the social reformist forces under Figueres did not win over an entrenched and reactionary oligarchy, but against a coalition of forces that by themselves had initiated a first phase of social reform. In the context of the political realignments brought about by World War II, after 1940 an alliance of conservative forces, the Catholic church and the Communist party (under the name of *Partido Vanguardia Popular*, PVP) supported the social-Christian government of Rafael Ángel Calderón Guardia; it was under this coalition that the project of social reform got underway. The reform process included the establishment of the public social security system, the *Caja Costarricense de Seguro Social* (CCSS), and the foundation of a public university, progressive labor laws and support for tenants' rights, and the promotion of national industry and support for the rural cooperative movement.

When Calderón Guardia resorted to fraud to win the presidential election of 1948, an armed conflict ensued, which the forces under Figueres quickly won. The changed international context was of indubitable importance: as the World War II alliance between the United States and the Soviet Union gave way to the beginning Cold War, it blew away the international cooperation that had been at the heart of the alliance between the conservatives and the Communists. The United States now favored the emphatically anti-Communist Figueres and his supporters.

After taking power, the *junta* under Figueres not only reaffirmed the social reforms of its predecessors: over its 18 months of rule, it also laid the political and economic foundations of modern Costa Rica. The chosen socio-economic model, often termed "social-democratic" (e.g. Wilson 1998: 81ff), found inspiration in the *cepalista* ideas of a 'developmental state' that, while maintaining a capitalist market system, would take on a central and active role in the nation's economy. Two landmark actions of the *junta* that signaled the new course were the nationalization of the banking system and the move to begin taxing the transnational banana companies. Among a whole series of state-administered or semi-autonomous public institutions that were to become pillars of the new development model were the National Production Council (*Consejo Nacional de Producción,* CNP), which regulated prices for agrarian products and supported peasant production, and the National Coffee Bureau (*Oficina Nacional de Café,* ONC), which was to protect the large number of small and medium-sized producers of Costa Rica's principal export crop from the fluctuations of world market prices.[1] A centerpiece amongst these 'founding institutions' of the Costa Rican development model has been the Costa Rican Electricity Institute (*Instituto Costarricense de Electricidad,* ICE). Created in 1949, the ICE not only was the state monopoly company for energy generation and electricity supply, but also for the nation's telecommunications system—a nearly unique combination that would have important implications for the country's telecommunications regime, as we shall see below.

The adoption of a *cepalista* development model also included efforts to establish import-substituting industries, primarily in consumer products, light industry and agro-industrial sectors. There was notable success in creating an industrial structure, which showed particularly high growth rates in the 1960s. Though foreign direct investment was significant, about 70 percent of industrial production came from domestic companies (Bulmer-Thomas 1987: 191). The extension of the domestic market through regional integration schemes such as the Central American Common Market (*Mercado Común de Centroamérica,* MCCA) was seen as vital to the ISI strategy, but by the end of the 1960s, the MCCA began to decline as a result of the exhaustion of the 'easy ISI' phase, as much as to political tensions between its members.[2]

Costa Rica always based its inward-oriented industrialization projects on the dominance of the two traditional agrarian exports: coffee, the backbone of the economy in most parts of the country, and bananas, produced in export enclaves by transnational companies. Given this division of the economy between agricultural exports and industry, Bulmer-Thomas (1987: 185–190) has spoken of a "hybrid model of industrialization" in Central America. However, as we have seen in chapter one, this hybridism is characteristic not only of the small countries of the Central American isthmus, but of the ISI strategy in general.

In addition to laying out the nation's economic and social development strategy, the Figueres-led *junta* of 1948/49 created the foundations for the country's political system, which have essentially persisted through today. The most spectacular step, enshrined in the new Constitution of 1949, was the abolition of the country's military. This was essential for the long-run viability of Figueres' reform project, since it eliminated precisely that institution to which anti-reformist elements across Latin America resorted when they felt their interests threatened. Although over time some police forces became para-military in nature and took over a number of functions typical for armed forces, these never became internal political actors of any relevance.

The political system that emerged under these conditions was highly successful; since its establishment in 1949, the institutional order has never been seriously disrupted by violence or electoral fraud. Costa Rica has accordingly been an island of political stability in a region that over decades suffered from oligarchic regimes, military dictatorships, weak and semi-authoritarian civilian regimes, and the armed revolutionary struggles which resulted largely from highly exclusionary and anti-democratic social and political conditions. In contrast, a stable two-party-system emerged in Costa Rica, alternating in power in almost every presidential election. Despite this alternation, Costa Rica is a 'two-party-system dominated by one party,' the *Partido Liberación Nacional* (PLN), which formed in 1951 under the leadership of Figueres. This party was the political heir of the founding *junta* and has dominated the political agenda even when its opponents held the presidency. The PLN conceived itself as a party with a social democratic program. The major party in opposition to the PLN has existed in different constellations and under different names, but since 1983 it has been the *Partido Unidad Social Cristiana* (PUSC), which tends to be (moderately) more conservative than the PLN.

In the context of the Cold War, when in Latin America initiatives for social reform were all too often regarded as subversive and 'pro-Communist'—an early example was the violent overthrow of the reformist government of Jacobo Arbenz in Guatemala in 1954–the PLN's deeply entrenched anti-Communism was crucial for the viability of the 'Costa Rican model' of reform. Figueres had not only defeated the Communist party militarily when it sided with Calderón Guardia in the 1948 civil war, but his reformist policies also won over large parts of the Communists' social base (cf. Ernst/Schmidt 1986: 112f, and Booth 1998: 68ff). Though the Communist PVP maintained some influence in the trade unions, neither they nor any other political force to the left of the PLN ever gained sufficient strength to seriously threaten the political or social order.

Precisely because the 'social democratic' project of the PLN had proven very efficient in immunizing the country against the radical left,

Costa Rica became ideologically important for the United States when the socialist course of the Cuban Revolution of 1959 inspired radical movements all across Latin America. John F. Kennedy sought close cooperation with Figueres while launching the 'Alliance for Progress' for Latin America.

While in many countries of the continent the United States did not hesitate, in the name of anti-Communism, to support military dictatorships, Costa Rica turned into the Latin American showcase model for what we might call 'capitalism with a human face.' In support the United States sent substantial financial assistance: the aid channeled through the official US-AID-program alone more than doubled, from US$253 million between 1946 and 1961 to US$499 million in the decade between 1961 to 1972 (Fox/Monge 1999: 70). In significant part enabled by US-AID money, Costa Rica initiated a far-reaching agrarian reform in 1961, passing the Land and Colonization Law and creating the *Instituto de Tierras y Colonización* (ITCO, later renamed to *Instituto de Desarrollo Agrario,* IDA)

CRISIS AND TRANSFORMATION OF THE MODEL: THE 1980s AND '90s

In the course of the 1970s, high world market prices for Costa Rica's principal export crop, coffee, combined with a growing reliance on external credit to ease the increasing exhaustion of the import-substituting industrialization process, the crisis of the Common Central American Market (*Mercado Común de Centroamérica,* MCCA), and other symptoms of economic crisis. In consequence, the accumulated foreign debt almost quadrupled between 1975 and 1978 from US$421 million to US$1,678 million (see table 1). The collapse of international coffee prices in 1978–79 and the rise of international interest rates made crisis inevitable. Per capita growth rates turned negative in 1980. The accumulated foreign debt soared, reaching US$4 billion in 1983, one of the highest per-capita-debts in Latin America. Inflation climbed to two-digit values and hit the 90 percent mark in 1982, when the crisis peaked.[3]

Although the political interpretations may vary, practically all observers agree that 1982 marks a central turning point in Costa Rican development. The largely incoherent crisis management of the Carazo administration (1978–82) was unable to stop the economic decline, negotiations with the IMF stalemated, and the economic crisis hit bottom. It was the newly elected PLN administration under Luis Alberto Monge that at the end of 1982 (relatively early, compared to other Latin American countries) signed the first stabilization agreement with the IMF. This marked the beginning of a long process of structural adjustment programs and liberalization

Table 1 Basic Macroeconomic Indicators of Costa Rica, 1975–1983

	GDP Rate (in percent)	GDP percapita (in percent)	Inflation p.a. in percent	External Debt in US$ million
1975	2.1	−0.4	17.4	421
1976	5.5	2.6	3.5	536
1977	8.9	5.8	4.2	735
1978	6.3	3.2	6.0	1,678
1979	4.9	1.9	9.2	2,109
1980	0.8	−2.11	8.1	2,735
1981	−2.3	−5.13	7.1	3,286
1982	−7.3	−10.0	90.1	3,627
1983	2.9	0.2	32.6	4,181

Source: Mesa-Lago (2000: 508, 520)

policies in which the Costa Rican development model underwent a gradual, but eventually profound transformation.

The structural adjustment programs begun in 1982 essentially aimed at liberalization and increased the market-orientation of the economy. Internally, this encompassed the reduction of state expenditures to lower the budget deficit, thereduction of subsidies for agrarian products, the privatization of state-owned companies (first and foremost the large holding company CODESA), the partial opening of the nationalized banking sector and the unification of the split exchange rates.

The external liberalization of the Costa Rican economy effectively ended import-substituting industrialization, as the new strategy called for export-led development on the basis of non-traditional exports. Its centerpiece in the 1980s was the so-called 'agricultura de cambio' (agriculture of change), which focused on the promotion of new agrarian export products such as ornamental plants and mango fruits, oil palms and spices, pineapples and palm hearts (Altenburg/Hein/Weller 1990). The dynamic growth of the country's tourism industry complemented this shift in the 1990s. The liberalization course, which also included the attraction of foreign direct investment and debt-for-equity-swaps, was reaffirmed by Costa Rica's adherence to the GATT agreement in 1989. Nevertheless, in Costa Rica the structural adjustment policies retained a far more heterodox approach than in most other countries. As Fürst (1990: 190) has argued, this was in part

> "due to the essentially unorthodox character of the political-institutional
> structures of social conflict regulation which has always been dominated
> by day-to-day negotiated compromises and gradual change."

As a consequence, a characteristic of the Costa Rican transformation process was the announcement of agreements, law initiatives or reform projects that were later postponed or only partially implemented. Explaining such a gradualist approach amidst acute internal and external economic crisis requires examining more than domestic factors. To a large extent, such gradualism depended on the resurrection of Costa Rica's role as an ideological 'front-line state' against left-wing revolutionary models in the wake of the Sandinista Revolution in Nicaragua in 1979. To forestall further destabilization of its prime ideological model, Washington exerted its influence on the international monetary institutions to call for moderation and the acceptance of gradualist approaches; direct U.S. aid flows also increased greatly, covering more than 50 percent of Costa Rica's current account deficit between 1982 and 1985 (Fürst 1990: 177).[4] Contrary to the experience of other countries, in Costa Rica U.S. aid aimed less at the acceleration of structural reforms and more at making them socially acceptable, resulting in the anomaly of an adjustment program accompanied by an actual increase in real social expenditures.[5]

The macro-economic data show a clear stabilization after 1982. After three years of negative rates, GDP per capita began to grow again in 1983. Led by the dynamic growth of exports, which surged from US$358 per capita in 1983 to US$1,558 in 1998 (Lizano/Zuñiga 1999: 19), GDP per capita more than doubled in these 15 years, growing from US$1,300 in 1983 to US$2,950 in 1998 (ibid). The external debt remained high, but became far more manageable, with the foreign debt/GDP ratio falling from 120 percent in 1983 to below 30 percent in 1997 (ibid). Inflation fell steadily to a one-digit-level in 1993 (Mesa-Lago 2000: 508). As for the distribution of national income, the increase in social expenditures contributed to an improvement of the Gini-coefficient from 0.47 in 1983 to 0.43 in 1988 (Mesa-Lago 2000: 494), though it did worsen to 0.46 by 1997 (UNDP 2002: 194). Social differences have increased, though less radically than in other countries, straining the social cohesion so central to the Costa Rican development model.

As a consequence of this gradualist, partial, socially cushioned and macro economically quite successful variant of structural adjustment, the Costa Rican model underwent a substantial transformation process but avoided radical ruptures in the country's socio-economic or political fabric.

Many authors in the 1980s discounted the resilience of the Costa Rican model, calling it an anachronism and predicting that the crisis would lead to a radicalization of society (e.g. Ernst/Schmidt 1986). Witness Trejos' (1986: 38) view:

> "Costa Rica is passing through a transition phase from a democratic system based on social consensus to a repressive and polarized society with a government ever more resorting to paramilitary organizations."

Such analyses greatly underestimated the capacity of the system to overcome the economic crisis and deal with the social tensions and conflicts within the established socio-political structures. For the political system of Costa Rica, it was crucially important that the Partido Liberación Nacional did not oppose but rather led the transformation of the country's development strategy all through the administrations of Luis Alberto Monge (1982–1986) and Oscar Arias (1986–1990).[6]

Another key person in the transformation of the Costa Rican model is the country's long-standing Central Bank President, Eduardo Lizano. Lizano, appointed head of the Central Bank by the Monge administration in 1982, has been the most vocal and influential advocate of liberalization policies. While a member of the Partido Liberación Nacional, he never made any bones about his intention to fundamentally transform the Costa Rican development model. He polemically decries the pre-1982 model as *"Gremialismo, Populismo y Paternalismo"* (Lizano 2000: 196f); the word *gremialismo* has no exact translation, but it means precisely the set of institutionalized interests (unions, independent associations, to public institutions etc.), that make up "the essentially unorthodox character of the political-institutional structures of social conflict regulation" of which Fürst has spoken (see above). Instead, Lizano explicitly argues for a "Model of Economic Liberalization," which he considers is on its way in Costa Rica, though still far from its target (cf. Lizano 2000; Lizano/Zuñiga 1999).

When in 1990 the PUSC won the presidential elections, the new strategy was already clearly established and that government and its successors essentially have continued its course. The personal continuity that surfaced in Costa Rican politics in the 1990s was remarkable: The PUSC candidate elected President in 1990 was none other than Rafael Ángel Calderón Fournier, the son of late Rafael Ángel Calderón Guardia, who had lost the 1948 civil war; the PLN candidate who won the 1994 presidential elections was no other than José María Figueres Olsen, the son of modern Costa Rica's 'founding father' José 'Pepe' Figueres. Thus, with the unspectacular interlude of Rafael Ángel Calderón Fournier's four-year-term from 1990–1994, the departure from the state-centered development model established by the Figueres-led *junta* in 1948/49 was not a project driven by its historic opponents, but rather by its 'organic heirs' in the PLN leadership. This helped the Partido Liberación Nacional to maintain its dominant position in Costa Rica's political system all through the crisis and transformation of its socio-economic model. The price the party paid for this was a considerable erosion of its political and ideological profile and a substantial loosening of the ties to its social basis. This was strikingly highlighted in the 2002 elections when a PLN dissident, Otton Solís, broke with the party to run as candidate of his own improvised party, the *Partido Acción Ciudadana,* PAC (Citizen Action Party). Solís, who had served as planning minister in the Arias administration but quit his office in open confrontation with Central Bank President Lizano, campaigned essentially on a platform that could be called 'Old PLN,' strongly opposing the liberalization course taken since 1982 (Solís 2001). In the elections, Solís gained a stunning 26.2 percent of the vote, only 5 percentage points less than the official PLN candidate.

The popular discontent that emerged in the 2002 elections concerns not the PLN alone, but the political establishment in general. In an opinion poll published in September 2000 a clear majority of Costa Ricans supported democracy as their preferred political system, while at the same time an equally clear majority showed discontent with 'the politicians,' sharply criticized the established parties and felt ill-represented by them.[7]

Political disenchantment, however, is not peculiar to Costa Rica, but is a common trend across Latin America—and in relative terms, Costa Rica still is among the few rather positive examples. A continent-wide poll by *Latinobarómetro* in 2001 showed that only in Uruguay (69 percent), Costa Rica (61 percent) and Venezuela (55 percent)—where 'democracy' has been recast in populist terms under Chávez—do a majority of interviewed persons report that they are 'very' or 'fairly' satisfied with the way that democracy works, as opposed to 'not very' satisfied or 'not at all' satisfied (Lagos 2001: 141). Nevertheless, in Costa Rica the profound erosion of support for the political establishment provided the background for the broad and successful social mobilization against the liberalization of the state electricity and telecommunications monopoly in 2000.

The PUSC managed to win the 2002 elections, although the preceding PUSC administration had been widely discredited. This victory was made possible by the great loss of votes by the PLN to its dissident Otton Solís, but also was due to the fact that the PUSC candidate, the well-known TV commentator Abel Pacheco, won the primaries against the candidate favored by the PUSC party leadership. This distance from the party apparatus helped him mobilize a lot of the discontent with the government and the political establishment.

THE SOCIAL STRUCTURE OF THE COSTA RICAN MODEL AND ITS IMPLICATIONS FOR NICT DIFFUSION

National pride in the Costa Rican development model comes from its social achievements; Costa Rica is one of the few non-OECD countries that continuously ranks in the category of 'high human development' in the UNDP's Human Development Index (UNDP 2003: 237). In place of 'halved Fordism' and its strong exclusionary characteristics, Costa Rica represents a Third World model of an incomplete but generally socially inclusive Fordism, albeit with lower income levels than the industrialized countries.

On the basis of the UNDP data, table 2 presents an overview of central social indicators, comparing Costa Rica with other countries of Central and South America; what is evident is Costa Rica's far above average performance in nearly all social, health and educational indicators. The average life expectancy of 77.9 years is not only higher than in the rest of Latin America, it is also above that of nations like Denmark or Portugal (UNDP 2003: 237). Costa Rica is also among Latin America's leaders, together with Chile and Uruguay, in measures like infant and maternal mortality. The country's extraordinary performance is even more evident when compared to the neighboring countries of Central America which, except for Panama, lag far behind in practically all indicators.

At 95.7 percent, adult literacy is very high in Costa Rica, and the share of public education expenses in the government's budget, 22.8 percent, is also much higher than in almost all other countries in the region, providing for relatively high levels of primary and secondary school enrollment. This educational record has been seen for a long time as a vital resource for the country's development, and it becomes an even more relevant factor with the need to integrate NICT into the national economy and development strategy.

It is interesting to note that, at under 60 percent, the degree of urbanization is far lower than levels in the South American countries ranking higher in the Human Development Index. The high public investments for fundamental infrastructure in rural areas, the continuous agrarian reform process and the support programs for peasant production have paid off, in that they have staved off a large-scale rural exodus typical of many Third World countries. At the same time, this geographical distribution implies that the development of the new information and communication technologies also has to follow an active strategy of diffusion to the rural areas if Costa Rica is to avoid excluding large parts of its population from the NICT.

Table 2 Selected Social Indicators of Costa Rica and Other Countries of Central and South America, 2001

Country	Human Development Index Rank	Life expectancy	Infant mortality rate[a]	Maternal mortality ratio[b]	Physicians per 100,000 people	Adult literacy rate[c]	Public education as percent of total gov't expenditure[d]	Urban population (percent of total)	GDP per capita (in US$)
Argentina	34.0	73.9	16	41	263	96.9	11.8	88.3	11,320
Uruguay	40.0	75.0	14	26	375	97.6	*15.5	92.1	8,400
Costa Rica	42.0	77.9	9	29	178	95.7	*22.8	59.5	9,460
Chile	43.0	75.8	10	23	115	95.9	17.5	86.0	9,190
Cuba	52.0	76.5	7	33	590	96.8	15.1	75.5	**5,259
Mexico	55.0	73.1	24	55	130	91.4	22.6	74.6	8,430
Panama	59.0	74.4	19	70	117	92.1	*16.3	56.6	5,750
Brazil	65.0	67.8	31	190	158	87.3	12.9	81.7	7,360
Peru	82.0	69.4	30	270	117	90.2	21.1	73.1	4,850
El Salvador	105.0	70.4	33	120	121	79.2	13.4	61.3	5,260
Bolivia	114.0	63.3	60	390	130	86.0	23.1	62.9	2,300
Honduras	115.0	68.8	31	110	83	75.6	*16.5	53.6	2,830
Guatemala	119.0	65.3	43	190	90	69.2	11.4	40.0	3,821
Nicaragua	121.0	69.1	36	150	61	66.8	13.8	56.5	2,450

[a] per 1,000 live births (2001); [b] per 100,000 live births (1985–2001); [c] percent aged 15 and above; [d] data for 1998–2000

* data from UNDP (2002) since n.d. in UNDP (2003)

** necessarily a rough estimate due to the inconvertibility of the Cuban peso

Source: UNDP (2003) and UNDP (2002)

A third important socio-structural aspect for NICT diffusion is language. In other countries in the region with large indigenous populations, Spanish and Portuguese can still be barriers to access in rural areas. In Costa Rica, indigenous communities maintaining native languages make up only about 11,000 people nationwide. However, even these communities are largely bilingual and on the whole integrated into the national school and health care system. In addition, the relatively high secondary school enrollment, the spin-off-effects from the tourism industry and the largely bilingual (Spanish and English) Afro-Costa Rican community along the Caribbean coast give the country a relatively large percentage of the population with a working knowledge of English, by far the dominant language in NICT content.

While migration frequently catalyzes NICT use, due to the increased communication needs, this factor has rather limited importance for Costa Rica. With 1.8 emigrants per 100 people, Costa Rica's emigration rate is the lowest in Central America and one of the lowest in Latin America (ibid). However, Costa Rica is host to large-scale immigration from Nicaragua: about 300,000 to 340,000 Nicaraguans are estimated to be living permanently in the country (Proyecto Estado de la Nación 2000: 112) and there is a continued flow of seasonal and temporary migrants. These immigrants do constitute a particular NICT user group, whether they seek to transfer remittances or maintain communications with their relatives in their country of origin—though these immigrants have limited effective demand for NICT use because of their often precarious social conditions and the highly deficient NICT infrastructure in Nicaragua.

The more than one million foreign tourists visiting Costa Rica each year make up a completely different type of clientele. In addition, the country has become a recipient of longer term visitors from the developed countries of the North, including students learning Spanish and entire communities of U.S. pensioners who see Costa Rica as a low-cost extension of the 'Sun Belt.' Due to the combination of their high purchasing power and high communication needs, these First World visitors or expatriates are a small but noteworthy clientele for NICT services: the blossoming of Internet cafés in tourist areas is their most visible effect.

THE STATE TELECOMMUNICATIONS MONOPOLY: A CORE INSTITUTION OF NATIONAL DEVELOPMENT

Whereas in the majority of Third World countries telecommunications was institutionally bound together with postal services (following the historical experience of most industrialized countries), the Costa Rican telecommunications regime has been different in that the state monopoly telecommunications carrier was institutionally tied to the state-owned electricity company, the *Instituto Costarricense de Electricidad* (ICE). The ICE, founded by the Figueres-led *junta* in 1949, became one of the most emblematic institutions of the Costa Rican development model. The institution's accomplishments were two-fold: on the one hand, it greatly increased national electricity generation, providing the necessary energy resources for economic growth and the state-led development and industrialization plans; on the other hand, it fulfilled a central social function by vigorously pushing for a geographically and socially inclusive system of electricity provision (ICE 1997a). Especially for Costa Rica's rural areas, connection to the country's electricity network became the symbol of material ascension and national integration.

Besides electricity generation and provision, the ICE also had sole responsibility for all telecommunications services and all uses of the country's radio electric spectrum (ICE 1997b). In telecommunications, too, the institution pushed forcefully for the nationwide extension of the network, following the same pattern of geographical and social inclusion as the power grid. The ICE subsidized rates for telephone access and domestic calls, and engaged in specific outreach programs for rural areas, including the rapid extension of telephone lines to provide nationwide coverage as well as, beginning in 1970,the installation of numerous public call offices on a concessionary basis in small towns and villages. [8]

In contrast to the dominant approach in the Third World through the late 1970s, telecommunications in Costa Rica was not seen as a low priority service or a luxury item for the urban elites. As a result, unlike other Third World countries, where the chronic deficiencies of the telephone system provided a strong argument for the dismantling of the state monopoly, in Costa Rica even outspoken advocates of liberalization like Tacsan (2001) cannot but pay respect to the historic achievements of the ICE:

> "No doubt, it accomplished an extraordinary job of networking the whole country with telephone lines, converting Costa Rica in[to] one of the Latin American countries with the most dense telephone line infrastructure. (. . .) In Costa Rica, one out of five persons has a telephone line which means that virtually every house owns one. The difference is certainly radical if compared to the neighbouring Central American countries where one out of 20 persons has access to a phone line. The data is significant in the sense that it shows the excellent performance of ICE in the provision of basic line communication in Costa Rica."

Even Ricardo Monge, the coordinator of the telecommunications liberalization bill in 2000 that was designed to open up the sector, underscores the extraordinary importance the state-owned energy and telecommunications company has for Costa Rica: "The ICE is the institution *par excellence* for Costa Rican national pride."[9]

MAIN LINE TELEPHONY AS THE CORNERSTONE OF AN INCLUSIVE TELECOMMUNICATIONS REGIME IN A THIRD WORLD COUNTRY

As a result of the policies of the state monopoly carrier outlined above and the high priority given to telecommunication development, Costa Rica enjoys one of the highest and socially most balanced teledensities of all developing countries. Table 3 ranks the teledensity, that is main line telephones per capita, in Latin America and the Caribbean and—while small Caribbean island states lead the statistics—it shows Costa Rica ranking second of all continental Latin American countries, behind only Uruguay. But while Uruguay, with its urbanization ratio of more than 90 percent and particular concentration in the capital, was easy territory for the deployment of telephone lines, Costa Rica faced the far less favorable conditions of a difficult and mountainous geography and a low rate of urbanization. If, as we noted in chapter three, the telecommunications regime in the Third World generally was typical of a 'peripheral Fordism' that failed to include large sectors of the population, then Costa Rica represents a clearly contrasting case. Here, the diffusion and access patterns instead resemble those of the inclusive Fordist telecommunications regime characteristic of the First World, though implemented in the conditions of a poor country.

In the last decade, the Costa Rican telephone net expanded from 478,900 main lines in 1995 to more than 1 million at the end of 2002 (see table 3). Despite the already high teledensity of the country, with the past years' average annual growth rate of 5.8 in the per capita supply of main line telephones, the Costa Rican state monopoly carrier outperformed the privatized telecommunications systems of countries like Argentina or Chile.

Also waiting lists for main line telephones are not a big issue anymore in Costa Rica. While at the end of 1999, the unsatisfied demand amounted to about 33,500 people who were applying for a telephone line and had to wait on average half a year (Mideplan 2000: 3f. and ITU 2002a), the ITU's data for 2002 see the waiting time at between two and three months which is even lower than in Chile and Argentina with their privatized system. This is all the more remarkable since Costa Rica has lower costs for telephone use than most countries. As table 3 shows, the ICE charges monthly subscription rates that are below the Latin American average. With US$0.03 per local call, also its telephone tariffs are among the lowest—and in Costa Rica, it must be noted, any call within the country is considered 'local.'

Table 3 Main Line Telephony, Basic Indicators Latin America and the Caribbean, December 2002 (ranked by telephone main lines per 100 inhabitants)

Country	Population absolute	Telephone main lines absolute	per 100 inhabit.	Annual Growth in percent 1997–2002* main lines per 100 inhabit.	Waiting time average (years)	Tariffs in US$ (2000) monthly subscrip-tion**	local call (3 min.)
Antigua & Barbuda	80,000	38,300	48.78	2.1	n.d.	11.1	0.06
Barbados	270,000	133,000	49.44	3.9	0.21	4.0	0.00
Bahamas	310,000	126,600	40.56	3.6	n.d.	n.d.	n.d.
Grenada	100,000	33,500	31.65	1.7	0.0	8.1	0.09
Uruguay	3,380,000	946,500	27.96	3.6	0.0	8.2	0.17
Costa Rica	4,140,000	1,038,000	25.05	5.8	0.2	4.6	0.03
Trinidad & Tobago	1,300,000	325,100	24.98	5.5	0.6	4.7	0.04
Chile	15,050,000	3,467,000	23.04	4.6	0.3	9.2	0.10
Brazil	173,880,000	38,810,000	22.32	15.9	0.0	6.5	0.03
Argentina	36,600,000	8,009,400	21.88	2.1	0.4	4.3	0.03
Colombia	43,290,000	7,766,000	17.94	5.9	3.2	2.7	0.03
Jamaica	2,620,000	444,400	16.97	0.5	1.2	0.0	0.01
Suriname	450,000	78,700	16.35	1.1	2.2	1.2	0.05
Mexico	101,880,000	14,941,600	14.67	8.6	n.d.	16.8	0.16
Belize	250,000	31,300	12.37	-2.2	n.d.	10.0	0.15
Panama	3,010,000	366,700	12.20	-1.9	n.d.	n.d.	n.d.
Venezuela	25,200,000	2,841,800	11.27	-0.3	n.d.	5.5	0.04
Dom. Rep.	8,710,000	909,000	11.04	-1.6	n.d.	12.1	0.06
Ecuador	12,940,000	1,426,200	11.02	7.9	0.1	6.2	0.03
El Salvador	6,460,000	667,700	10.34	11.2	0.7	8.7	0.07
Guyana	880,000	80,400	9.15	7.0	>10.0	2.6	0.00
Guatemala	12,000,000	846,000	7.05	11.5	n.d.	—	0.00
Bolivia	8,340,000	563,900	6.76	6.5	0.4	1.6	0.09
Peru	26,750,000	1,766,100	6.60	-0.5	1.3	14.5	0.08
Cuba	11,280,000	574,400	5.11	11.1	n.d.	6.3	0.09
Paraguay	5,780,000	273,200	4.73	2.0	n.d.	3.0	0.09
Honduras	6,710,000	322,500	4.81	5.0	>10.0	2.4	0.06
Nicaragua	5,370,000	171,600	3.20	3.0	>10.0	7.0	0.08
Haiti	8,300,000	130,000	1.57	14.4	n.d.	n.d.	n.d.
Latin America total	529,200,000	86,263,100					

Note: Of the Caribbean island states only a selection has been listed.
* Compound Annual Growth Rate (CAGR)
** for residential users
Figures in italics are estimates or refer to previous years; data for Cuba are for end of 2001
Source: ITU (2003)

Despite already high coverage rates, growth in main line telephony is still a principal priority of the Costa Rican telecommunications regime. In 2002 the state carrier ICE signed contracts with four international telecom companies to provide the country with an additional 445.000 main line telephone connections during 2003, with particular emphasis on the zones outside the major urban concentration in the Central Valley around the capital. As the national audit office has approved this deal, in the coming years Costa Rica will likely show one of the highest main line telephone growth rates in Latin America.

NICT DEVELOPMENT IN A 'PRE-LIBERALIZATION REGIME': PIONEERING ACADEMICS AND TRAILING STATE MONOPOLY

Costa Rica certainly can claim to have been the Internet pioneer in Central America. As in many countries, initially academic institutions were the driving force in Internet use and development.[10] The *Universidad de Costa Rica* (UCR) established the country's first international connection in 1990, when it introduced a node of the Bitnet network, which connected via satellite to the United States. In the same year, the so-called *'Huracán'* Project was launched. From Costa Rica, it established a Central America-wide network of academic and civil society institutions, making e-mail and other services available to its members. All extra-regional traffic had to pass through the central server in Costa Rica, which relayed data packages daily via satellite to the United States Two years later, on January 26, 1993, the *Universidad de Costa Rica* established Costa Rica's direct access to the Internet. Parallel to these efforts, the UCR, in cooperation with the Ministry of Science and Technology, created the national science network *Red Nacional de Investigación de Costa Rica* (CRNet). This network linked all major domestic centers of higher education and investigation.[11]

When Costa Rica's CRNet established an Internet connection to Nicaragua, this was the first such connection between any two countries in Latin America; in the following years, the CRNet also led the development of a Latin American university network, the *Red Hemisférica Inter-Universitaria de Información Científica y Tecnológica* (RedHUCyT).

In this initial phase, support from international donor institutions as well as donations of hard and software from transnational companies was essential for the NICT development in the country. The *Universidad de Costa Rica*'s early Bitnet connection was only possible through financing from the Inter-American Development Bank (IDB) and a major equipment donation by IBM. The *Huracán* project became possible through the support of the United Nations Development Program (UNDP) and the Canadian International Development Agency (CIDA). The country's first Internet connection and the CRNet were largely financed through a cooperation of US-AID and the Organization of American States (OAS).

Domestically, one man played a key role in all these developments: Guy de Téramond, the undisputed 'father of the Internet' in Costa Rica. As director of the informatics department at the *Universidad de Costa Rica,* he directed the institution's pioneering work, and he also became founding president of the CRNet. Additionally, Guy de Téramond was the first director of the national Network Information Center (NIC; cf. www.nic.cr), the institution charged with the administration and assignment of domainnames below the country's top level domain '.cr.' He held this post until he became Minister of Science and Technology in 2000. There is no commercial market for domain names under the '.cr' ending, but the Costa Rican NIC is offering domain name registration on a monopoly basis as a public service institution. Since 1996, it charges US$50 annually for private or commercially used addresses, whereas the government, academic, educational, and health addresses are assigned free of charge. [12]

A specific regulatory body for the NICT sector does not exist so far, nor does it exist for traditional telecommunications. Instead, the Regulation Authority for Public Services (*Autoridad Reguladora de Servicios Públicos,* ARESEP) deals with both, along with water and electricity, public transport and garbage collection (www.aresep.go.cr).

Although in the early stages Costa Rican academic institutions pioneered the regional use of the Internet, NICT development in the country since then has lost much of its dynamism. In an interview in early 2000, Guy de Téramond (2000: 84) underscored the ICE's historic accomplishments but went on to criticize it sharply for its failures to adapt to the new technologies:

> "Until 1980 the ICE had been an exemplary institution. (. . .) Ten years ago we asked the ICE to offer Internet services and there was no will to do so. The ICE was no longer the great driving force in telecommunications. The ICE failed to go along with the technological advances, and it never gave much importance to the Internet, and what turned out was a disaster."

Indeed, the early 1980s mark a turning point for the state telecommunications carrier. Since the economic crisis and structural adjustment programs greatly increased the pressure on the state budget, the ICE was obliged to spend much of its revenues on government bonds to finance the public deficit. As a consequence, between 1982 and 1986 the investment volume for the ICE's telecommunications branch sank dramatically from 16 percent to 4 percent of annual expenditures (Monge 2000: 285) and did not recover in the following years. Slowly but steadily the country turned into what can be described as a 'pre-liberalization telecommunications regime': A state monopoly on the defensive, chronically under-capitalized and losing its capacity to invest and modernize, with the resulting deficiencies providing ammunition for arguments to liberalize and privatize the sector.

The institution was still able to expand main line telephony since this required a relatively low level of technological innovation. However, there was little room for the high-cost investments needed for the new technologies, be they mobile telephony or Internet. Failure to invest was not simply due to the fact that the state monopoly company did not recognize the development potential of the Internet, as its critics argue, but was also due to the chronic undercapitalization of the ICE given the drain on its financial resources. Even political advocates of dismantling the state monopoly, like the PUSC parliamentarian Vanessa Castro (2000: 71), recognize this: "The ICE's incapacity to realize big investments results from the fact that its resources are bound to finance the country's internal debt." It is remarkable, however, that she (and many others) does not conclude that the company's financial resources need to be freed, but rather that transnational companies should take over its job.

It was not until October 1994, a year after the CRNet had begun to offer e-mail and other services to its affiliates on a non-profit-basis, that the ICE began to offer commercial Internet services to business users and private clients. From the beginning, the ICE delegated all Internet activities to a subsidiary it fully owned, *Radiográfica Costarricense S. A.* (RACSA), which became the country's monopoly Internet service provider (ISP) for all commercial uses. However, CRNet still remained active beyond the academic sector. When the first network linking central government and administrative institutions was established in July 1995 it was not done via RACSA but through the CRNet. Openly departing from what should have been its proper boundaries, the academic CRNet served for years as provider for many administrative offices and reserved part of its capacities for this purpose. In return, the government paid roughly a third of CRNet's connectivity costs (Krennerich 2000: 49).

The data on Internet users and ISP accounts also reflect the strong role of the academic net and the less dynamic development of the state monopoly as Internet service provider. In 2000 RACSA's company data shows that it served 33,000 accounts nation-wide, fewer than the approximately 35,000 accounts served by the cost-free, non-profit CRNet.[13]

THE LONG ROAD LEADING UP TO TELECOMMUNICATIONS LIBERALIZATION

In Costa Rica, as elsewhere in Latin America, it was the strong pressures on state finances, stemming from the profound economic and debt crisis of the early 1980s, which prompted the push for a major transformation of the country's telecommunications regime and the first initiatives to liberalize the sector. The Monge administration (1982–86) greatly accelerated the drain of financial resources from the ICE in order to cover the growing public deficit. As a consequence, the company began to lose the capacity for carrying out the large investments needed for the digitization and modernization of its infrastructure. The following government, under Oscar Arias (1986–1990), launched the first plan to privatize some of the services offered by the ICE. In line with the gradualism of Costa Rican politics, the plan proposed not a comprehensive sale, but rather the partial sale of company shares and the participation of private companies through strategic alliances. Protests from the unions of the ICE and other groups of the PLN's social base quickly halted this particular initiative.

A second way the Arias government tried to break up the state telecom monopoly focused on mobile telephony. Under the government's guidance, an agency of the Ministry of the Interior, the *Oficina de Control de Radio,* issued a license for the U.S. company Millicom to operate mobile telephony in Costa Rica. The country's audit office, the *Contraloría General de la República,* quickly stepped in to block the arrangement after the ICE's unions began a formal law suit against the license and Costa Rica's supreme constitutional court, the so-called *Sala Cuarta,* declared it the deal unconstitutional. In return, Millicom sued the Costa Rican state before a U.S. court; when this failed, the company completely withdrew from operations in the country (Monge 2000: 286ff.).

In the early 1990s, the ICE saw its earnings from long-distance calls decline as call-back mechanisms began to circumvent the company's monopoly. Although such mechanisms are illegal in Costa Rica, enforcement proved difficult: in 1999, call-back mechanisms were responsible for an estimated share of 22 percent of total long distance calls to the United States (La Nación, 15.3.2001).

A specific aspect in the process of the Costa Rican telecommunications liberalization results from the fact that telecommunications and electricity are bound up in the same institution. The (expensive) long-distance calls subsidized not only local calls, but also the low rates for electricity provision. In the light of this complex scheme of cross-subsidization, one of the first demands on the liberalization agenda calling for business efficiency was the administrative, financial and organizational separation of the two sectors. The Inter-American Development Bank in particular stressed this issue in its dealings with Costa Rica. In addition, the national audit office demanded an end to the cross-subsidization of electricity through telephone earnings. Under these external and internal pressures, the ICE began in 1996 to cut down on its subsidization practices, gradual raising rates for electricity and local calls, and lowering tariffs for long-distance calls, a process continuing all through the 1970s.

In the 1980s, the foreign debt was seen as the crucial economic problem; in the 1990s attention increasingly turned to the state's domestic debt and the resultant fiscal deficit. To deal with this issue, the administration of José María Figueres (1994–98) set up a national commission headed by Eduardo Lizano, the country's most vocal advocate of liberalization. Thus it was not surprising that the solution this commission recommended was essentially the sale of state assets, including the state electricity and telecommunications monopoly (Lizano 1997: 59). This immediately provoked the protest of unions in the ICE.

With this background, the Figueres government launched a liberalization project that, rather than focus on privatization via the (partial) sale of the ICE, instead sought to break up its monopoly and open up the market for private competition. The project encompassed three bills that were jointly introduced in the National Assembly in September 1996: 1) a new 'organic law' concerning the ICE that called for a full separation of the telecommunications and electricity sectors, which would operate as separate companies under the ICE, recast as a rather loose umbrella to be called *Grupo ICE*; 2) a new law for the energy sector; and 3) a new general law for telecommunications. In both sectors, the 1996 draft law opened the ICE to private capital in the form of joint investments. As for the opening of the telecommunications market to private competition, a gradualist schedule proposed stretching the process over five years (Monge 2000: 289f).

These bills, however, never really got political life. They were introduced in the second half of Figueres' term, when considerations for the 1998 elections dominated the calculations of the parliamentarians of both parties—and it was widely felt that the ICE liberalization would be unpopular. Ultimately, both parties opted to avoid conflict, and the initiative died a slow and silent death in parliament. In first dealing with the issue, parliamentarians voted for a commission to analyze the bill within 120 days. Then, in a tacit agreement of both parties, the initial deadline was extended for another 120 days, and another; eventually, the Figueres administration ended without the commission ever producing a report on the bill (ibid: 291).

LIBERALIZATION TAKES SHAPE:
THE '*COMBO DEL ICE*' LAW OF 2000

After winning the 1998 elections, the government of Miguel Angel Rodríguez (PUSC) inherited the pending liberalization bills. The pressure from the external and internal advocates of liberalization mounted quickly, as proponents of liberalization wanted to use the post-electoral setting to push through the reform project that they regarded as long overdue. This time, anticipating the unpopularity of the reform and learning from past failures, the government launched a broad public relations campaign (which could count on significant financial support from the U.S. government and international financial institutions) to prepare the ground for the liberalization initiative (Monge 2000: 294).

A practical example of such support was US-AID's funding of trips, by which politicians, union leaders and journalists were taken to see the benefits of liberalization experiences in other countries. Vanessa Castro, the PUSC parliamentarian most active on the issue, openly acknowledged the impact this initiative made on her:

> "In May 1998 I had a very different view on the state of telecommunications in Costa Rica. I thought everything was wonderful. Today, as a result of my travels to Latin America and Europe, I have a wholly different opinion."[14]

Another one of these agenda-setting efforts was the workshop on 'Costa Rica in the XXI Century—Information Technology for Development' organized by the Inter-American Development Bank in October 1999. Under the guise of an independent expert conference, the workshop outlined a 'National Policy on Information Technology,' which concluded:

> "In the coming 18 months the political agreements have to be consolidated that allow the opening of the telecommunications sector before the next change in government. (. . .) A gradual opening beginning in 2002 would not have the sufficient aggressiveness to turn the vision developed here [in the course of the seminar—B. H.] into reality" (CENAT 1999).

A third example that is characteristic of the approach and tone of the campaign was a study elaborated by the Alvin Toffler Institute in the United States, commissioned by CINDE and financed through US-AID. This text, entitled "Telecommunications—the Key to Development" (Instituto Alvin Toffler 1999), was promoted with press conferences and distributed free of charge to journalists and public actors of all sort. According to the dichotomous view of the study, "Costa Rica is at a crossroads" (as the title of the first chapter proclaimed): either Costa Rica would maintain the state telecommunications monopoly, leading inevitably to technological backwardness and a loss of international competitiveness (Instituto Alvin Toffler 1999: 10), or it would opt for the path of liberalization, which would empower the country to become nothing less than "a leader on the global level" (ibid: 5).

Besides the increased public relations efforts, the government of President Rodríguez found new ways to facilitate the political implementation of the liberalization project.[15] He did present the electricity bill directly to the National Assembly, but prior to submitting the other two bills, concerning telecommunications and the restructuring of the ICE, the administration established a consultation process, the *Foro de Concertación Nacional*. This forum reached a partial agreement, which the government (and the representatives of the governing party, PUSC), business sectors and trade unions all signed. Although the PLN and the small left-wing party *Fuerza Democrática* did not support the agreement, acceptance by unions was seen as an important step towards the viability of the project. The government then introduced both bills into Parliament on the basis of these accords.

The bills sped up the gradualist approach to liberalization, envisaging full competition in the sector as early as 2002. At the same time, it promised to continue the traditional social commitment through the establishment of a universal access fund called FOSUTEL (*Fondo Universal para los Servicios de Telecomunicaciones*) and its new regulatory body, ARETEL. Unlike hidden cross-subsidies, this method would establish transparent forms of subsidization. The financing of the fund would come from obligatory contributions from all companies operating on the telecom market, amounting to 15 percent of their earnings in the first four years and 40 percent for the following years.

In contrast to the process in other Latin American countries and the recommendations of the commission headed by Lizano, this liberalization law did not include the privatization of the state carrier. Instead, the ICE was supposed to stay in the hands of the state, though it would lose its monopoly position. This rejection of privatization, despite fiscal pressures, was clearly a response to the high political costs associated with such a measure, since the ICE, as described above, was still held in high regard as an institution worthy of national pride. Advocates of the liberalization project tried to portray the opening of the sector as a means to modernize, not weaken, the ICE through competition and private capital.

In this sense the law concerning the ICE itself aimed at corporatizing the company, streamlining its administrative and bureaucratic functions and strengthening its business orientation. This approach also allowed strategic alliances with private capital, thus addressing the capital shortage of the ICE not by freeing funds that had been frozen in government bonds, but by gradual 'joint-venturization.' Thus we can in fact speak of a strategy of privatization in the sense that private sector competitors were supposed to drive the ICE out of part of the market; that private capital would play an increasing role within the operation of the state company; and that the ICE would be forced to adapt its structures, services and tariffs in order to compete with private business. Such a 'privatization,' if not by sale, still implies a lot of pressure to follow a commercial business logic that is inherently in conflict with the developmental function of overcoming social disparities.

An additional factor contributing to the decision not to sell the state company is the fact that the Costa Rican liberalization initiative came much later than in most of the continent. As a result, policy makers could take into account the sobering results in many Latin American countries where the privatization often meant the replacement of a public monopoly with a private one (see chapter three). In a market as small as Costa Rica's, it seemed rather doubtful that more than one company really would enter into competition in general telephone services.

The experience of the privatization processes in other countries also affected expectations for telephone service rates. Whereas mainstream arguments hold that liberalization brings lower prices, the Latin American experience casts doubt on this claim. As discussed earlier, prices indeed went down for international calls but rates for monthly subscription and domestic calls went up. Though the pro-liberalization publicity campaign and much of the Costa Rican media focused criticism of the ICE on the state monopoly's shortcomings regarding the new information and communication technologies, these distant issues were not priorities for the majority of the population. What mattered to them was the real social accessibility of telephone service. Given the wide geographical diffusion of the telephone network and the extremely low tariffs for monthly subscription and domestic calls, proponents of liberalization could not promise much improvement on issues that did matter to the populace. Their remaining inducement was a reduction of the waiting list for main line telephones, but at half a year, the waiting list was still within a manageable dimension. Finally, the pro-liberalization forces turned to a fairly unattractive argument: admitting that domestic telephone tariffs would rise with liberalization, but arguing that they would even were the monopoly maintained (e.g. EKA 1999:13).

The governing PUSC did not have a parliamentary majority, so the refusal of the PLN (which supported more gradual liberalization) and the left-wing Fuerza Democrática to sign the national consultation accord led to a new round of negotiations in the Legislative Assembly. Both the PUSC and PLN pushed for rapid progress, since each wanted to finish with the issue before the next long pre-electoral period could hamper implementation. In consequence, in November 1999 the National Assembly decided to modify its procedures to allow for a fast-track approach to the bill.

A complex negotiation process between the two dominant parties ensued, producing an agreement relatively quickly (cf. Asamblea Legislativa 2000). Modifications to the bill essentially affected two dimensions: the time frame for the stepped opening of the telecommunications market to competition, and the composition of the law's new regulatory body, ARETEL. The time frame was extended, while negotiators threw out the agreement of the consultation forum to establish a board for ARETEL with seven members: two from the government, two from business organizations, two from the trade unions and one from the national commission for the protection of consumer rights. The new version proposed a board of five members appointed by the government, since both parties were interested in a clear subordination of the new body to the executive branch of government. As a result, the trade unions vehemently protested their exclusion and withdrew their support for the liberalization law: the parties ruined the greatest success of the consultation process, union support for the liberalization bill.

In December 1999, another decision by the leaders of the two dominant parties in Parliament, who merely intended to accelerate the process, proved to be even more decisive in its final failure. The parliamentary commission that oversaw negotiations decided to bundle the three separate bills—the law restructuring the ICE, the law on energy and the law on telecommunications—into a single project that the media and public called the *Combo del ICE*.[16] The political rationale was that without such a package deal, the three projects could not pass before the regular end of parliamentary sessions in April 2000.

In terms of parliamentary process, fast-track proceedings and bill combination achieved their aim: on March 20, 2000 the National Assembly easily passed the comprehensive reform law of the *Combo del ICE:* all PUSC and all but three PLN representatives voted in favor, while only the three PLN dissidents and the seven representatives of the small parties voted against it.[17] Liberalization had become law—but not for long.

THE FORMATION OF THE ANTI-LIBERALIZATION ALLIANCE

The passage of the *Combo del ICE* law by the Legislative Assembly soon proved to be a pyrrhic victory. The leadership of both parties had obviously underestimated the depth of the popular disenchantment with 'the political class' and, by brushing aside the consultation accords in the National Assembly negotiations, the party leaderships confirmed widespread negative views of 'the politicians.' While the combination of the three laws into one package deal served to accelerate the parliamentary process, it also was essential for rallying very diverse social forces into a heterogeneous alliance of opposition to the *Combo del ICE.*

The least surprising opposition came from the unions of the ICE, whose workers feared the loss of employment and many of the particular benefits they had obtained through the years. The government never succeeded in neutralizing these fears, though it promised the opening of the telecommunications sector would generate more than 6,500 new jobs.[18] Other unions of public employees quickly closed ranks, fearing similar initiatives in their own sectors if the country's liberalization course, for which the *Combo del ICE* law was increasingly symbolic, were pushed forward. Unions from that of the teachers to that of the oil workers staged solidarity strikes and called for members to participate in the anti-*Combo*-demonstrations. With hindsight, what actually might deserve more consideration than the unions' opposition to the project was their initial acceptance of the bill in the consultation process. It may be that the unions' leadership at that point in time had come to see the ICE liberalization as inevitable, since practically all dominant political forces supported it, and so union leaders participated in order to negotiate at least some favorable conditions for unions and the workers in the process.

More surprising than the unions' protests was the strong mobilization of the country's university students against the *Combo* law. These students mainly came from upper or middle class sectors of society, and for them the 'heroic' days of the ICE's role in national development should have been distant history. In addition, since the students certainly have disproportionately high interest in Internet use, they seemed a prime target group for the argument that the state telecommunications monopoly is an obstacle for a more dynamic development of the new information and technologies in the country. However, the majority of the students did not perceive the *Combo* law as a means to improve these services but instead as part of a neo-liberal 'sale of national interests.' Also important in the students' mobilization was the overall disenchantment with the political establishment. Beyond the telecommunications and electrical issues, the *Combo del ICE* law presented a concrete case in which often diffuse frustrations were publicly articulated. In the wake of the students' demonstrations and teacher strikes, many secondary students also joined the street protests.[19] Practically all the ecological associations in the country joined against the *Combo del ICE,* particularly galvanized by a passage in the law that facilitated the building by private companies of new hydroelectric plants, even in nature reserves (article 119).

Another unexpected, but highly important, sector participating in the protests against the liberalization law was the countries' peasantry. Pressured by the fall of world market prices for a number of important agricultural products, peasant organizations in all parts of the country had been mobilizing their bases in order to demand more support from the government's agrarian policy. However, peasants quickly joined the coinciding anti-*Combo* demonstrations in the cities. Particularly in the countryside, appreciation of the ICE's role as a cherished institution of development and social integration was still very much present; also in the countryside, the commercialization and privatization of services posed particularly strong threats to the prevailing developmental and social function of the ICE.

For the peasants, the electricity sector was a more pressing concern than telecommunications. And in electricity, the government had not deemed it necessary either to initiate a social consultation process or to accompany the project with much communicative effort.[20] As a consequence, it remained very much unclear what—if any—benefits the break-up of the ICE electricity monopoly would have for for the population at large; in contrast, the danger that the liberalization would lead to rising tariffs was sufficiently clear. The peasants' principal form of protest was the blocking of important overland roads, a measure that greatly disrupted the country's transportation and many of its economic activities, exerting a strong pressure on the government.

The general loss of confidence in the political establishment also greatly affected the public perception of the universal access fund, FOSU-TEL, that was included in the law. Many saw this as a mere paper-promise, without much assurance that it would be fully realized—whereas the past actions of the ICE vouched for its continued developmental commitment.

Among the political parties, the small left-wing *Fuerza Democrática* gained a high national profile as a political force that strongly supported the protests on the streets and tried to voice the movements' concerns in the formal political arena. However, the 'defense of the ICE' was an issue that drew support from prominent dissidents in the two dominant parties. In the case of the PLN, the former planning minister Ottón Solís led the dissidents, as he broke with the party to form his own *Partido Acción Ciudadana* (which, as we noted earlier, garnered a remarkably high 26.2 percent of the vote in the 2002 presidential elections). Within the governing PUSC, it was ex-President Rodrigo Carazo (1978–1982) who a few years earlier had formed a 'National Committee against the Sale of State Assets' (*Comité Nacional contra la Venta de Activos del Estado*). As the protest gained strength, he led a 'National Citizens' Front' (*Frente Cívico Nacional*) against the liberalization law and was supported by, amongst others, the social-Christian trade union *Confederación de Trabajadores Rerum Novarum*.

The coalition against the liberalization law encompassed groups that loudly articulated their specific interests, but—and this was important for its success—the coalition could count on broad, less visible support from the population at large. Opinion polls have left little doubt that the country's 'silent majority' viewed the demonstrations favorably. A survey conducted in May 2000 showed a solid two-thirds majority of those interviewed supporting the protests against the *Combo* (Unimer-La Nación 2000b). More telling still is that no less than 86.9 percent agreed to the statement that "protests organized by some groups received spontaneous support by the population" and 93.7 percent agreed that "the protests were a message to the PLN and PUSC that the Costa Ricans are tired of what they have done to the country" (ibid). In the political arena,' the country's ombudswoman (*Defensora de los Habitantes*—Defender of the Population), Sandra Piszk, spoke out against the *Combo* law in the name of this 'silent majority, and played an active and prominent role in the conflicts over the ICE's liberalization.

The conflict certainly can be seen as ideological in the sense that "Defending the ICE!" as the main slogan of the anti-*Combo* coalition was about more than just a specific institution. To a great extent, it was a call for "Defending the Costa Rican model!" with a strong economic role and social commitment by the state, and a vote against ongoing erosion of the state's role through liberalization policies that a political elite pushed forward.

ESCALATION OF THE CONFLICT AND WITHDRAWAL OF THE LIBERALIZATION LAW

When the parliamentary majority of PLN and PUSC voted the *Combo del ICE*-package into law on March 20, 2000, the social protests escalated. Massive strikes, demonstrations and road blockades paralyzed the nation. Compared to many other Latin American countries with long records of far more intense and violent social conflict, the anti-*Combo* demonstrations might not have seemed all that spectacular. However, in Costa Rica they were almost incredible. The scope and aggressiveness of these protests reached a level of social conflict that was unknown in the country's recent history. After two weeks of social unrest that showed no signs of subsiding, the government withdrew the law on April 3, 2000.

To save face, the government did not cancel the law altogether, but called for a 60-day freeze to open up a national dialogue. A 'Mixed Commission' (*Comisión Mixta*) was established and a broad spectrum of political and social actors were invited to discuss and reformulate the law bill. However, this was in fact merely a conciliatory measure and no longer a political attempt to produce and implement a transformation of the country's telecommunications regime. Virtually everybody, including the representatives of the government, now distanced themselves from the initial project. Eliseo Vargas, who headed the PUSC delegation in the mixed commission, openly declared that the break-up of the monopoly was no longer option, since "there are political and social realities we have to respect" (La Nación, October 27, 2000).

The political price the government paid for the retreat certainly was high, since the Rodríguez administration had staked a lot, nationally and internationally, on the reform. On the condition of anonymity, one of the President's closest advisors recounted the political dynamics of the law's withdrawal from the executive's perspective.[21] According to this source, in the face of the street protests the President at first remained committed to pushing the law through "no matter what." It was additional pressure from three different directions that finally forced him to give up.

Before delving into the three sources of political pressure that changed the President's calculations, it is important to note that one of the most striking features of the political crisis was the lack of an option for 'restoring order' that likely would have been employed in many other Latin American countries: the use of sufficient repressive force, whether through police or military action. In Costa Rica, the use of the country's police for a large-scale repressive action against non-violent protest with broad public support would have shaken the foundations of the country's political and social system so profoundly that virtually all national actors shied away from this option. At the same time, the government was also well aware of how limited its practical ability would have been to implement such a repressive strategy. In the words of the unnamed Presidential advisor:

> "In any other country the government would declare a state of emergency, and if strikes would endanger the national electricity generation, the army would take over the power plants and guarantee at least a minimum of electricity supply. In Costa Rica, however, if the ICE staff refuses to produce electricity, there simply does not exist any other institution which simply technically would be able to operate a company like the ICE."

In this situation, one source of increasing pressure for the government to back down were business sectors that suffered millions in losses through the ongoing road blockades. This not only affected agricultural products, which rotted due to lack of transport, but also companies such as Intel, whose microprocessor plant (see chapter four) suffered high losses due to the disruption of its electricity supply and the blocked access to the airport. These business sectors now called on the government to restore order. While the business community largely supported the liberalization law in principle, such a long-term change in the telecommunications regime was a far lower priority to them when their concrete earnings were affected. As a consequence, the government lost a lot of support from one of its principal domestic allies in the project.

A second source of pressure on the government came from the President's party itself. When the protests did not, as expected, subside after the passing of the law, fears grew that the conflict might put at risk the foundations of the two-party-system. These 'structural' fears mixed with the more concrete and short-term worries of delegates and party members who, with their eyes on the next elections, saw their personal situation getting increasingly uncomfortable. They recognized that the longer the government kept to its unpopular project, the deeper the rift between the party and the electorate would be.

The third element pressuring the government into retreat came from the institutional structures, namely the judiciary system. About ten days after the law's passing, the word began to spread that the country's highest constitutional court, the *Sala Constitucional de la Corte Suprema de Justicia* (also known in short as *Sala Cuarta)*, would suspend the law due to procedural errors. At this point, even for the government's hardliners who had maintained their position of persevering "no matter what," it no longer made sense to stand by the bill. As the senior Presidential advisor described the situation:

> "The question could not be answered anymore, why we should keep on fighting at a truly high political price when two weeks later the *Sala Cuarta* would turn back the whole process anyway."

Indeed the constitutional court revoked the law package on April 18, less than a month after it was passed. In its judgement it avoided any statement on the content of the law, ruling strictly on the grounds of serious procedural mistakes (cf. Sala Constitucional de la Corte Suprema de Justicia 2000). Nevertheless, the decision has widely been seen as an expression of how deeply the rejection of the law was entrenched in the country's institutional establishment. At the same time, the decision of the constitutional court also helped the government to save some face: rather than cede to 'pressure from the streets,' the government could officially bow to the limits on executive and legislative authority resulting from the democratic separation of powers.

THE NICT ISSUE: LOST IN THE MAELSTROM OF THE *COMBO*

Although, as we noted above, the ICE commanded respect from large sectors of the population, the picture is more complex when its different services are evaluated separately. Polls show an extremely high degree of satisfaction, 90.9 percent of those interviewed regarding its electricity services (Unimer-La Nación 2000a). A still high 72.6 percent expressed satisfaction with residential main line telephony. However, satisfaction rates decreased significantly in regard to the newer technologies, to 36.5 percent for Internet services and 33.2 percent for mobile telephone services. In both of the latter cases, however a high percentage of those interviewed—49.5 percent in the case of Internet services and 33.6 percent in the case of mobile telephony—did not respond either way since they had no experience with these services (see figure 6).

In many countries, the issue of telecommunications liberalization has been divided up, frequently leading to the maintenance of the monopoly for main line telephony while opening the market in mobile phones and Internet services to private competition. Of the world's 201 countries listed in the ITU statistics, only 43 percent have fully opened main line telecommunications, whereas 78 percent have done so in the case of mobile telephony and 86 percent in the case of Internet services (ITU 2002a: 5). An essential reason for this disparity is that, in the sector of the new technologies, liberalization generally encounters less resistance since it affects far fewer vested interests and established structures.

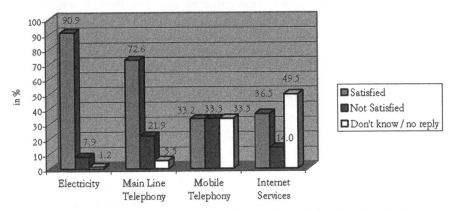

Figure 6 Satisfaction with the ICE According to Type of Service: Results from an Opinion Poll, 2000

Source: Unimer-La Nación (2000a)

In the statistics above, the high percentage of 'don't know / no reply' regarding the state monopoly on mobile telephony and Internet services suggests that in Costa Rica, a selective liberalization in the field of the NICT could hardly have raised passionate popular resistance. Therefore it seems likely that the conflict that arose did so because the break-up of the state monopoly for mobile telephony and Internet services was additionally bound up with the politically much more controversial hotter topics of doing the same for main-line telephony and electricity. As a consequence, all the public relations activities concerning the importance of NICT liberalization for the country's development perspectives proved worthless and the package deal of the *Combo* provoked a 'package answer' in favor of the state monopoly.

A STATE MONOPOLY DEFENDED: COSTA RICA'S TELECOMMUNICATIONS REGIME AT THE BEGINNING OF THE 21ST CENTURY

The political confrontation triggered by the *Combo del ICE* law was a major earthquake in Costa Rica's political and social development. Thanks to the government's virtually unconditional withdrawal of the bill, its negative effects on the political system and on the position of the dominant parties could be limited. Its effect on the issue of telecommunications liberalization, however, was profound: in Costa Rica the break-up of the state telecommunications monopoly has become a highly symbolic issue with extraordinary political stakes involved. The leader of the PUSC parliamentarians, Eliseo Vargas, summed up the 'lessons learned' as follows:

> "The Costa Rican populace has been very clear, and it has said 'No' to the opening up of telecommunications and electricity. (. . .) We are no masochists, we will not insist on something to which 'No' has been said. Any such rumors are malicious. Let me be very clear about this: The *Combo* has died, and it has died forever" (El Semanario, 2001: 7).

Even President Rodríguez acknowledged: "If the Costa Ricans do not agree with the proposed option—which I personally would have considered to be the best—then we have to look for other ways" (La Nación, December 1, 2000).

As the government changed its course, first to disappear was the alarmist discourse that linked the nation's fate to the liberalization law. The coordinator of the telecommunications part of the *Combo* bill, Ricardo Monge, called for self-criticism at this point: "We drew a picture that this reform was super-super-urgent and absolutely indispensable. This dramatist approach was wrong. Eventually, the reform didn't come, and life goes on."[22] Hernando Pantigoso, responsible for NICT matters in the President's office, confirms:

> "The failure of the *Combo* has led us to adopt a different approach. Before, we thought it to be better to hand over things to private business, and right before the opening of the sector major state investments in the telecommunications sector certainly were not a priority. Now we have adopted a 'plan B,' a 'second best option,' and we have to go forward with the modernization of the sector with much involvement from the government and with the state-monopoly of the ICE. This will be slower, but it must be pushed forward—this sector is way too important for the country to afford a standstill."[23]

The government decided on two changes in personnel that were to symbolically signal its political turnaround. One was the replacement of the ICE's President Rafael Sequeira, who had regarded the opening of the sector as inevitable and had supported the *Combo,* with Pablo Cob, who had the support of the ICE's unions. The other decision was the appointment of Guy de Téramond, the highly respected 'father of the Internet in Costa Rica,' as the new Minister for Science and Technology; whereas before there hardly was any governmental NICT policy worthy of the name, this appointment promised the government's commitment to dynamically expand and modernize the country's NICT infrastructure and services. The government followed up these appointments with the announcement of a program called 'The Digital Agenda' (*La Agenda Digital*). This included numerous state-led initiatives for infrastructure investments, diffusion of access, improved regulation and the integration of the new technologies in public institutions, the education and health sectors, and in small and medium businesses (cf. Presidencia de la República 2001).

Before the *Combo* initiative, Costa Rica had been characterized by what we have called a 'pre-liberalization telecommunications regime': a state monopoly on the defensive, producing deficiencies, which would serve the arguments of those in favor of liberalization. The *Combo* law then meant to implement a moderate variant of the dominant neo-liberal model. What resulted from the failure of this project, however, was not a return to the 'pre-liberalization' regime, but something substantially different. Costa Rica entered the 21st century with a telecommunications regime centered on a state monopoly which could claim broad popular support and which, after the *Combo* failure, received the financial resources indispensable to introducing the NICT dynamically while seeking to continue the socially inclusive patterns of Costa Rican telecommunications. If in the coming years this regime is given the chance to consolidate, instead of bowing to new initiatives to erode it, Costa Rica would come to represent a model of telecommunications and NICT development that is lacking in the Third World. Such a model would sharply contrast with the dominant paradigm of neo-liberal transformation and its polarizing social effects, and therefore it should be of great interest to those engaged in the international debate on the NICT and Third World development.

Chapter Four
Active NICT Development by State Monopoly: A New Costa Rican Model?

In the preceding chapter we outlined the particular conditions of the Costa Rican telecommunications regime and the failure of the liberalization initiative. We shall now take a closer look at what has become a rare species: NICT development in the context of an ongoing state monopoly, not only in mainline telephony but also in mobile telephony and Internet services.

MODERNIZING NATIONAL NETWORKS AND EXPANDING INTERNATIONAL BANDWIDTH

An integral part of national NICT development is the connection to international telecommunications and data networks. While some authors go so far as to claim that bandwidth is the single most important indicator for NICT development (e.g. Monge 2000: 278), this is far too schematic and ignores all aspects of social distribution and use. Still, limitations in the capacity of international connectivity certainly are an important element of a country's NICT development and, particularly for Third World countries, they tend to elevate the costs and reduce the quality of available services.

It is noteworthy that antagonists in the wars and other conflicts in Central America during the 1970s and 1980s largely respected the regional telecommunications network, which started as an analogue system before migrating to a digital microwave one. In addition to these terrestrial connections, the increasing demand for international traffic in Costa Rica led to the establishment of the first earth station for satellite connections in 1981; its capacity grew from an initial 240 channels to 504 in 1989. Reflecting the steady, but still relatively modest, increases in demand, a second, now digitized station with a capacity of 1,400 channels was established in the early 1990s and included Costa Rica's first fiber optic ring.[1]

However, with the rapid expansion of the new information and communication technologies, transnational data flows increased far out of proportions to earlier projections, soon absorbing the existing capacities. In addition, undersea fiber optic connections, with far higher capacities and greater reliability, began to compete with satellite transmission in the 1990s. Once the initially very high costs of such connections had decreased substantially, most countries in the region chose to connect to the submarine fiber optic cables and made this a high priority issue in national telecommunications development in the second half of the 1990s.

The first project in which the Costa Rican ICE participated was the Columbus II cable which, since 1994, linked Cancún on Mexico's Caribbean coast with Florida, thereby connecting to the U.S. networks and, via the transatlantic cables, to Europe. In Central America, all traffic had to pass through the regional terrestrial digital short wave relay system to reach Cancún, which meant increased costs and lower quality than a direct connection could offer. The Panamericano cable that went into operation in November 1998, leading from Chile to the Caribbean, did not solve the problem. The consortium dropped the initially planned landing point in the Costa Rican coast town of Puerto Quepos and Costa Rica had to use terrestrial micro-wave connections to reach the fiber optic cable at its landing-point in Panama City.

A direct access point to the international fiber optic cables was not available until the arrival of the Maya-1 cable, which spread out from Florida in a 4,400 kilometer-long arch, connecting Mexico (Cancún), Honduras (Puerto Cortés), the Cayman Islands, Panama (Colón) and Columbia (Tolú)—and which also included a direct landing-point near Costa Rica's Atlantic port town of Puerto Limón. As part of a larger consortium of telecom and other companies, the ICE had signed the contract for the building and operation of the Maya-1 cable 1998. The ICE's financial share amounted to US$10 million, out of a total project cost of US$207 million; when the cable first became operational in Costa Rica in December 2000, its capacity was 45 Mbps (Pratt 2000), which greatly expanded the international bandwidth available.

For political analysis, it is important that national politicians have limited influence on the medium-term investment projects that cover the nation's connection to undersea fiber optic cables. In addition, there is a notable time-lag between the decision to participate in such a project and the practical reaping of its benefits. In the case of Costa Rica, this lag had implications for the outcome of the fight to liberalize telecommunications. Had the government's liberalization initiative in 2000 succeeded, its implementation would have neatly coincided with the connection to the Maya-1 cable. The resultant major improvement in quality and costs of Internet use certainly could have been passed off as a first practical benefit of liberalization. Instead, with the failure of the initiative, the political effect was quite the opposite: as a result of Maya-1 overcoming the bandwidth bottleneck, the state-owned monopoly provider could pass along the benefits of infrastructure improvements to its clients in the form of greatly reduced rates.[2]

After the arrival of the Maya-1 cable in 2000, a high capacity 2.5 Gbps fiber optic cable between the landing point in Puerto Limón and the capital of San José went into operation, establishing a high-bandwidth connection between the country's urban centers and the transnational cable net. In a second step, a project called Frontera a Frontera, serving a North-South axis, upgraded the existing connections between Nicaragua in the North and Panama in the South, as part of the land-based Central American fiber optic network administered by the *Comisión para las Telecomunicaciones en Centro América* (COMTELCA).

As newly appointed Minister of Science and Technology after the failure of the *Combo,* Guy de Téramond, who devoted his Ministry's resources largely to modernizing Costa Rica's Internet infrastructure,[3] could build upon the qualitative leap in international connectivity that Maya-1 represented. In his new position, Téramond focused on two projects. One was a new high-capacity academic network, the *Red Nacional de Investigación Avanzada,* established by legal decree in April 2001 (cf. MICIT 2001: 14); Téramond had to recognize, however, that the idea of linking up with the 'Internet 2' project of leading universities in the United States failed to materialize (La Nación, April 26, 2002). The second and more important project was the push for a nationwide broadband network of DSL quality, the so-called Advanced Internet Network (*Red Internet Avanzada,* RIA). Since such a network would use the existing telecommunications copper wires, to which high-end fiber optic rings would have to be added only at the nodal points, this leap to a second generation network could occur in a relatively short time and with modest investments.[4]

Though these modernization projects may sound purely technical, they play an important part in the political disputes about the telecommunications regime and the new information and communication technologies. One of the central arguments for liberalization had been the inevitable technological backwardness and high costs for users if the state monopoly were maintained. Yet with the new infrastructure projects, the government not only could promise a rapid expansion of high-capacity broadband access, but from the very announcement of the modernization, it offered massive price cuts for all types of users, from leased lines to residential access (MICIT 2001: 9).

In consequence, when on April 18, 2001 the Advanced Internet's backbone around the capital was inaugurated, in Costa Rica this was seen less as a technological and more as a political milestone.[5] It was, in the words of the leading newspaper, a "symbolic exorcism of the recent national nightmare that had become the *Combo del ICE*" (La Nación, April 19, 2001). President Rodríguez himself explicitly linked the two events, declaring that he looked back on the failed ICE reform "with republican humility" and was proud that the new project showed the ability of the Costa Ricans to find common solutions, even after such a bitter conflict (ibid). As to his political opponents, even the prominent PLN dissident and outspoken *Combo* critic Otton Solís congratulated the government's Science Minister Téramond for this achievement. By celebrating the new network, politicians underscored the symbolic "national reunion" this project represented; at the same time, Solís did emphasize that the project refuted all those that "for years have tried to tell us, that without liberalization the ICE would end up a heap of scrap" (Solís 2002).

EXTENDING THE REACH OF INTERNET SERVICES

In a preceding section we have seen how, after the pioneering role of academic institutions, the dynamics of NICT development in Costa Rica slowed down. In 2000, the state monopoly's provider RACSA still served only about the same number of accounts as the academic, non-profit CRNet. Since then, however growth in the number of Internet users has been high in Costa Rica and RACSA, which in two years doubled its number of clients, has been responsible for most of the increase. By the the end of 2002 user numbers had soared to 800,000, 19.3 percent of the total population, putting Costa Rica among the top three Latin American countries (see table 4 on the following page). Using a different methodology, a recent CID-Gallup poll in Costa Rica carried out in September 2003, even counted as many as 950,000 persons who access internet regularly at home, at their work-place or through internet cafés and public access centers; this number would represent 23 percent of the population (RACSA 2003). In another important indicator, computer density, Costa Rica is the undisputed leader of continental Latin America, with 19.7 computers per 100 inhabitants (see table 4).

The strong increase in Internet use Costa Rica has experienced comes in part as the result of RACSA's rate reductions. Whereas RACSA's Internet access tariffs before were extraordinarily high, the new tariffs offered Internet access without time-limit for a flat US$15 per month (excluding

Table 4 Internet Indicators (Hosts, Users, PC diffusion); Latin America and the Caribbean, December 2002 (ranked from highest to lowest share of Internet users per 100 inhabitants)

Country	Internet Startup date	InternetUser absolute	per 100 inhabitants	InternetHosts absolute	per 100 inhabitants	Estimated PCs per 100 inhabitants
Chile	Jan 92	3,575,000	23.75	135,155	0.90	11.93
Jamaica	Aug 94	*600,000	*22.91	1,276	0.05	5.39
Costa Rica	Jan 93	800,000	19.31	7,725	0.19	19.72
Guyana	Oct 96	125,000	14.22	63	0.01	2.73
Antigua & Barbuda	1995	7,000	12.82	622	0.80	n.d.
Uruguay	Aug 93	**400,000	**11.90	78,660	2.32	11.01
Argentina	1989	4,100,000	11.20	495,920	1.35	8.20
Trinidad & Tobago	Sep 95	138,000	10.60	7,209	0.55	7.95
Peru	Feb 94	2,500,000	9.34	19,447	0.07	5.56
Belize	Aug 95	22,000	8.69	1,498	0.59	13.83
Brazil	1994	14,300,000	8.22	2,237,527	1.29	7.48
Bahamas	Sep 95	21,200	6.79	32	0.01	n.d.
Grenada	Oct 9	66,500	6.13	14	0.01	13.21
Barbados	n.d.	30,000	11.15	160	0.06	10.41
Venezuela	1994	1,274,400	5.05	24,138	0.10	6.09
El Salvador	Jan 96	300,000	4.64	269	—	2.52
Mexico	1989	4,663,400	4.57	1,107,795	1.09	8.20
Colombia	May 94	1,982,000	4.57	55,626	0.13	4.93
Panama	1995	120,000	4.13	7,393	0.25	3.83
Ecuador	Jan 93	503,300	3.88	2,648	0.02	3.11
Dominican Rep.	Jun 95	300,000	3.44	45,508	0.55	n.d.
Guatemala	Dec 95	400,000	3.33	9,789	0.08	1.44
Suriname	Oct 95	**14,500	**3.30	24	—	4.55
Bolivia	Jul 95	270,000	3.23	1,413	0.02	2.28
Honduras	Jan 96	200,000	2.97	160	—	1.36
Paraguay	1996	100,000	1.70	24,351	0.08	3.46
Nicaragua	Feb 9	490,000	1.60	73,370	0.06	2.79
Cuba	Mar 97	**120,000	**1.06	1,133	0.01	3.18
Haiti	Nov 96	80,000	0.96	—	—	n.d.
Ø Latin America			6.70		0.80	6.60
Latin America Total		35,670,800		4.249.796		

Note: Data for host computers and Internet users are only estimates. Of the Caribbean states with a population of under one million only a selection has been listed.
* Jamaica's spectacular increase of user data from 100,000 users (or 3.85 percent of the population) in 2001 (ITU 2002a) to 600,000 (or 22,91 percent of the population) in 2002 (ITU 2003) certainly reflects more changes in measurement than real developments on the island in that year.
** data for end of 2001(cf. ITU 2002a).
Source: ITU (2003); except for start-up date: ITU (2000: 100)

telephone costs), which made residential Internet accessible for small businesses, as well as for a significant section of Costa Rica's middle class. Price reductions of 50 to 70 percent also came to services used by bigger businesses and institutions, such as leased lines. With these rates, Costa Rica ranks well below the average ISP tariffs in lower and upper middle income countries worldwide (ITU2002a: A-69, A-70).

The advocates of liberalization have pointed to the 'unmet demand' for main line and mobile telephony, signified by the waiting lines of several months to get access, as proof of the 'inertia' of the state telecommunications carrier. In fact, this 'unmet demand' has had at least as much to do with keeping rates below what would be the equilibrium price. Thanks to the increased international bandwidth and RACSA's modernization offensive after the *Combo* failure, the state company now boasts that it can fully supply the national demand for Internet services, despite its low new rates.

Though the ICE subsidiary RACSA holds the state monopoly on ISP services in the country, it should be noted that one form of broadband access—Internet via the cable TV net—is offered through joint-venture cooperation with private business, namely the two private cable TV providers operating in Costa Rica, 'Amnet' and 'Cable Tica.' With just over 11,000 clients by September 2003, this service, which provides high speed service comparable to a DSL connection, so far is limited to small and medium-size business and a small number of high-income residential users (RACSA 2003). This joint-venture deal was initiated, if hardly noticed by the public, well before the government's liberalization initiative of 2000. However after the failure of the *Combo* and the heightened attention that came to this arrangement as a result, this cooperation with the cable TV companies brought criticism as a violation of the state monopoly condition. Although the critics announced a law suit before the nation's constitutional court, so far the service continues unchanged.

CRNet and RACSA are the only Internet service providers in the country, but only two weeks after the withdrawal of the *Combo* law, on April 17, 2000, the government tried another form of market opening: it tried to oblige the ICE to enter the market as Internet service provider in direct competition with its own subsidiary, RACSA. Though the ICE repeatedly rejected this idea, it remained on the agenda for some months (cf. La República, September 28, 2000; La Nación, December 3, 2000 and February 3, 2001; La Nación, March 3, 2001). However, over time this project did not materialize. At any rate, it seems rather doubtful that such competition within the state monopoly, with one regulatory body overseeing both services and rates, would make much economic sense.

To complete the picture, we should mention the existence of illegal Internet service providers. For obvious reasons, there are no sure figures on these 'pirate ISPs,' but an approximate indicator is the number of them closed down by the police; in 1999, that was 17 (they generally reappeared soon afterwards under different names), with an estimated number of 1,000 to 2,000 accounts (La Nación, September 27, 2000). These typically sprang up around legal software companies, using the company's leased line provided by Racsa in order to resell its use. This illegal competition was particularly profitable when RACSA's Internet tariffs for residential users were rather high. Since RACSA introduced its reduced rates for residential access, the market share of illegal ISPs has dropped significantly.

The state monopoly is limited to the simple ISP function and the company does allow private enterprises, using RACSA as a provider, to offer value-added Internet services to the public. Such enterprises may offer the hosting of websites or intra-nets for businesses, or e-commerce applications. The country's roughly 300 Internet-cafés, which rent their RACSA access to users, are another legal source of Internet services, in a form reminiscent of older public call offices that offered phone service on a concessionary basis.[6] In fact, the government's NICT strategy explicitly counts on the dynamics of the private sector. For instance, in the case of the 'Advanced Internet Network' project, the government explicitly argued that the prospect of greatly increased capacities and reduced rates

> "will allow the boom of numerous Internet-cafés which, together with other initiatives in schools, post-offices or municipal offices will achieve a nationwide coverage and the universalization of access and use of e-mail and Internet applications" (MICIT 2001: 9).

COSTARRICENSE.COM: A STATE-LED INITIATIVE TO PROVIDE UNIVERSAL ACCESS

Although we have seen that the government's commitment to providing 'universal access' does not mean that the state monopoly has to be responsible for all the access points, there have certainly been direct state-led initiatives in the promotion of the social diffusion of NICT use in the country. The flagship program was the 'Costarricense.com' initiative, which made Costa Rica the first country in the world to provide all its citizens with a free e-mail account, assigning a total of 4.5 million e-mail accounts and homepages to both citizens and permanent residents based on their identity card or registration number (cf. www.costarricense.com). The name *Costarricense* ('Costa Rican') and the resulting e-mail addresses '*name*@costarricense. com' implicitly convey the program's message that NICT access is as much part of 'being Costa Rican' as an identity card.

The web-based e-mail program was integrated in the homepage 'www.costarricense.com,' which was designed as the central web portal for Costa Rican citizens, with well-structured links to the main public institutions and their online-services. Besides Costarricense.com's free e-mail and website, the project made an important commitment to provide free Internet access through the establishment of public access centers in communal centers, administrative offices or public libraries in each of Costa Rica's 81 municipalities. In these municipal telecenters, every citizen was supposed to be able to use the Internet free of any charge for at least 15 minutes daily.

This initiative certainly appeared promising for the social diffusion of the NICT, since it would have extended basic Internet and e-mail access to those Costa Ricans that cannot afford the prices of commercial Internet-cafés or residential access, or those who live in places where no such option exists. In addition, since Costarricense.com uses established public buildings, socio-cultural access barriers tend to be lower for any who do not feel at ease in the Internet cafés usually dominated by young people and an informal atmosphere.

Although the program was criticized as 'too slow and too little' (e.g. La Nación, March 16, 2001), its impact should not be underestimated. In little more than one year, some form of public access centers were established in 54 of the 81 municipalities (La Nación, August 25, 2002) and the homepage became the second most frequently visited Costa Rican website, after the site of the leading newspaper *La Nación*.[7] Also, with 170,000 users, Costarricense.com was the most widely used e-mail service in the country (La Nación, September 18, 2002). However, as we noted earlier that access is a necessary but insufficient condition for meaningful NICT use, the Costarricense.com effort still failed to address the need to provide a coherent social context in which the new technologies could be embedded. In order to have a greater developmental impact, the initiative should have been accompanied by educational efforts at different levels and by initiatives to integrate it with existing social groups, programs or activities.

It soon became evident that the most fundamental problems of the initiative were due to its organizational structure and financing. Costarricense.com had been initiated as a public-private cooperation; the private company was called 'Power Access' and it operated the e-mail service and donated the hardware to the municipal telecenters. On the state side, Costarricense.com was directly subordinate to the President's office;

in addition, the so-called 'Triangle of Solidarity' scheme, directed by the country's Vice President,[8] and RACSA were involved in the project.

The first manifestation of this problem to surface was the issue of technically maintaining the municipal telecenters, which none of the institutions engaged in the project covered. Since the communal centers themselves lacked the know-how and financial resources to assume this task, this rapidly led to the 'rusty tractor syndrome,' and a high percentage of computers were soon out of service due to technical problems (La Nación, August 25, 2002). From a more medium-term perspective, other shortcomings in the program's financial structure were foreseeable. For one, RACSA had committed itself only to providing Internet connections free of charge for an initial period; after this, either the municipalities would have had to assume these costs— an unlikely prospect—or RACSA would have had to continue this service as a permanent subsidy to the state-led program. In addition, the project faced administrative and security costs, and investments would have been necessary to keep the telecenters technologically up to date in the coming years.

In 2002, after only two years of operation, the initial structure of the Costarricense.com initiative collapsed, as the crisis of the dot com companies in the United States affected Power Access, which had to declare insolvency (La Nación, August 25, 2002). To cope with this setback, the state monopoly company RACSA decided to assume the initiative entirely; the Costarricense.com site was closed and the service re-opened under the national domain name ending (www.costarricense.cr), with all e-mail accounts transferred to the new address. The new website has given up the ambitions of being Costa Rica's leading web portal and has essentially been reduced to operating the e-mail service. By assuming full responsibility for the initiative, RACSA obviously incurs higher costs than originally planned, but the company argues that this is compensated by the positive effect this service has in attracting future RACSA clients (La Nación, September 18, 2002).

A very similar initiative to provide free Internet access was undertaken by the Costa Rican postal system, Correos de Costa Rica, in cooperation with the dotcom company 'Intermedia.' With a total of 12,000 clients at its peak, this initiative operated on a much smaller scale than Costarricense.com. In 2002, this service also ran into insolvency problems with its private sector partner; Intermedia's withdrawal meant service was discontinued altogether (ibid).

As to the municipal telecenters, the new administration elected in 2002 announced its intention to relaunch the universal access program under the name 'Comunicación sin Fronteras' (Communication without Borders). Under the direction of the Ministry of Science and Technology, this program plans to establish a total of 187 public access centers in the country. Postal offices equipped with the necessary hardware, from the postal service's own free-Internet-access initiative, will serve as public access centers in this new endeavour. Now, as the initiative does not rest on the cooperation with private companies, it will be entirely up to the government to give it sufficient priority and financial resources to ensure its success. The interaction between the state, the educational system, public institutions and civil society groups will be crucial in determining how well these municipal telecenters will be embedded in a social context that helps make these centers meaningful instruments for development.

LINCOS: TELECENTERS WITHOUT SOCIAL VISION— AN EXCURSUS ON A PROMINENT MODEL

One telecenter project operating in Costa Rica, called LINCOS (short for 'Little Intelligent Communities'), deserves a separate analysis in this work because it has become one of the most prominent models worldwide for NICT diffusion in developing countries. The international profile of this project received a boost when its initiator, the former Costa Rican President José María Figueres, was named Special Representative of the United Nations for the Information and Communication Technologies by Kofi Annan in 2000. Although LINCOS centers in Costa Rica have little practical importance, the extraordinary degree of international attention that has followed them makes an excursus worthwhile: we will critically examine the developmental implications of this project so widely hailed as exemplary.

The LINCOS concept was developed at the Media Lab of the Massachusetts Institute of Technology (MIT) and it was implemented in Costa Rica by the *Fundación Costa Rica para el Desarrollo Sostenible* (*Costa Rican Foundation for Sustainable Development*), which was founded and run by José María Figueres after his presidency ended in 1998. Although we spoke previously of an NGO-based initiative, this case demands qualification: the Costa Rican Foundation for Sustainable Development is definitely no grassroots organization but, to the contrary, is an entity established and directed by a political leader in a highly personal style.[9] Besides the MIT and Figueres' Foundation, leading computer companies such as Intel, Microsoft, and Hewlett Packard have supported the LINCOS project by donating the equipment for the initial pilot projects, with the reward of much publicity.

Although the first LINCOS were implemented in Costa Rica, these are merely conceived as 'pilot projects' within a broader system conceived for worldwide use. The very physical appearance of the LINCOS telecenters, which are housed in recycled shipping containers, symbolizes their envisioned role as autonomous technological units that can be transported to the most remote places on earth (for photos of the containers, see www.lincos.net). The LINCOS containers, operated by solar batteries and equipped with satelite Internet connections, are to be multi-purpose centers, concentrating a maximum of functions on the smallest space possible, taking its cue from the Swiss army knife. The prototype has six Internet-connected computers, which can be used individually or for training in small groups, with printer and scanner, satellite fax and phone connection, a video recorder and giant TV screen, a digital radio station with its own antenna and a digital camera, a laboratory for soil and water analysis, and complete telemedical equipment—and all utilize state-of-the-art technology, thanks to the sponsorship of the leading international companies in these fields.

In Costa Rica currently no more than a handful of LINCOS containers are in operation, with one in San Marcos de Tarrazú and one in San Joaquín de Cutris in the San Carlos region; in 2003, a third one opened in Río Frío as a project sponsered by the United States-Costa Rica Foundation. Another ten LINCOS have been deployed in the Dominican Republic and, according to government sources there, a total of 60 are planned (La Nación, March 19, 2001).

While LINCOS has received enormous international applause and praise in a very short time, it is amazing that none of this approval was based on any serious evaluation of the project's concrete experience, but solely on its technological possibilities and their promised merits. When actually looking at the practical impact of the Costa Rican LINCOS projects, they instead seem to be a prototypical example of inappropriate technology because they are not sufficiently embedded in the social context. These shortcomings begin with the container itself, which was a stroke of genius in terms of marketing the project to the international media and donor organizations.

Though the 'all-that-in-one-little-box!' effect of the container has fascinated outside observers, in practice it simply means very little room in which to work. When the computers are used for teaching purposes, two people could share one screen in any ordinary school room, but in a container that is only 2.40 m wide, there is practically no room left for the instructor to move—even with just one person at each of the monitors. In the case of telemedicine and soil and water analysis, the use of these facilities requires a specialist's know-how; the laboratory would make much more sense in a local agronomist's office than in a communal telecenter designated for the population at large. Moreover, telemedicine facilities only function properly if they are integrated into the local medical structures, if the personnel are appropriately educated and trained and if the poor population has effective access to its benefits. Even so, in most remote rural areas of the Third World, medical posts equipped with the most essential equipment, medicine and decently paid public health workers are the most pressing needs—not the additional services telemedicine facilities provide. The author's visit to the LINCOS site in San Marcos de Tarrazú in April 2001 could not have brought more sobering results: since the site was inaugurated two years ago, the staff admitted that neither telemedicine nor the laboratory has been used a single time.

In addition, as a place to work, a metal container is one of the most unsuitable environments in tropical countries, where the sun quickly heats up the metal roof. This makes it necessary to cover the LINCOS container with a tented roof three times as large as the container—and such an addition certainly exceeds the costs of building a regular stone house with local means. In practice, not much is left over from supposedly money-saving idea of recycling shipping containers.

What is most used in the LINCOS site in San Marcos are the six computer-workplaces, for courses as well as for individual use. As to the latter, according to the local staff the most commonly used services are e-mail, to maintain contact with relatives that have emigrated, and Internet navigation regarding sports, cooking recipes and games. The giant TV screen has its moments of glory, when important soccer matches are shown on national television. All this is not wrong, but it certainly does not maximize the usefulness of the new information and communication technologies for development. The same practical purpose could have been achieved with a technologically far more modest communal telecenter at a fraction of the cost. And since one LINCOS staff member in San Marcos concluded that "what has benefited most from LINCOS has been the local school," a serious question is whether it would have been more coherent, cheaper and in the long run perhaps even more sustainable to have installed a computer classroom in the school building. Pointedly, one could say that rather than 'Little Intelligent Communities' LINCOS might stand for 'Largely Inadequate Containers.'

As for the aesthetic effect of the tented roof, LINCOS' initiator Figueres (2000: 28) himself once remarked: "it looks like it just landed from Mars." This phrase may be fitting for more than just the physical appearance: the people of San Marcos did not seem to perceive the LINCOS as something 'of their own,' but rather something 'put there' by someone 'from above.' The project was not based on any communal structure nor was it coherently inserted into the social context. Moreover, San Marcos de Tarrazú seems an odd choice for the project, since it certainly is not a remote and underdeveloped community, but a town not far from Costa Rica's central valley, perfectly accessible by paved road. Since 2000, San Marcos has even had its first commercial internet access center. 'Allen's Internet Café' is operated by a local youngster—and it has been successful despite the existence of LINCOS, which, as a subsidized demonstration project, offers all its services for free. (This mode of operation, of course, is only applicable to a limited number of demonstration plants—begging the question of what the financial basis for the projected extension of the model could be.)

The great impression of the LINCOS initiative on the international public hardly results from either its concrete experiences or a sober cost-benefit analysis. From its very beginning, the project was accompanied by an intensive and well-planned public relations strategy, including the invitation of numerous functionaries and journalists to the country, who have been spreading the good word of the project. Probably more important than any real experiences is the symbolic effect of the LINCOS containers. The complex problem of the integration of the NICT in rural Third World communities now has a concrete technological 'answer' in a box measuring 8 by 2.40 meters, which is white, clean and perfect, shipped as a ready-to-use package from the North to the South, a kind of high-tech version of CARE packages.

Finally, the institutions and people representing LINCOS have played a crucial role for the project's international success: they are extraordinarily well positioned within the development establishment, combining the MIT Media Lab's academic reputation and business connections with the person of José María Figueres, in the prestigious post of the supreme UN Representative for New Technologies.

Nevertheless, NGOs and activists who work for the developmental use of the NICT with a social vision have sharply criticized the LINCOS project:

> "The project is extremely costly (. . .) the concept is completely oriented on connectivity and does not include a community vision (. . .) The impressive apparatus was not put near a school, community center or church, but to the contrary, far away from what the community symbolizes. From all parts of the village one sees this object, symbolizing technology in its most distant form from the people (. . .) It is a model project—but a model of how not to do it!" (Pimienta 2000b)

THE 'DIGITAL AGENDA': GOVERNMENT INITIATIVES FOR NICT INTEGRATION

Beyond the Costarricense.com initiative described above, the administration of Miguel Angel Rodríguez also initiated the 'Digital Agenda' program, which brought together a wide range of measures for NICT integration in diverse fields of society, politics and economy. Under the title 'Gobierno Digital,' this included an e-government program in which different government institutions improved their Internet presence and offered online versions of a number of administrative services. The social security system, the *Caja Costarricense del Seguro Social* (CCSS), assumed a lead role in this process, establishing a centralized system to collect the employers' contributions.

The 'Digital Agenda' gives much importance to active NICT use by the public institutions, since these are seen as having a strong pull effect on the rest of society:

> "Since the state has evident influence on the economy and social organization, if state institutions are more and more present on the Internet this will lead to the private sector and the citizens using it more, too." (Presidencia de la República 2001: 9).

However, with the elections of 2002, the transition from one administration to the other did much to disrupt the government's web presence. For more than half a year, the central portal of Costa Rica's 'digital government' (www.go.cr) was discontinued, leaving it with vintage "News" and showing the outgoing President's photo as if he were still in office. The new government now has assigned new personnel within the Ministry of Science and Technology to coordinate the national e-government initiatives, and it certainly is too early to judge their efforts. However, the case of the www.go.cr portal is typical of the problems of political discontinuity that many e-government initiatives in the Third World encounter—and in the Costa Rican case, it was after regular elections in which the governing party retained power!

The 'Digital Agenda' also stressed the lead role of the government in e-commerce development, which so far has found little acceptance in the population. It is estimated that only about 4 percent of residential Internet clients have used it for buying online (La Nación, July 25, 2001; cf Farah Calderón 2001: 3–6). In mid-2001 the government's own system for buying online became operational, and the hope was that this would stimulate other sectors. In addition, the initiative encompassed a set of programs to support small and medium-sized enterprises integrating NICT and establishing on-line-business activities (Presidencia de la República 2001). These programs included the creation of a state-financed Internet portal mainly for export-oriented businesses, named 'www.marketplacecostarica.com,' as well as direct support measures in the areas of training and financing. Cooperation between RACSA and the—also state-owned—Banco Nacional de Costa Rica led to a special credit program with preferential conditions for small and medium-sized businesses acquiring computer equipment, software and

Internet access. A special program targeted the promotion of the Internet presence of Costa Rica's tourist sector.

As for regulatory measures, the Legislative Assembly has adopted a number of laws protecting intellectual property, in accordance with the international norms promoted by the WIPO and WTO. A bill providing a legal framework for the so-called 'digital signature,' securing authorship and identity in business transactions, has been introduced in the parliamentary body, though at the time of writing it had not yet passed. A 'hábeas data' bill limiting the intrusion on the citizens' privacy rights through NICT use and prohibiting the publication of personal data without the person's consent is in preparation (Hess Araya 2002a). A case in point for the necessity of such protection is the Costa Rican company 'Datum,' which accesses online public information regarding ongoing legal cases, posted on the Internet by the judicial system, to provide this information—often without even waiting for the cases results—to financial institutions inquiring about a person's credit-worthy-ness (Gregorio 2002; see also Carvajal 2002). In addition, the bill also contains protective clauses against content defamatory to a citizen.

Finally, in the education sector a broad program for the promotion of NICT and computer use has been announced. As is the case in all these programs, it has yet to be seen how many of these high-flying plans will be implemented effectively in the coming years. However it should be noted that as early as 1988, the *Programas de Informática Educativa* began to implement new technologies in primary and secondary education by introducing computers. In addition to the Ministry for Education, this long-standing project receives important support from the *Fundación Omar Dengo,* a non-profit foundation started in 1987 with the declared goal of promoting the quality of the country's educational system through the introduction of new technologies and innovative pedagogical concepts (cf. www.fod.ac.cr, Anfossi/Fonseca 1999). A specific legal framework provides for close cooperation between the autonomous Fundación Omar Dengo and the public school system.

There has been little governmental interest in the promotion of open source software, which has essentially failed to become a public issue in Costa Rica. Farah Calderón (2001: 38), though certainly sympathetic to the movement, concludes: "The Linux Association is small and it seems almost like an 'underground' manifestation. (. . .) The open source software movement is still in its beginning and of little significance."

DIFFUSION AND STRUCTURE OF MOBILE TELEPHONY

In mobile telephony growth has been less dynamic than in mainline telecommunications. With 460,000 mobile phones (that is 11.1 per 100 inhabitants) operating in the country at the end of 2002, Costa Rica ranks rather low in Latin America (see table 5). A particularly telling indicator is the percentage of mobile phones as percent of total telephone subscriptions; at 30.7 percent Costa Rica's level is the lowest in mainland Latin America. This reflects as much the modest development of mobile telephony as the extraordinarily broad diffusion of main line telephony. Some have pointed to this ratio as an indicator of the country's poor ability to adapt to new technologies relative to its success adapting to the old. Such a conclusion, however, is rather misleading: as we noted earlier, it is the 'old' main line telephony that opens up the gateway to more complex NICT uses such as the Internet, whereas the current generation of mobile phones essentially is limited to classic (though now mobile) one-to-one voice communication with significantly less developmental potential.

The shortcomings of the mobile telephony sector in Costa Rica have provided a central argument for the forces pushing for liberalization. One of the deficiencies criticized is the average waiting time of several months for service after applying for a mobile phone. Another point of criticism concerns the technological standards or 'generations' of the existing mobile telephones in Costa Rica, which lag behind other countries on the continent. In addition, some authors (e.g. Tacsan 2001) point to the rather high initial cost for mobile phone access (about US$200) though this argument is incomplete because it fails to take into consideration the very low rates for mobile phone use. Monthly subscription rates are less than half the continent's average and the tariffs per unit for mobile phone calls are the very lowest of all Latin American countries, as figure 7 shows (for the underlying data see table 5).

The Costa Rican government has not been inactive on the issue, however. Shortly after the failure of the *Combo,* the Costa Rican state monopoly not only improved Internet services, it also responded to the unmet demand in mobile telephony. In a first step, the ICE acquired 214,000 digital mobile phone lines from the Ericsson and Lucent companies. The lines could be quickly implemented, since these companies' technology would make the lines available through the expansion of already existing plants. ICE struck a similar deal with Alcatel to buy a further 160,000 mobile connections based on the extension of existing capacities.

Such investments could take off some of the pressure on the ICE regarding mobile telephony. However, the ICE ran into problems: whereas the first direct deal passed without problems, the deal with the French company Alcatel was legally challenged by other companies—amongst them Nortel, Motorola and Lucent—and ICE had to withdraw from the Alcatel deal in

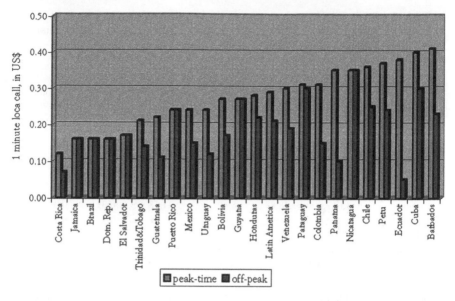

Figure 7 Mobile Telephony Tariffs, Latin America and the Caribbean, 2000 (ranked from lowest to highest rates)

Source: ITU 2000

July 2001 to pre-empt a negative report from the country's audit office. Also, the U.S. embassy publicly spoke out against the direct deal with Alcatel, calling instead for a new invitation of competing offers (La Nación, March 22, 2001).

Other sectors also criticized these deals since they are visibly based on "donations that work," as La Nación (March 26, 2001) titled an article on the issue. Alcatel had donated a mobile phone plant in 1998 that was greatly underused, leaving a lot of free capacity for expansion—supporting the argument that now buying Alcatel connections would bring results faster than buying from any other company. Although a parliamentary commission was set up to deal with these kind of self-interested donations, it is likely that this practice will continue: in March 2001, Siemens donated a similarly oversized plant to Costa Rica, clearly hoping to thus attract future orders (La Nación, March 22, 2001).

The public sale of the 214,000 Ericsson/Lucent mobile phones began in April 2001, one year after the failed *Combo*. The waiting list at this time was at 239,000, people; with an estimated 17,000 new applicants each month (La Nación, April 4, 2001) and the block of the Alcatel deal, the gap was still far from closing. In mid-2002, however, a major US$150 million

Table 5 Mobile Telephony Indicators Latin America and the Caribbean, December 2002 (ranked by mobile telephone subscribers per 100 inhabitants)

	Mobile telephone subscribers			tariffs in US$*		
Country	Absolute	per 100 inhabitants	as percent of total telephone subscribers	monthly subscription rate	per minute peak-time	per minute off-peak
Jamaica	1,400,000	53.48	75.9	19.46	0.16	0.16
Antigua & Barbuda	38,000	48.98	50.1	n.d.	n.d.	n.d.
Chile	6,446,000	42.83	65.0	28.01	0.36	0.25
Bahamas	122,000	39.03	49.0	n.d.	n.d.	n.d.
Paraguay	1,667,000	28.83	85.9	31.58	0.31	0.30
Trinidad & Tobago	362,000	27.81	52.7	32.68	0.21	0.14
Venezuela	6,464,900	25.64	69.5	35.67	0.30	0.19
Mexico	25,928,000	25.45	63.4	37.67	0.24	0.15
Suriname	108,000	22.52	57.9	18.70	0.56	0.56
Brazil	34,881,000	20.06	47.3	14.81	0.16	0.16
Dominican Rep.	1,701,000	20.66	65.2	25.16	0.16	0.16
Belize	52,000	20.54	62.3	—	0.43	0.25
Barbados	53,000	19.80	29.2	45.00	0.41	0.23
Uruguay	652,000	19.26	40.8	26.58	0.24	0.12
Panama	570,000	18.94	60.8	29.95	0.35	0.10
Argentina	6,500,000	17.76	44.8	43.00	n.d.	n.d.
El Salvador	889,000	13.76	57.1	19.94	0.17	0.17
Guatemala	1,577,000	13.15	65.1	30.81	0.22	0.11
Ecuador	1,561,000	12.06	52.3	27.00	0.38	0.05
Costa Rica	460,000	11.10	30.7	13.18	0.12	0.07
Colombia	4,597,000	10.62	37.2	44.16	0.31	0.15
Bolivia	873,000	10.46	60.7	15.91	0.27	0.17
Guyana	87,000	9.93	52.0	44.98	0.27	0.27
Peru	2,307,000	8.62	56.6	16.62	0.37	0.24
Grenada	8,000	7.13	18.4	n.d.	n.d.	n.d.
Honduras	327,000	4.87	50.3	20.00	0.28	0.22
Nicaragua	203,000	3.78	54.2	24.95	0.35	0.35
Haiti	140,000	1.69	51.9	n.d.	n.d.	n.d.
Cuba	18,000	0.16	1.5	40.00	0.40	0.30
Latin America Total	263,641,000					

* Data for 2000 according to ITU 2000: 95, since ITU 2002a as well as ITU 2003 only provide highly incomplete data which are not usable for continent-wide comparison.

Source: ITU (2003) and ITU (2000)

deal to acquire 400,000 second generation mobile phone connections from Alcatel could be completed. While in this deal, the 'European' GSM standard was chosen over the 'American' TDMA technology, later in the same year a major contract was awarded to Lucent to provide wireless equipment that will increase the capacity and coverage of ICE's existing TDMA network. Additionally, the country began major investments in improved mobile service coverage, with a projected total of 242 antennae sites for the GSM network (La Nación, August 17, 2002). According to ICE's President Pablo Cob, though, the entry of third generation phones (such as UMTS) is not foreseen until 2007 (La Nación, March 6, 2002).

FROM THE *COMBO* TO CAFTA: COSTA RICA'S TELECOMMUNICATIONS REGIME UNDER NEW PRESSURE

As we noted earlier, the state telecommunications monopoly was strengthened by the successful protest against the *Combo* law, making a re-launch of comprehensive liberalization unlikely to appear on the political agenda for some time to come. Instead, the forces pushing for liberalization have turned their efforts to projects of partial liberalization. Such an approach has found success in Costa Rica's commitment to liberalizing core parts of its telecommunications sector as part of the agreement on a Central American Free Trade Area (CAFTA), involving the United States, in January 2004.

The government had taken a similar line of retreat right after the failure of the *Combo* by initiating the 'mixed commission,' thus dropping the general liberalization project whilst maintaining the call to open the ICE for 'strategic alliances' with private capital and to open the end-user markets in mobile telephony and Internet services for private competition (La Nación, November 15, 2000).

Only four months after the withdrawal of the *Combo* law, a PUSC parliamentarian launched the first such bill, advocating a complete liberalization of Internet service providers under the name 'Law on the Right to Internet Access.'[10] Though a driving force in the anti-liberalization mobilizations was the desire to guarantee the social and geographical accessibility of the telecom and NICT services, this bill tries through its very name to use the right-to-access argument in favor of liberalization. If, in the logic of the traditional Costa Rican development model, the ICE's monopoly was the expression of the 'public interest' in the issue, the bill explicitly argues the contrary:

> "The present bill declares access to Internet services to be a matter of public interest; in consequence, any person or enterprise, be it public or private, can offer Internet access services" (Proyecto de Ley 2000).

To make this change viable, meaning that it not violate the telecommunications monopoly enshrined in the country's constitution, the bill argues that Internet services are not to be considered as part of telecommunications, but as part of the information and media sector. It seemed unlikely that the country's constitutional court would follow this unconventional interpretation. In addition, the dominant parties shied away from embarking on an initiative that might revive the memories of the *Combo*. Though the Legislative Assembly's economic commission accepted the bill in May 2001, since then it has been placed on the bottom of the parliamentary agenda, meaning that it can take years until it is introduced into plenary discussions.

In fact, since the failure of the *Combo* the practical implementation of the ICE's planned modernization projects has been hindered by the ICE's limited autonomy in investments decisions. Since it has to follow the formal procedures of public entities and not of private companies, all major projects have to pass through an invitation for bids—and time and time again competing companies and/or the national audit office have challenged the results and caused long delays. Though an independent control on spending by a state monopoly company doubtlessly is necessary, the current practice represents an obstacle to the effective and speedy process of modernization of the country's telecommunications and NICT infrastructure, leading to an overall impression of bureaucratization and the inertia of the state company—a situation easily exploited by those advocating liberalization. A reform giving the ICE greater autonomy in its investment decisions certainly is needed to improve the development of the telecom sector and to make it more predictable and reliable for the economic and social actors. The social forces that resisted the *Combo* recognize this and have repeatedly demanded such a reform, regarding it as an indispensable part of strengthening the institution and defending the public monopoly.[11]

Though the PUSC retained control of the government in the 2002 elections, Guy de Téramond's term as Science and Technology Minister was cut short after only two years in office. He was replaced by Rogelio Pardo, a medical doctor with no particular expertise in the NICT or the Internet modernization plans of his predecessor (see the interview with him in La Nación, April 26, 2002). Pardo stated that he was determined to continue these projects to strengthen the ICE in applying the NICT and that he wanted to make Téramond his advisor in NICT matters. However, it seems very doubtful if these plans have the same priority for the Ministry as they enjoyed during the brief leadership of 'the father of the Internet in Costa Rica' in the Ministry.

In addition, the new government's decision in August 2002 that the ICE would have to contribute US$27 million to the state budget for the following year marked a significant turn away from the active NICT and telecommunications strategy initially adopted after the *Combo* failure. Negotiations between the company and the government halved this sum to US$13.5 million (La Nación, September 13, 2002, and La Nación, September 26, 2002). According to the company, this shall be implemented through the delay of ongoing construction of hydroelectric plants, the slowdown of investments and delays in the installation of new main line telephones for one or two months; critics fear that even more programs and services will be affected (ibid).

This return to the old practice of diverting ICE's resources to finance the general deficit of the state budget is a highly worrisome aspect of current telecommunications development in Costa Rica. At the same time, in the negotiations over the ICE's contribution to the 2002 budget, President Pacheco pledged to support a law project establishing that, in the future, the state will not be able to touch the ICE's financial resources. Since this openly contradicts the government's actions at this very moment, it is doubtful how strong this verbal commitment will be.

While domestic political forces, still traumatized by the *Combo* experience, were keeping a low profile regarding any new initiative to liberalize or privatize the ICE, the government of the United States became the key actor in putting the issue once again high on the Costa Rican agenda. The instrument for raising the issue has been the project of a Central American Free Trade Area (CAFTA), announced by President Bush on January 16, 2002 (U.S. Presidency 2002). It is worth noting that in 2000, the United States exported $8.8 billion to the Central American states—more than to Russia, Indonesia, and India combined (ibid). Following in the footsteps of the North American Free Trade Agreement (NAFTA) between the United States, Canada and Mexico, CAFTA promises improved access for Central American products to the U.S. market in exchange for further opening the Central American economies to U.S. companies. Such an arrangement implied tariff reductions, change in trade regulations and opening clauses for strategic sectors. Amongst these, the initial draft of the agreement included provisions that oblige all signatories to fully open the telecommunications sector to private competition.[12]

In December 2003, the governments of Honduras, El Salvador, Nicaragua, Guatemala and the United States signed the CAFTA agreement. Costa Rica held back its approval and opted to continue negotiations. In such an important area as insurances, where Costa Rica so far had resisted any liberalization of the public monopoly, now the government committed the country to totally opening up the sector to private competition, including foreign companies. Regarding telecommunications, however, the government of Abel Pacheco showed its "lessons learned" from the failure of the package-deal of the *Combo* law: it conceded to opening the sector in the areas of Internet services, mobile telephony, and data-network services, but it resisted liberalization of the ICE's monopoly on main line telephony and electricity. U.S. negotiators accepted this attitude.

On January 25, 2004, the Costa Rican government finalized negotiations with the United States and joined CAFTA—pending ratification of the agreement by the U.S. Congress and Costa Rica's Legislativa Assembly. According to the document signed by both countries, Costa Rica has to open Internet services and private data networks to free market competition by January 1, 2006, and mobile telephony by January 1, 2007. A new regulatory law governing the activities of the ICE, adjusting it to the partial liberalization, must be in effect no later than December 31, 2004; by January 1, 2006, new legislation for the entire telecommunications sector must be in place. This legislation needs to include the establishment of a regulating authority typical of a liberalized telecom regime.

The Costa Rican government hailed the CAFTA accord as a milestone achievement, with President Pacheco announcing: "[With CAFTA] we have in our hands a tool that will help us overcome underdevelopment and poverty" (The Tico Times, January 30, 2004). Minister of Foreign Commerce, Alberto Trejos, argues that "the ICE will not be adversely affected—it will be strengthened by a modern legislation and it maintains its monopoly in what it does best: the generation of electricity and the provision of main line telephony" (La Nación, January 28, 2004). However, the public sector unions and the students' organizations so active in the anti-*Combo* mobilizations reject the idea that singling out Internet services, data networks and mobile telephony for private competition is likely to help concentrate the ICE on "what it does best." Instead, they point to the fact that these activities generate much of the revenues the company uses for its socially and geographically inclusive pattern of mainline provision and for maintaining low user tariffs (Semanario Universidad 2004).

Denouncing the CAFTA accord as a sell-out of national interests, the political and social forces opposed to liberalization have announced massive protests and a general strike once the treaty gets to the Legislative Assembly. Recalling the profound social unrest in the wake of the *Combo* law, Fabio Chaves, President of the Association of Costa Rican Electricity and Telecom Institute Employees (ASDEICE), threatened: "Deputies must understand: If CAFTA passes, this country will break down" (The Tico Times, January 30, 2004).

Although the Costa Rican government has learned from the Combo failure not to insist on a package deal to open up the entire telecom sector at once, with CAFTA the issue of the privatization of NICT services once again will get lost in the maelstrom of a much broader debate over the transformation of the country's economic model. Since the anti-*Combo* mobilizations gained much political momentum when the country's *campesinos* joined the protests, it will be critically important whether or not the CAFTA agreement will provoke a similar alliance of opposition. UPANACIONAL, the National Union of Small and Medium Farmers, has already spoken out strongly against CAFTA. Whereas the exporting sectors of Costa Rica's agriculture may benefit, all those producing for the domestic market stand to lose through increased competition from U.S. imports—especially as it happens that the US$19 billion of annual agriculture subsidies in the United States were an issue entirely left out of the CAFTA negotiations. The Costa Rican government was aware of this pressure. The extraordinary negotiating efforts the Costa Rican delegation dispensed on exempting a number of 'sensitive' agricultural products—such as potatoes and onions, which are produced by small farmers in highland Costa Rica—from trade liberalization measures can be seen as a cornerstone in the government's efforts to limit the potential oppositon to the CAFTA accord from the country's peasantry. It remains to be seen if these exemptions will be enough.

In a sharp attack on the CAFTA opponents, President Pacheco has already turned to the baseline argument in favor of CAFTA: the horror of being "left out in the cold." President Pacheco did not hesitate to say: "I want to ask those that go around threatening with strikes and conflicts to prevent the ratification of CAFTA: What are you going to say to those half a million of workers that will lose their jobs if CAFTA is not ratified?" (La Nación, January 29, 2004). There is not the slightest rationale for this statement. It may be an acceptable estimate that half a million jobs in Costa Rica in one way or the other depend on exports to the United States. But CAFTA is not about having or not having access to the U.S. market: Costa Rican exports already have preferential treatment by U.S. customs in the context of the Caribbean Basin Initiative (CBI) and CAFTA essentially would extend and deepen these arrangements—no less, no more. Pacheco's turn to such an alarmist and grossly exaggerated discourse greatly resembles the attitude of the preceding government towards the *Combo del ICE* law. Rather than be a show of strength, this approach seems to demonstrate the government's nervousness about committing the country to a liberalization course which, if past experience is any indicator, is likely to be highly unpopular with the populace at large.

At the time of writing, it still is unclear if CAFTA will materialize as depicted. Not only is ratification by the Costa Rican legislature an open issue. in the United States; passage in Congress is uncertain during the 2004 election year. But whatever the outcome, the CAFTA project definitely has succeeded in putting the issue of telecom liberalization back on Costa Rica's political agenda.

The resolution of the *Combo* conflict in 2000 created an opportunity to establish a 'new Costa Rican model': a telecommunications regime centered on a dynamic, socially backed state monopoly that could stimulate an evolution of the inclusive telecommunications regime that had characterized Costa Rican development. The first two years following the failure of the *Combo* certainly were promising for such a prospect, as the modernization and extension of the NICT infrastructure pushed forward, strong main line telephony expansion continued and a relatively inclusive social approach was taken in regard to the diffusion of the new information and communication technologies. However, the return to a policy of resource-draining from the state telecommunications company and, even more importantly, the government's commitment to open the NICT and mobile telephony sectors to private companies in the context of the CAFTA agreement, both signal the government's unwillingness to follow such a model. Instead, they point to a return to the situation that we described earlier as a 'pre-liberalization telecommunications regime.' Should this trend persist, the country would sacrifice what could be a vital asset in its development potential.

INTEL INSIDE: A HIGH-TECH CLUSTER AS DEVELOPMENT MOTOR?

Under the administration of President José María Figueres (1994–1998), the government was not only concerned with the adaptation and use of the new information and communication technologies, it also began to actively promote the production of the new technologies as a centerpiece of a new development strategy. The much repeated phrase that the NICT would be "the coffee beans of the 21st century" (e.g. Bate 1999) announces the intended shift in the simplest of terms: the transformation from an export-led model based on agricultural products to one based on a diversified export and services structure with an increasing participation of high technology products that would make use of one of the country's strategic assets, the high standards of its educational system. As Figueres explained:

> "We wanted to incorporate Costa Rica into the global economy in an intelligent way. Globalization was more than simply opening the country

to foreign trade. We needed a national strategy not based on cheap labor or the exploitation of our natural resources. We wanted to compete based on productivity, efficiency and technology (. . .) Many textile firms [had] left the country, and the government received severe criticism for not trying to sustain the *maquila* industry (. . .) but our foreign investment attraction strategy had changed. We wanted to attract industries with higher value-added, that would allow Costa Ricans to increase their standard of living" (cited in Nelson 1999: 5).

With this turn to a new strategy Figueres followed the recommendations of a World Bank study that proposed concentrating on attracting foreign companies in the electronics sector (World Bank 1996). The conceptual base of the adopted high-tech strategy was the development of industrial clusters along the lines of the approach developed and popularized by Michael Porter (see chapter two). This argued for spatially concentrated, sectorally defined growth and innovation centers, which would act as a dynamic element for the country's economy as a whole. In fact, President Figueres and the director of the INCAE business school in Costa Rica, Brizio Biondi-Morra, sought direct contact with Porter in the design of the Costa Rican strategy. The strategy passed from the drawing board to the implementation stage when Costa Rica managed to attract Intel, the world's leading microprocessor company, which built a huge production plant in Costa Rica in 1997.

After President Figueres left office in 1998, his successor, Miguel Angel Rodríguez, took up the new development strategy:

"We are in the transition from a country whose development was based on agrarian products, then on import-substituting industrialization, then on non-traditional exports, and now we are in a new stage, which focuses on technological innovation, on the 'new economy' (. . .)" (cited in: La Nación, February 19, 2002).

President Abel Pacheco, elected in 2002, promised to follow this approach and underscored his intentions by going to Taiwan on one of his first foreign visits, with the declared aim of attracting investments from Taiwanese high-tech companies (La Nación, October 7, 2002).

The transformation of Costa Rica's economy resulting from the Intel production site certainly has been considerable. The scale of the Intel plant is such that currently about one quarter of all Intel chips worldwide are being shipped from Costa Rica; from this export volume alone Costa Rica now ranks among the world's 30 leading export nations of high-tech products (UNDP 2001: 42). The following sections will examine these developments more closely, along with the factors leading to Intel's decision to select

Costa Rica for its first production plant in Latin America, the conditions the company demanded and how much developmental potential the new high-tech sector holds for the country.

THE WHALE IN A SWIMMING-POOL: FROM STRUCTURAL CHANGE TO A DUALIST ECONOMY

The structural changes in the Costa Rican economy, which in the 1980s were initiated with the promotion of non-traditional agrarian exports, took on impressive proportions in the 1990s. In 1999, a mere 15 percent of Costa Rica's exports came from the two traditional mainstays of the country's economy, coffee and bananas, and each was surpassed by textiles (12 percent) and other non-traditional exports. Moreover, no less than 38 percent of all national exports fell under the category of 'electronic components' (see figure 8), all of which were the products of a single company: microprocessor chips from the Intel site that had started operations in Costa Rica in 1997.

Rather than express a general structural change, the above data reflect the duality that has begun to characterize the Costa Rican economy. As a consequence, to adequately reflect the state of the country's economy, all macro-economic data have to be divided into separate sets of data: one referring to the economy including Intel and one excluding Intel. For instance, the astonishing fact that Costa Rica's chronic foreign trade deficit (of minus US$497.6 million in 1997) almost over night turned into a surplus (of US$632 million in 1999–see Larraín et al. 2000: 15) is simply and solely explained by the 'Intel effect.'

Given the enormous weight Intel as a single company has within the entire Costa Rican economy, Intel's Vice-President Bob Perlman has spoken of the Costa Rican site as "putting a whale in a swimming-pool" (cited in:

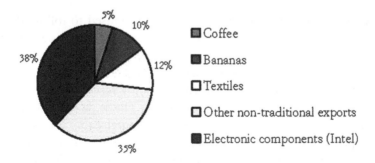

Figure 8 Costa Rica: Composition of Exports, 1999

Source: Banco Central de Costa Rica

Nelson 1999: 6). The dimension of the 'Intel effect' on Costa Rica's economy is also illustrated by a closer look on the composition of GDP growth.
In 1998 and 1999 Costa Rican GDP growth was at 8 percent per year, the
highest rate of all Latin American countries. Of this 8 percent growth, the
growth due to Intel alone accounted for more than 2 percent of GDP in
1998–and in 1999 it was even responsible for 5 percent (Larraín et al. 2000:
13–see figure 9).

The impact of Intel on the development of exports paints a similar picture: Intel produced more than one third of Costa Rica's record US$6.8 bil-
lion in exports in 1999. Figure 10 clearly shows that the strong growth in
Costa Rica's total exports between 1996 and 1999 was due almost entirely
to the 'Intel effect,' as exports in the rest of the economy showed only minor
fluctuations.

With net exports of US$1.5 billion in 1999, Intel's impact on Costa
Rica's foreign trade balance is impressive. However, this should not lead to
the assumption that Intel's net exports translate into net foreign currency
earnings for the country. According to Intel's own data, in 1999 US$1.2 bil-
lion—80 percent of its export earnings—immediately left the country as
transfers to the mother company based in the United States. The remaining
US$300 million still are a sizable amount for the Costa Rican economy;
however, it is not certain that Intel's official data are necessarily exact, nor
is it sure that the future relationship between export earnings and transfers
will continue to be as favorable for Costa Rica.

The dualist character of the Costa Rican economy is underlined by the
system of free production zones (*zonas francas*) under which Intel operates
and which exempts the company from all import and export duties, as well
as tax payments, for eight years. Since 1982, nine such free zones were established in the country; they currently house more than 200 companies,

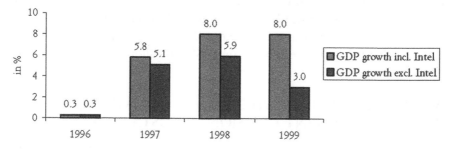

Figure 9 Costa Rica: The 'Intel Effect' in GDP Growth, 1996–1999

Sources: Banco Central de Costa Rica, Procomer

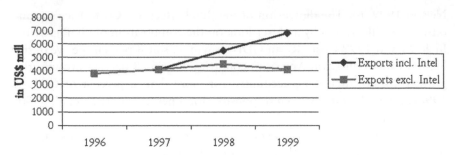

Figure 10 Costa Rica: The 'Intel Effect' in Exports, 1996–1999

Sources: Banco Central de Costa Rica, Procomer

most of them foreign-owned. In 1997, before the arrival of Intel, they made up 25 percent of the country's total exports; in 1999, including Intel, they accounted for 54 percent (Minkner 1999: 166f., Jiménez 2000: 61 and Delgado 2000: 5–11).

Such figures have led to the argument that the dividing line for the stated dualism in Costa Rica's economy should cut along a different line. That is, it should not consider 'Intel vs. rest of the economy,' as described above, but a broader distinction of 'free zones vs. rest of the economy.' The official data of the country's Central Bank indirectly follow this method when, under the heading "exports by principal products," they list coffee, bananas, sugar etc. and end with "free zones"—which certainly are not a 'product,' but a specific regime of business operation (cf. www.bccr.fi.cr).

However, for the present analytical problem, this approach offers no convincing solution, since the weight of Intel then would continue to distort the data—not that of Costa Rica's economy in general, but that of its free zone production. For instance, when Intel's exports sharply dropped after 1999, the data for 'exports from free zones' show a fall by one third, from US$3.6 billion in 1999 to US$2.3 billion in 2001. However, this apparent decline is completely misleading, since the production under the free zone regime, discounting Intel, actually expanded strongly in this time-span: it increased by more than one third, from US$0.9 billion in 1999 to almost US$1.5 billion in 2001.[13]

ATTRACTING THE CLUSTER'S CORE: FACTORS FOR INTEL'S INVESTMENT DECISION IN COSTA RICA

For Third World countries, the strategy of promoting high technology clusters generally relies on the attraction of a sufficiently strong investment from one or more dynamic foreign companies, well integrated into the world market, which will act as the core of the hoped-for future cluster. While the production of computer equipment was limited to the largest Third World countries in the import substituting industrialization scheme, this changed with the world market orientation of the export-led model that became dominant in the Third World in the course of the 1980s and '90s. Still, Intel's choice of such a small country for such a large production plant caused much surprise. Indeed, Intel's initial list of possible sites concentrated on the medium-to-large Third World countries that already had a considerable industrial base: in Latin America these were Argentina, Brazil, Chile, and Mexico. Costa Rica was not even on the list, as the head of the 'Intel Site Selection Team,' Ted Telford, acknowledged (cited in Nelson 1999: 3).

A central role in attracting the company's interest went to the *Coalición Costarricense de Iniciativas de Desarrollo* (CINDE), a business-led private association founded in 1982 and largely funded by US-AID. While CINDE's initial mission had been to promote private sector interests in the process of turning away from the ISI model, in the 1990s its work focussed on lobbying for foreign direct investments. When the Figueres government made the bid for the Intel site a top priority, it handed responsibility to CINDE, which was as central to contacting the company as it was to the Costa Rica visit by Intel's site selection team. It is precisely this transfer of a task, traditionally performed by the Ministry of Exterior Commerce or a similar governmental institution, to a private, business-led association that both textbook stories for business schools (e.g. Nelson 1999) and the United Nations Development Program (UNDP 2001: 81) have celebrated. Not bound by bureaucratic regulations that apply to public budget spending and the public service's salary structure, CINDE could freely spend substantial resources in the bid for the Intel site. As Larraín et al. (2000: 4) conclude: "The process of making Intel executives aware of the advantages that Costa Rica represented for the company was neither easy nor cheap in financial terms."[14]

However, all this marketing activity would have been of no use if the conditions offered by the country had failed to impress the company. There are essentially two different motivations for a manufacturing company to invest in a Third World country: either to obtain access to that country's domestic market or to benefit from the differential in production costs. As Intel produces entirely for the world market, the motivation doubtlessly lies in the latter consideration.

For the Intel investment, a *conditio sine qua non* was favorable fiscal and tariff regulations. In Costa Rica, a number of free production zones had been created since 1982, in the course of the country's gradual liberalization process; these zones provided a duty- and tax-free environment, without the need to create these conditions especially for the company. The positive examples of other transnational companies, such as Motorola, DSC Communications and Baxter Healthcare, which had years of experience producing in the country's free zones, carried significant weight in Intel's decision for Costa Rica.

For a production site with 2,000 employees, such as Intel's, the most important differential in production costs between a Third World location and production in the United States is certainly in labor costs. The case of Intel's Costa Rican plant, which covers the last stage of microprocessor production (assembly and testing), is somewhat atypical in the Third World, as it requires a rather high share of qualified personnel with a background in information technologies and engineering. Whereas Costa Rica' wage level for low-skilled labor is considerably higher than in other Third World countries (causing industries like textiles to move to nearby Honduras), it does hold a competitive advantage in the segment of medium to highly qualified personnel, due to the country's high educational level and the relatively broad experiences with NICT. In addition, Intel explicitly sought a country without combative labor unions or any tradition of strikes in the private sector—which Costa Rica could deliver (Nelson 1999: 9).

Nevertheless, the issue of labor brought up what seemed Costa Rica's most crucial disadvantage: its small domestic market. Though this was no obstacle in terms of demand for the product (as it would have been within the ISI scheme), it did raise strong concerns regarding the country's ability to supply the high number of qualified workers that the company needed. On this aspect, the central element in allaying Intel's fears was the *Instituto Tecnológico de Costa Rica* (ITCR) in San José, one of the most prestigious centers of higher education in Central America. Not only was such an institution present, but the Costa Rican government and the ITCR's directorate were also willing to restructure the institution according to the needs of the transnational company.

A third key reason for Intel's decision to invest in Costa Rica stems from the particularities of Costa Rica's political and social development: the extraordinarily high political stability, the relatively high social cohesion and the very low levels of social conflict over a long time, in addition to—as the company saw it—a high level of predictability in the country's institutions, a low level of corruption and a smoothly functioning legal system. In fact, a number of accounts point out that Intel shied away from special conditions other governments were willing to grant, since they feared that any such preferential treatment or 'Lex Intel' could either trigger political strife or be undermined after a change in government. For instance, Larraín et al. (2000: 6) conclude:

"A fundamental point is that the Costa Rican government did not promise Intel any special benefits, fiscal or other; rather it offered existing advantages that any other foreign investment under similar conditions could obtain. The latter is a key factor in making FDI [foreign direct investments—B. H.] policy credible and reducing the perceived risk of policy change."

This version has come to be accepted standard wisdom, even by some outspoken critics of the neo-liberal discourse such as Hanson (2001), who argues that the lack of specific subsidies to Intel is the most striking positive difference to other cases studied in his critical discussion of the developmental impact of foreign direct investments (ibid: 30). However, such praise paints a rather too idyllic picture. In fact, Intel did demand specific conditions for its investment, and the Costa Rican government acquiesced. What was different about these arrangements was their form: the government, rather than creating particular laws, subsidies or regulations for the Intel investment, tailored (where needed) the general regulations of the country to the interests of the transnational company, as we shall see in the following section.

THE POLITICAL SPILL-OVER: THE TRANSNATIONAL COMPANY'S PRESSURE FOR LIBERALIZATION— AND ITS LIMITS

Before and during the set up of Intel's production plant in Costa Rica, the company pressed for a number of conditions. One of these referred to the incentive and tax exemption system of the free production zones, which Intel found insufficient. The Costa Rican government heeded the call—but it did not seek preferential treatment for Intel, rather it launched a bill improving the conditions for all companies in the free zone regime. In 1998, this bill passed in the Legislative Assembly and became law.

A second point referred to air traffic. Since practically all of Intel's export and import shipments are done by air, in view of the company's tight time schedule and its globalized production chain, the company put a high priority on securing a sufficiently high number of international flights to the United States and other destinations. The government acquiesced by launching an 'open skies' program, which eliminated all restrictions on air traffic, and attracting a number of additional international airlines to serve the San José airport (located less than 10 kilometers from the Intel production site). A similar concession involved the customs clearance office at the airport which, as Intel discovered, did not work on Sundays—a situation the microprocessor company found completely unacceptable and the government immediately ordered that it be changed.[15]

Another point of contention was electricity. ICE's tariffs for industrial users were between US$0.07 and US$0.09 per kilowatt-hour (kwh); Costa Rica's strongest competitor in the bid for the Intel investment, the Mexican city of Guadalajara, offered US$0.02 per kwh. Intel saw a need to negotiate with Costa Rica while avoiding 'special treatment.' After discussing of the issue with the government, the solution was that the ICE announced a new, reduced rate of US$0.05 per kwh for especially heavy users of electricity, which would apply to any company using more than 12 megawatt-hours (Nelson 1999:10). Curiously enough, Intel was the only company in the country with this level of consumption.

As some aspects of the Costa Rican system required negotiation and adjustment, what had seemed to be Costa Rica's principal disadvantage, its small size, turned out to be quite useful: "Executives of the company seem to have valued the fact that Intel's bargaining power would be greater in a smaller country, as opposed to a larger one like Mexico" (Larraín et al. 2000: 4). This is the other side of the 'Intel effect': the 'whale' did not need special treatment. Because it was so huge in relation to the 'swimming-pool,' general conditions were tailored to meet its particular needs.

Notably, authors like Larraín et al. (2000) do not take this 'bargaining power' as an uncomfortable fact that has to be accepted, rather they welcome it as a highly desirable effect of foreign direct investments: "Large FDI firms like Intel may help bring about needed institutional reforms by influencing the political balance through a new arrangement of stakeholders" (ibid: 2). Typical examples are the intensification of air traffic through the 'open skies' program or the seven-day-opening of airport custom services, which advocates of this line of thought describe as measures that also benefit other business sectors and the tourist industry. (Obviously, this argument considers this type of reform beneficial for the country, a view that particular social actors, shouldered aside by the "new arrangement of stakeholders," probably would not share.)

Larraín et al. (2000: 2) describe this type of pressure for structural change somewhat awkwardly as a "backward-forward linkage." Their reference to the Costa Rican conflict over the state telecommunications and electricity monopoly is more than obvious when they write:

> "Consider a case where there is opposition to private participation in some type of essential infrastructure, like electricity generation and distribution. (. . .) The arrival of the multinational corporation, and its demand for the service, in terms of both quantity and quality, may become an important force to weaken opposition to the reforms, by making evident the insufficiency of the existing service and the incapacity of the government to invest the required amounts to satisfy demand. (. . .)

Multinationals would in this indirect way play an important political role in pushing for the required structural reforms" (Larraín et al. 2000:8f.).

When Larraín et al. (2000) wrote their text, they took it for granted that the *Combo del ICE* law soon would pass and they attributed this success in part to the Intel investment:

"The political balance among stakeholders in this realm [telecommunications] seems to have changed after Intel's arrival, and the reforms are more likely to take place, benefiting not only Intel, but also all other companies in the Costa Rican economy, and even the consumers themselves" (ibid: 30).

However, on this point the political spill-over from the 'Intel effect' showed its limits. First, the country's telecommunications regime had *not* been a primary concern for Intel's investment decision. Since RACSA could provide sufficient bandwidth for Intel's worldwide 'virtual factory' at an acceptable price, the transnational company did not see the state telecom monopoly as negatively affecting their operations in the country. As to electricity, the state monopoly offered the advantage that the government could respond directly to Intel's call for cheaper tariffs.

Second, as we saw above, the "political balance among stakeholders" did not change in the way Larraín et al. (2000) anticipated and the "consumers themselves" had different thoughts about what would be beneficial for them. Intel was intelligent enough to restrain itself from public comments on the *Combo* conflict, and when the conflict escalated and road access to the airport was blocked, these activities affected the company's business interests far more severely than the state monopoly could have. The reaction from the Intel leadership was prompt: it pressured the government to restore normal production and export conditions—and if no other alternatives were viable, then it approved of withdrawing the liberalization law package.

A year later, when Intel announced cut-backs on the expansion plans for the plant in Costa Rica, the company explained this step with its "worries about the strain such developments would put on Costa Rica's already overloaded electricity and telecommunications infrastructure" (Latin American Economy & Business 01–03, March 2001: 28). This, however, was little more than a weak pretext: since the global economic downturn had affected the demand for computer chips, Intel's plants produced far below capacity and the company announced that it would have to lay off 5,000 employees worldwide (Pinzler/Tenbrock 2001). In this situation, the most radical liberalization law in the world would not have led Intel to expand its investment in Costa Rica.

NICT AS 'COSTA RICA'S NEW COFFEE BEANS'? LINKAGES AND DEVELOPMENT IMPLICATIONS OF THE INTEL INVESTMENT

So far we have analyzed the factors for Intel's decision to invest in Costa Rica and its pressure for structural reform measures in the country; now we shall take up the issue of Intel's impact on the country's economy and its development perspectives in more detail. Is the country really moving "from Coffee to Semi-conductors" (Hershberg/Monge/Pérez 2003)? How far has the Intel investment shown positive linkage effects to the rest of the economy? Has it indeed led to the development of a dynamic cluster, in which transnational and national, private and public enterprises and institutions build an innovative network?

One of the most immediate effects of the attraction of foreign direct investments by Third World countries is the creation of employment. Higher employment in turn raises national income, contributions to the social systems and consumer demand. In the case of a high technology investment such as Intel, additional motivations for attracting such companies are the improvement of the country's technological competence, the knowledge transfer and the impulses the investments might create for a more general process of industrial upgrading. Between 1990 and 1997, an estimated 21,300 jobs were created by companies operating under the free zone regimes in Costa Rica, with the largest part in *maquiladora* industries with low salaries and few skills needed (Minkner 1999: 166). The case of Intel is different. The company employs a work force of about 2,000 people and more than half of these jobs require skilled and highly skilled labor, with salaries that on average are more than 50 percent above those at other industrial companies operating in the country (Larraín et al. 2000: 11). The sum of salaries and social welfare contributions amounted to US$25.3 million (ibid: 9) in one year.

In contrast to the typical sweatshop-style *maquiladoras* of the textile sector, which can easily be transferred to production sites in other countries, the Intel plant is a long-term strategic investment. The invested capital of US$ 420 million is equivalent to about 2.6 percent of Costa Rica's annual GDP. This is a sizable sum even for a company like Intel, so the transfer to another production site would be an emergency option, rather than a means to seek benefits from salary differentials or similar advantages in other locations. Due to the complexity of much of the work in the plant and the time needed to train the employees, Intel seeks a low fluctuation rate of its employees and offers relatively good labor conditions for its qualified personnel.

It is noteworthy that, with a few exceptions, Intel's work force comes entirely from the national labor market. At the beginning of 2000, only six leading employees were from the company's U.S. headquarters and the rest were Costa Ricans, according to information from Costa Rica's former planning minister, Juan Manuel Villasuso (author's interview, April 24, 2001 in San José). More than 40 Costa Rican employees have come to occupy high level positions in Intel sites in other parts of the world. In addition, a number of Costa Rican engineers and informatics experts have been grouped together in a new 'Latin America Engineering Services Group' within the company, with the task of developing new methods in the design of microprocessors and software development. After 18 months in the Intel headquarters in the United States, they now operate in the Costa Rican plant which, with the addition of this group, includes not only assembly and testing, but also a genuine research & development branch (La Nación, June 2, 2001, and information provided by Intel Costa Rica through www.intel.com/costarica). Thus, if critics have spoken of the Intel site as 'high tech maquiladora,' they have a point in reference to the exploitation of the wage differential and the attraction of the free zone regime as the prime investment motivations; however, it is highly misleading in regard to the type of labor employed and the working conditions, and it also ignores the development of technological competence in the country.

In relation to the whole of Costa Rica's labor force, the Intel plant, of course, has created such attractive employment opportunities only for a tiny number of people. However, the public effect has been immense; enrollment in the engineering fields in the two most important higher education institutions of the country almost doubled in the two years after Intel took up operations (Larraín et al. 2000:23). For the Costa Rican society, the Intel investment has provided a tangible illustration for the emphatic political discourse that has pinned the developmental hopes of the country on the 'information society,' in which Costa Rica's crucial assets are its high educational standards and its human capital. On this background, education in general has received a remarkable stimulus, as much on the microlevel within families as on the macro-level of state budget considerations.

It is precisely in the education system that Intel's presence has shown some of the most noticeable linkage effects. As noted earlier, the close cooperation with the *Instituto Tecnológico de Costa Rica* (ITCR) was an important part of the country's bid for the investment. After the plant was set up, this cooperation continuously expanded. The ITCR was promoted to the status of an official 'Intel Associate' and, as such, it is integrated into a worldwide network of prominent academic institutions cooperating with the company. Besides international recognition, this brings a number of benefits to the ITCR: a series of exchange programs with the other universities of the network have been established; Intel has equipped the Institute with state-of-the-art computer technology; additional courses and certificates can be offered; Intel finances English classes; and the ITCR can apply to a US$300 million a year fund established by Intel exclusively for research projects by its 'Associates.'

While some have praised these changes, the effect is certainly rather ambiguous. One price for the material benefits from Intel is a very visible subordination of the ITCR to the company's needs, both in research projects and curricula. For the Institute's students, this orientation toward Intel may be useful in career terms, since many aspects of the 'Intelization' of the ITCR tend to be well received by other business sectors (cf. Nelson 1999: 8). Nevertheless, there is no doubt that the 'new ITCR' lacks the autonomy of higher education and research which had been a guiding principle in Costa Rica's educational system. In fact, it spearheads a transformation of public education institutions into appendices of private business. Moreover, while most international attention focuses on the *Instituto Tecnológico,* the early pioneering work—and much of the current work—in the field of computer networking has been done by the *Universidad de Costa Rica,* which is far less business-oriented and operates without any similar corporate attachments.

As to direct fiscal linkage effects from the Intel investment, in the short and medium term these are limited to the income taxes from the employees' salaries. While the state's internal debt has become a pressing macroeconomic problem and has led to a series of federal spending cuts, Intel and all other companies operating under the free zones regime are entirely tax-exempt for eight years. Currently, the company pays only a minor fee to the community in which it is located. Obviously, beyond fiscal considerations, this also creates a general sense of injustice and unequal treatment felt by many in the country, since the biggest and richest company in the country does not pay taxes while any local small or medium-sized business has to.

Additionally, for a development strategy focusing on industrial clusters, the linkages between the foreign company and domestic companies and institutions are of special importance. However, tax and tariff exemptions under the free zone regime distort costs leading to a form of inverted protectionism that is structurally biased against national production, in favor of imported production inputs. Due to this effect, in their definition of clusters, Altenburg / Meyer-Stamer (1999: 3) specifically excluded any agglomerations within export processing zones, since these by their very structure do not build upon intensive linkages.

In addition, in the case of Intel it is particularly difficult for any local company to insert its products into the company's production process, due to its high degree of technological sophistication and due to Intel's complex global logistics. Even the little stickers with the 'Intel inside' logo are not produced locally, but in one single section of the company's global production chain, to be shipped from there to wherever needed.

In the case of Intel Costa Rica, an estimated 200 local companies work in some capacity as suppliers. However, low value-added products and low-tech services, including the cafeteria, transportation, security services, garbage disposal and insect control make up the majority of theses (Larraín et al. 2000: 24). Therefore, a program organized by CINDE and sponsored by the Inter-American Development Bank, which aims at strengthening national companies' capacity to become suppliers to high-tech-companies in the country's free zones (ibid: 28), faces long odds.

The promotion of Intel investment by the Figueres government included the argument that this would have a demonstration effect on other companies, drawing an avalanche of high-tech firms to Costa Rica. These high hopes have not been fulfilled. The Intel investment, while having a certain positive demonstration effect, at the same time created a major problem for the attraction of other high-tech companies, since it absorbed much of the qualified labor market in these technical fields; due to the increased competition, a number of companies in the country felt it necessary to raise the salaries in order to keep their technical staff (Larraín et al. 2000: 21f.). With an increasing number of graduates from technical majors, this situation is likely to change in the coming years. Nevertheless, only a handful of companies have followed Intel. The largest of these is a U.S. company, *Photocircuits,* which established a 210 employee plant producing printed circuits as a direct supplier to neighboring Intel's production of microprocessors.

From a development perspective, probably more important than the attraction of other transnational companies as Intel suppliers is the development of a national software industry. Though long overshadowed by the Intel investment, a recent UNCTAD report highlighted Costa Rica as "a rising star in computer-related services exports" (UNCTAD 2002: 236), taking the country's software sector as a prime example of Third World countries successfully exporting NICT services (ibid: 235–238). It is noteworthy that this sector began to emerge well before the arrival of Intel: 39 percent of the currently operating companies were set up between 1991 and 1995

(Caprosoft 2001). Contrasting it with the Intel investment, Stamm (2002: 79) characterizes the Costa Rican software sector as a "bottom-up innovation cluster," and he argues that so far it has developed rather independently from the transnational chip producer (ibid: 81). Since quite a few of the software companies have emerged from or are linked to the ITCR, Stamm emphasizes that the *Instituto Tecnológico* serves as the main hinge between the software sector and Intel (ibid: 80).

These arguments are a well-founded counterpoint to the dominant perception that—in somewhat blunt terms—whatever high tech has developed in Costa Rica is there thanks to Intel. Nevertheless, there is a specific 'Intel effect' that should not be underestimated: the chip company's decision to set up shop in Costa Rica put the little country, over night, onto the global map of high-tech producers. For the international marketing of "software made in Costa Rica," this image gain certainly is of great value. Costa Rica's largest and most successful software producer, ArtinSoft, which was founded in 1993 as a spin-off of the Instituto Tecnológico de Costa Rica and quickly became one of the world's leaders in automatic software migration and conversion tools, has benefited directly from the Intel investment. In February 2001, Intel announced an investment in ArtinSoft, its first in any Latin American company, and a few months later, Microsoft also announced an alliance with ArtinSoft (cf. www.artinsoft.com; Latin American Economy & Business 01–03, March 2001: 28; and La Nación, July 24, 2001).

Beyond ArtinSoft, which today has more than 250 employees, Costa Rica's software sector consists of about 150 small or medium-sized companies. 80 percent of these are nationally owned and employ 3,500 to 4,000 mostly highly skilled employees, according to the Association of Costa Rican Software Producers, Caprosoft (cf. www.caprosoft.org). Even though the sector's share of national GDP is still rather small, Costa Rica can boast the world's highest rate of software-producers per capita (Bortagaray/Tiffin 2000: 22). While initially providing tailor-made software mainly for national companies and institutions moving into e-commerce, e-government or e-tourism activities, the limited domestic market has made the expansion of the sector dependent upon exports. Since 1998, the export value has surged from below US$5,000 to more than US$60 million in 2000, rising in just three years from close to 0 to 3.2 percent of total exports (UNCTAD: 2002). Very much in contrast to the far more voluminous exports by Intel, Costa Rican software has 90 to 95 percent of its value added domestically (ibid and Caprosoft 2001).

Any export-led development model is highly dependent on the ups and downs of the world market. This is particularly true if exports are highly concentrated on a single product, even if that product is sophisticated microprocessors. This became all too clear in 2000, when the downturn in the U.S. and Asian economies sharply reduced the demand for Intel's product, leading to a decrease in its exports of 34.5 percent. Due to the enormous weight of the company, this immediately appeared in the macro-economic indicators of the country in general: export data showed a steep decline (see figure 11), and the GDP growth rate fell from 8.0 percent in 1999, the highest in Latin America, to 1.5 percent in 2000 (figure 13), one of the lowest rates on the continent. With the low international demand for computer chips continuing in 2001, Intel's exports were halved once again, reducing the national growth rate to 0.9 percent. Suddenly, the spreading news was the 'negative Intel effect.'

However, the above analysis has also shown that the supposed 'hyper dependency' of the country on Intel is far greater in the statistics than in the real economy, as the impact of the sharp decline of Intel's exports was actually quite limited for Costa Rica's economy. GDP excluding Intel remained rather stable, with a growth rate of about 3 percent between 1999 and 2001 (figure 12) and, again not counting Intel, the country's exports actually increased slightly (figure 11). In the good as in the bad, the inter-connection between the Intel plant and the rest of the Costa Rican economy is still rather low.

In the final analysis, the developmental impact of the Intel investment remains ambivalent. What is clear is that the talk about the NICT being

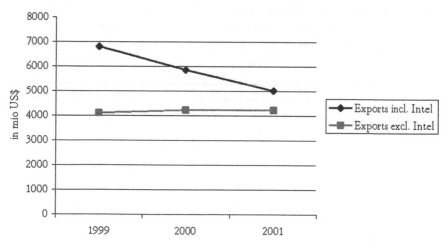

Figure 11 Costa Rica: The 'Negative Intel Effect' in Exports, 1999–2001

Sources: Banco Central de Costa Rica, Procomer

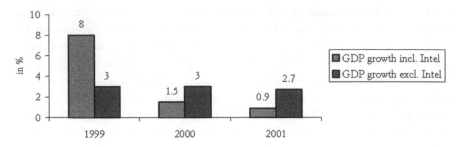

Figure 12 Costa Rica: The 'Negative Intel Effect' in GDP Growth, 1999–2001

Sources: Banco Central de Costa Rica, Procomer

'Costa Rica's new coffee beans' is highly misleading. Unlike the current NICT economy, the Costa Rican coffee economy historically has been a paradigmatic example of a narrowly woven net of economic structures that became the vertebral column of the nation, with numerous linkage effects and the integration of large social sectors. While Bortagaray/Tiffin (2000: 21f.) characterize Costa Rica's Central Valley as a promising 'proto cluster' with Intel as its anchor point, a sober assessment has to note that, until now, Intel has not sparked wide cross-fertilization effects or other cluster dynamics. Therefore, Lindegaard/Vargas (2002), who contrast the Intel experience in Costa Rica with the densely integrated high tech region in North Jutland (Denmark), underscore the persistence of the chip company's enclave character, referring to "Intel island" (ibid: 1). They conclude that the links to the local environment would have to be greatly strengthened in order to consider a local system of innovation. Though the government and other national and international institutions are making efforts in this direction, they face considerable structural obstacles.

In fact, the Intel plant illustrates the typical problems of modernization approaches that promote a modern sector as a 'motor of development.' It has shown that Third World countries can become production platforms for foreign direct investment not only in agricultural, mineral or *maquiladora* sectors, but in high-technology products. But the much thornier question is whether, and to what degree, this production can be integrated into the national economy and perhaps serve as a 'motor' that would stimulate the country's economy and society as a whole, with more than minor 'trickle down' effects. If it cannot, an investment like Intel's in Costa Rica ends up deepening the structural heterogeneity and economic dualism of the country instead of helping overcome it.

Part III
Latin America's 'Socialist Model': Cuba

Chapter Five
Cuba's State-Socialist Development Model and Its Telecommunications Regime

For more than four decades, Cuba has been the most prominent model of socialist development in the Third World. [1] It was the Revolution of 1959, led by Fidel Castro that laid the foundations for the political system that prevails today. This situation makes for a highly specific context for the political and developmental debate about the new information and communication technologies in the Third World.

As noted in the introduction, it has become a leitmotif in the political debate about Internet & co. that, due to their intrinsically transnational and pluralist character, they have an erosive effect on non-pluralist forms of government. "In the new century liberty will spread by cell phones and cable modem!" U.S. President Bill Clinton declared in March 2000 (Clinton 2000). This type of belief has been specifically applied to the Cuban case: for instance, returning from an official visit to the island, Hans-Olaf Henkel, then Chairman of the Association of German Industry (BDI) and earlier the head of IBM Germany concluded: "Internet will topple Fidel Castro!"[2] Will it really?

THE POLITICAL REGIME: CHARISMATIC STATE-SOCIALISM AND ITS ENEMIES

When the Revolution led by Fidel Castro triumphed on January 1, 1959, it was not yet a socialist revolution. Instead, its character was essentially one of radical nationalism with an ambitious social reform program. However, the implementation of this agenda quickly led to a radicalization of the revolutionary project and to a sharp polarization of the political conflict, within the country as much as between Cuba and the United States. This process culminated in the official proclamation of the 'socialist character' of the Cuban Revolution on April 16, 1961, on the eve of the 'Bay of Pigs' invasion by forces of Cuban exiles trained and supported by the U.S. government. This invasion sought to overthrow the Castro government, but in its failure it actually marked the definite consolidation of the revolutionary regime's power.

Before the triumph of the Revolution, one of the central political banners of the 'M-26' movement led by Fidel Castro had been the restitution of the progressive, if essentially liberal, 1940 Constitution that had been suspended by the Batista regime after its putsch in 1952. The rebels' promise was the establishment of a sovereign and non-oligarchic political regime with a strong social commitment, but they made no mention of Marxist or Leninist lines of thought.[3] However, this mission was soon transformed. Rebel leaders reinterpreted the Revolution's commitment to democracy, measuring the 'democratic character' of the regime by the economic and social transformations that were to favor the objective interests of the masses. On the basis of the corrupt, oligarchic and repressive regime of the pre-revolutionary period, party-pluralism was cast as a system inherently serving the interests of an elite and descriptively termed *pluriporquería* (roughly a 'plural-pigs-system'). The Revolution would ratify its democratic character not by elections, but through the direct communication between 'the people' and the Revolution's charismatic leader; the most prominent expression of this communication was mass mobilization on Havana's central *Plaza de la Revolución*.

Although we have spoken of the binomial rule of party and state as typical of state-socialism in the Soviet Union and Eastern Europe, the Cuban case is somewhat different. In Cuba, socialism neither came as a product of the country's pre-revolution Communist Party—to which Fidel Castro did not belong, and which for a long time rejected his 'adventurism'—nor in the wake of the tanks of the Soviet Red Army. When the Revolution triumphed, Fidel Castro's leadership essentially relied on two supporting factors: the guerrilla army and the mobilization of the masses. While the latter became institutionalized in new revolutionary 'mass organizations,'[4] the rebel army became the national military and the backbone of the revolutionary state. What today is the Cuban Communist Party (*Partido Comunista de Cuba*, PCC) was a creation of the revolutionary state, not vice versa. Fidel Castro's olive-green army fatigues are a prominent reminder of this order of events.[5]

The different titles of Fidel Castro reflect the resulting parallel systems of power: of course, he is the First Secretary of the Central Committee of the Communist Party and the President of both the Council of Ministers and the Council of State, just like any other leader of a state-socialist country. Fidel Castro's principal title, however, is *Comandante en Jefe de la Revolución*

Cubana (Commander-in-Chief of the Cuban Revolution), although neither the country's Constitution nor the Party's statutes mentions such a position.

Much in contrast to the bureaucratic character of government typical of state-socialism in Eastern Europe, Cuba's brand of state-socialism prominently bears the personalist stamp of its charismatic leader. It was not until the 1970s that, in the so-called 'Process of Institutionalization,' the nation's political structures were formalized in the mould of the Soviet model. It was not until 1975–16 full years after the Revolution—that the Cuban Communist Party celebrated its first national convention. The next year, the country passed a new socialist Constitution, which closely followed Soviet patterns. Article 5 defined the role of the Communist Party in the same wording as anywhere between Vladivostok and East-Berlin: "The Communist Party of Cuba, the organized Marxist-Leninist vanguard of the working class, is the leading force in society and state (...)" (República de Cuba 1976).

However, this ideological alliance with the socialist camp should not hide important differences between Cuba and the state-socialist countries of Eastern Europe. For a long time, the Cuban Revolution certainly could count on the authentic support of a large majority of the population. Moreover, whereas in Eastern Europe's state-socialist countries nationalism was a key part of the opposition, the Revolution in Cuba was so deeply anchored in the historic struggle for national independence that nationalism has always been a trump card of the socialist government.

Ever since the early days of the Revolution, the conflict with the United States has provided the fundamental backdrop for the dichotomous military logic dominating Cuban politics, based on an 'us vs. them' view. The external conflict, however, translated into domestic politics. In the context of a 'besieged fortress mentality,' the call to 'close ranks against the enemy' became an overriding justification for rejecting not only political pluralism, but also deviant positions within the system. Any sign of dissent or friction became a threat to the 'unity of the nation' that was regarded as vital to the fight against the imperialist designs of the United States.

On the flip side, in the context of the Cold War, the United States saw the Cuban Revolution not only as a violation of its long-standing political and economic interests on the island, but also as an unacceptable Soviet incursion in the Western Hemisphere. In addition, in the *longue durée* of international politics, since the United States had banned European powers from any expansion in the American hemisphere in 1823, the Soviet alliance with Cuba seemed an intolerable breach of the Monroe doctrine (Smith 1996: 20ff.). Furthermore, Castro's Cuba also was seen as a dangerous example for other Latin American or Third World countries.

The prospect of an outright military confrontation between Cuba and the United States faded after the failure of the Bay of Pigs invasion and the negotiated solution of the missile crisis of 1962. In the latter, the United States guaranteed to abstain from any direct military intervention against the island in return for the Soviet Union's withdrawal of the missiles stationed in Cuba.[6] Although covert actions continued for some time, the focus of U.S. Cuba policy essentially shifted to economic pressure through an embargo on all trade between the two countries, which was enacted in the early 1960s. This policy also included pressure on third parties with trade relations with Cuba and a U.S. veto against any cooperation with Cuba in the major international institutions. Parallel to this, the United States continuously stepped up media efforts regarding Cuba, as part of their efforts to 'win the hearts and minds' of the people on the island.

THE SOCIO-ECONOMIC TRANSFORMATION: STATIZATION AND INCLUSIVE DEVELOPMENT

Before 1959, Cuba had followed a model of dependent capitalist development. The island's insertion in the world economy essentially was based on a single traditional agrarian product, sugar, which made up no less than 90 percent of Cuban exports. Cuba's dependency on the United States was ever-present, in the economy as much as in politics. With the rapid radicalization of the Revolution, the domestic economy underwent a fundamental transformation, essentially liquidating the capitalist order within the first two years. The island's international status also changed drastically, as the Castro government openly confronted the hegemonic claims of the United States and turned to the Soviet Union for support. Both the internal upheaval and the new external alliance laid the foundation for the state-socialist development model that Cuba pursued in the following decades.

At the time of Revolution, Cuba was not so much a 'poor' country as a highly dependent one, with enormous economic, social and geographic disparities. In the cities, above all in Havana, a relatively broad urban middle class had emerged with a life-style and consumption patterns similar to the standards in the United States. In 1957, Cuba had the second-highest income per capita of all Latin American countries and it ranked first in the number of private motorized vehicles and similar indicators (Pérez 1995: 296–297).

However, the chasm between social groups was profound. The relative welfare of the upper and middle classes went hand in hand with the dramatic economic and social exclusion of large parts of the population, which paralleled the striking disparities between the capital city and the rest of the country. In rural areas, semi-feudal labor structures persisted and living conditions were miserable. Cuba's overall literacy rate was 76 percent, but it was only 58 percent in the rural areas—and an even lower 50 percent in the marginalized *Oriente* province in the east of the island. Here, too, medical institutions were largely inaccessible to the majority of the population. If Havana counted one medical doctor for every 227 inhabitants, *Oriente* had 1 for every 2,423. The capital city, with 20 percent of the country's population, concentrated 62 percent of the nation's salaries, 70 percent of electricity use, and 73 percent of telephones (Pérez 1995: 295–303). Within Havana, too, the dividing lines between rich and poor were deeply entrenched and they were reinforced by open racial discrimination against 'mulattoes' and 'blacks.'

In the wake of the Revolution, not only were foreign companies nationalized, but also all major domestic business. The state expropriated the holdings of all large and medium-size landowners: the state was going to be not only the guiding force for national development, as in the case of the '*desarrollista* state' in the *cepalista* ISI strategy, it was to be the center of all economic planning and the owner of all major means of production. Based on this large-scale statization of the economy, rapid industrialization was seen as the primary vehicle by which to overcome economic dependency and backwardness. In its structure, the industrialization effort was essentially one of import-substitution. In pursuing this strategy, Cuba could draw as much on the Soviet experience as on concepts developed for Latin America's capitalist countries within the context of the ISI strategy—in fact, Cuba's first Minister of Economy had worked many years for the CEPAL (Fabian 1981: 393). As a result, Cuban industrialization plans never concentrated as strongly on the production of machinery and capital inputs as the Soviet development model had done. Instead, arguing that pre-revolutionary consumption levels were unacceptable and had to be raised immediately, the government also focused much attention on light industry and consumer products typical of the 'easy phase' of ISI (ibid: 389–41).

When the early industrialization plans failed, Cuba underwent a massive re-orientation towards the sugar sector. The Soviet Union had not only stepped in after the U.S. market closed to Cuban sugar, it also turned the island's sugar exports into the centerpiece of its trade relations and, through highly preferential prices, its economic support. Under these conditions, the sugar industry, which had been synonymous with dependency and underdevelopment, was declared to be the motor of development. This strategic turn culminated in the *Gran Zafra*, the project of a gigantic 10-million-ton sugar harvest in 1970.

The failure of this *Gran Zafra* sealed the end of a decade in which Cuba's economic development took on a character that was quite autonomous, experimental and 'utopian,' though often highly voluntarist.[7] After the 'push for Communism,' came the 'retreat to socialism,' as Eckstein (1994: 31) put it. The 'Process of Institutionalization,' initiated in the 1970s, not only streamlined the political structures along the patterns of 'really existing socialism' in the Soviet Union and Eastern Europe, it also introduced the economic planning system and bureaucratic administrative rationality of the Soviet model. These adaptations included the integration of Cuba into the Council of Mutual Economic Assistance (Comecon) in 1972, a step which institutionalized and extended the high trade subsidization the USSR had come to provide for Cuba.[8] Following a model of extensive growth, the Cuban economy showed high growth rates for most years between 1971 and 1985. Sugar remained the principal product in Cuba's international trade, and it helped finance a primarily import-substituting industrialization.

The break with the economic and political system of pre-revolutionary Cuba opened the doors to an equally fundamental restructuring of society. While political polarization led to a strong current of emigration, the development model proved highly inclusive for those who stayed on the island and were integrated into the socialist process. The strong push for a redistribution of the nation's wealth levelled the profound social discrepancies within short time, making Cuba one of the most egalitarian societies in the Third World (even if a culture emerged that privileged the 'nomenclatura,' the functionaries of the state, party, army and security apparatus; cf. Stahl 1987). The statization of the economy eliminated unemployment, the range in salaries was low and far-reaching social security provisions, as well as a high quality health and educational system equally accessible to all, were established. These social accomplishments of historic proportions became landmark symbols of the Cuban revolution and were important elements in the appeal Cuban socialism had across the Third World. In addition, a central focus of Cuban socialism was the levelling of the disparities between urban and rural areas. It should be noted, however, that underneath the official structures of the socialist society and economy, family ties and other informal social relations remained important elements; for this contradiction, the ironic term '*socio*lismo'—from *socio* (buddy)—was coined.

It is the great weight of the social achievements that earned Cuba, despite its low figures for national income, and even after the crisis decade of the 1990s, a rather high ranking in the Human Development Index elaborated by the United Nations Development Program (UNDP 2003; see also table 2). With an average of 76.5 years, life expectancy matches the level of developed nations, as do the adult literacy rate (96.8 percent), school attendance (99.9 percent of the children in the indicator's age group) and infant mortality (7 deaths per 1,000 live births) (ibid). With 590 physicians per 100,000 people, Cuba is the leading country worldwide in the quantitative provision of medical personnel (ibid).

Cuba's high educational standards are particularly important for the issue of the new information and communication technologies. Within the first years of the triumph of the Revolution, a large-scale literacy campaign was initiated. All private and religious schools were expropriated by the state and all educational facilities became free of charge. Prior to the Revolution, Cuba counted 12 institutions of higher education, with a total of 15,000 students; in 1985 the country had 46 such institutions with a total of 200,000 students (CEDISAC/Prensa Latina 1997). In addition, a considerable amount of state resources went to scientific activities: according to official data, in 1997 more than 60,000 specialists worked in 422 academic centers of diverse disciplines (ibid).

Although these accomplishments certainly are impressive, the contrasting downsides must be mentioned. Particularly in the areas of education and culture, two of the most prominent of the Revolution's accomplishments, heavy-handed political ideologization was widespread, reaching its most dogmatic phase in the 1970s.[9] Access to scholarships for study abroad and responsible positions in academic facilities often were dependent upon active participation in the official organizations. Early in the Revolution, political control extended to the media and, though less rigidly, to the cultural sphere. After a documentary film about Havana's night life—Sabá Cabrera's 'PM'—had been prohibited Fidel Castro explicitly defended the government's right to censor the cultural production with his famous "Words to the Intellectuals" in 1961:

> "Could one, in the midst of the Revolution, deny the government the right to evaluate, examine and confiscate the films to be shown to the people? (. . .) To deny the revolutionary government this right, would mean to deny it its function and responsibility to lead the people and to lead this Revolution" (Castro 1972: 356f.).

More than any formal law, the slogan Castro coined in this speech— "Within the Revolution: Everything! Against the Revolution: Nothing!" (Castro 1972: 357)—became the standard for what was allowed and what not. The precise boundaries of what was acceptable were marked by 'cases' of writers or artists who had passed the line of the permissible and received exemplary sanctions.

Of particular importance is the massive emigration that resulted from the Revolution. The polarization of the political conflict after 1959 led to an increasing stream of emigrants, mainly from the upper and middle classes. Between 1959 and 1962 more than 225,000 Cubans left the country; by 1985 the number was up to more than half a million; and by 1997 almost 900,000 people born in Cuba lived in the United States, mostly settling in and around Miami (Pérez 1999: 19–21). Today's Cuban community in the United States, in addition to this first-generation group, includes about 400,000 more people of Cuban origin, but born in the United States (ibid). After 1959, this mass emigration served as the principal escape valve for the social conflicts that accompanied the profound disruption of the old economic and social order. At the same time, however, this solution externalized and internationalized the conflicts of Cuba's society: the turn to an inclusive development model on the island was based on the exclusion and geographic separation of the former elites and of those who opposed the new course.

As the revolutionary regime consolidated its power, the Cuban exiles' political organizations centered their efforts less on activities directed towards the island and more on gaining influence on U.S. policy towards Cuba, to assure an uncompromising, anti-Castro attitude. In the context of the Cold War, this attitude was largely congruent with mainstream U.S. foreign policy, with Jimmy Carter's *détente* policy towards Cuba between 1976 and 1980 being the only major exception (cf. Smith 1987).

Over time, most Cuban emigrants to the United States became U.S. citizens. In 1980, the Cuban-American National Foundation (CANF) was founded and since then it has stood out as the single most powerful political organization of the Cuban emigrant community.[10] Based on their considerable economic power, reliable support from right-wing sectors in the United States and, additionally, the specificities of electoral arithmetic, which give the state of Florida strategic importance for presidential elections (cf. Hoffmann 2002), the Cuban-Americans, with the CANF as their dominant political lobby group, managed to exert enormous influence on U.S. policy towards the island. Although the emigrated Cubans may hold U.S. passports and most of them probably would not return to live on the island if Castro were gone, they still are a factor of great importance for Cuban politics— and even more so regarding the new applications of transnational media and communication that have come with the Internet.

CUBA'S WORLD MARKET CRISIS: CRISIS AND TRANSFORMATION SINCE 1986

Through its alliance with the Soviet Union and its integration into the Comecon, Cuba's socialist regime for a long time not only was able to resist the economic pressures and sanctions from the United States, it also was dissociated from the world market. However, this dissociation had been a partial one and in Cuba, as in other state-socialist countries, the exhaustion of the model of extensive growth in the 1980s motivated the government to seek hard currency credits. While many Cubans remember the first half of the 1980s as a 'golden age,' in which consumption levels rose markedly, it was this period that saw Cuba's debt towards Western creditors more than double within four years, from US$2,800 million in 1983 to US$6,100 million in 1987 (Carranza 1992: 132f.). In 1986, Cuba could not continue servicing its debt, and the country de facto declared insolvency towards its hard currency creditors. This crisis in Cuba's relations to Western creditors and, in consequence, to the capitalist world market, proved to be a prelude of the profound economic crisis that the country suffered after 1989, when world market integration became inevitable.

Other Third World countries in similar situations faced harsh negotiations with the international creditors and the IMF that inevitably led to 'structural adjustment programs.' Cuba, however, in 1986 still had an alternative to the international credit institutions: as a consequence of the unresolved debt crisis, the government decided to reduce trade with the West to a minimum and to reconcentrate foreign trade still more than before on the countries of the Comecon (ibid: 133). The government made this decision, however, precisely when Gorbachev's *perestroika* policies began to strain political and economic relations and, between 1986 and 1989, the Soviet Union gradually reduced its subsidies towards the island. In open contrast to Gorbachev's reform project, the Cuban government announced the so-called *Proceso de Rectificación* ('Process of Correction'). In this "unique case of anti-market reforms" (Mesa-Lago 1993) the government abolished peasant markets, re-centralized the economy and, invoking the idealistic visions of Che Guevara (cf. Tablada 1987), increasingly relied on moral appeals to substitute for material incentives and economic mechanisms. Ideologically, the *Rectificación* returned to a strong emphasis on the nationalist character of the Cuban Revolution, a turn which also implied a certain dose of self-criticism regarding the rigid adaptation of Soviet models since the 1970s.

In striking contrast to the deteriorating political relations, Cuban trade dependency on the Comecon countries reached an all-time high, at more than 85 percent of trade volume, in these same years (Comité Estatal de Estadística 1989). When these trade relations virtually imploded after 1989, Cuba fell into an economic crisis of dramatic proportions. Total imports fell from more than US$8 billion in 1989 to below US$2 billion in 1994 (see table 6). Practically all sectors of the economy suffered severely from the crisis and harsh shortages began to dominate life in Cuba.

Politically, Fidel Castro bitterly blamed Gorbachev for having destroyed socialism in the Soviet Union and for "handing over the rule of the world to the Yankee imperialism without having fired a single shot" (Castro 1993a: 7). As to the impact on Cuba, the government began to speak of "two blockades."[11] Russia and other post-Soviet states did in fact remain important trading partners for Cuba, but now, even if most arrangements were barter deals, trade with Cuba followed world market prices. It became increasingly clear that perhaps the most fundamental change facing Cuba was the need to integrate into a capitalist world market. This was a daunting task, as the country's economy was not at all prepared for the exigencies of such a market and—beyond agrarian products like sugar and mineral resources like nickel—most of its tradable output, particularly of manufactured products, simply was not competitive. Considering the challenges facing Cuba, Brundenius/Monreal (1998: 5) conclude: "In fact, the trade shock associated to [sic] those events [the breakdown of European socialism] should be better considered as the most immediate cause but not the only, and surely not the most important, cause of the crisis of the Cuban economy."

WAR ECONOMY WITH DOLLAR ENCLAVES: A SURVIVAL STRATEGY OF ECONOMIC DUALISM

The Cuban government responded to the external shocks with a dualist economic strategy. In the domestic economy, it proclaimed a strict emergency and austerity program. Its official name, *Período Especial en Tiempos de Paz* (Special Period in Peace-Time), immediately indicates that it was the adaptation of a concept developed for wartime. To cope with severe supply shortages, the government began to ration almost all goods. Fidel Castro himself declared: "At this moment we have virtually a war economy" (Castro 1991: 57). Though the de-facto war economy resulting from the government's strategy could not prevent very high social costs, it did manage to distribute these costs relatively equally across all sectors of society. This ability to spread out the social effects of the crisis doubtlessly was a crucial factor in safeguarding social cohesion and political stability in the first years after 1989.

In an earlier part of this work, we analyzed the state-socialist system of the Soviet Union and Eastern Europe as a sort of 'halved Fordism,' which implemented the rationalization of industrial work but lacked the counterpart of mass production of consumer goods and mass consumption. In Cuba, the perspective is somewhat different. Here, too, the government demanded material sacrifices from the population for the sake of future well-being. Nevertheless, the Revolution brought a substantial rise in living standards for a large part of the population. Yet with the crisis in the late 1980s and early 1990s, the fall in living standards was sharp. Public transport nearly collapsed, power was cut up to eight or even 12 hours daily and the population's consumption fell to painful levels; central 'achievements of the Revolution,' such as the free education and health system, also suffered greatly from the economic crisis. In both areas, a wide array of material inputs, from medicine to school books, became scarce or were simply lacking. The decline in real salaries led to a drain of qualified personnel and created great material and motivational problems for those remaining. Even basic food provision became so problematic that in 1992, 50,000 Cubans suffered from an epidemic of a rare eye disease, optic neuritis, which essentially resulted from insufficient nutrition.[12]

Despite the accompanying severe suppression of national consumption, reintegration into the changed world economy remained imperative for an economy structurally dependent on foreign trade. Under pressure to obtain foreign exchange, Cuba worked to develop a new world-market oriented sector, based on the massive expansion of international tourism and an opening-up of the economy to joint-venture enterprises with foreign capital. In addition, around these dollarized world market enclaves, quasi-state Cuban enterprises were established, the so-called *Sociedades Anónimas* (stock companies), which operate on a hard currency basis and openly imitate capitalist organizational forms.

Finding new sources of hard currency revenues became all the more urgent as sugar production fell by half within four years, from 8.4 million tons in 1989 to a mere 4.2 million in 1993, when the country's economic crisis peaked.[13] According to official data, Cuban GDP fell by 34.8 percent between 1989 and 1993 (see figure 13 and table 6).

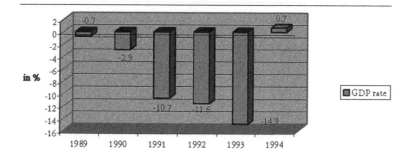

Figure 13 Cuba's GDP Rate, 1989–1994.

Source: Banco Nacional de Cuba

Table 6 Cuba's World Market Shock: Selected Macroeconomic Indicators, 1989–1994

Year	1989	1990	1991	1992	1993	1994
Accumulated change in GDP over 1989 (in %)	—	–2.9	–13.3	–23.4	–34.8	–34.3
Imports (cif, in US$ mill)	8,139.8	7,416.5	4,233.8	2,314.9	2,036.8	1,956.1
Accumulated change in imports over 1989 (in %)	—	–8.9	-48.0	–71.6	–75.0	–76.0
Exports (fob, in US$ mill)	5,399.9	5,414.9	2,979.5	1,779.4	1,136.5	1,314.2
Accumulated change in exports over 1989 (in %)	—	0.3	–44.8	–67.1	–79.0	–75.7
Sugar production (in mill tons)	8.4	8.0	7.6	7.0	4.2	4.0
Cuban peso (extra-official exchange rate to the US$)*	—	7:1	20:1	45:1	100:1	60:1
Gross Domestic Investment (in % over preceding year)	10.1	–2.9	–45.9	–58.3	–39.7	5.8
Net current transfers** (in US$ mill)	–48.0	–13.0	18.0	43.0	263.0	470.0
Tourism (Visitors in millions)	n.d.	0.340	0.424	0.460	0.546	0.619
Tourism / Gross Revenues (in US$ mill)	n.d.	243	387	567	720	850

* data by CEPAL 1997; for 1989 no data are given; however, a rate of 5:1 is considered to be the 'historic' black market rate in pre-'89 socialist Cuba; data given are for end of year, but CEPAL 1997 Cuadro A.14 also provides data for 'middle of the year' which reflects the peak of the peso's devaluation in mid-1994 reaching 130:1.
** largely remittances since 1993
Source: Banco Nacional de Cuba and CEPAL

In 1993, the combined hard currency revenues from exports and tourism sank below US$2 billion, a level which the government's economic planners had considered to be the lowest 'survival margin' for the Cuban economy.[14] Under this pressure, the government took a spectacular step with far-reaching economic and social consequences: the legalization of the U.S. dollar on the island (Castro 1993a). Far from the days of insulation from the world market, international currency competition now takes place in Cuba itself, visible to all and in drastic forms.

While much of the Cuban productive structure had virtually collapsed, the government had held prices and salaries artificially constant.[15]

The consequence was a dramatic devaluation of the Cuban currency, reaching a black market rate of 130 pesos per U.S. dollar in summer 1994, while officially a 1:1 parity was maintained (see table 6). In this situation, the legalization of the U.S. dollar did not just mean the entry of a second currency: the U.S. dollar came to be the only currency of value in the country. Moreover, when the average wage in the state economy of 180 pesos corresponded to one and a half dollars, this distortion of monetary relations not only disrupted e conomic mechanisms, it also opened deep social discrepancies. When a steelworker or a medical doctor earns the same in a month as a waiter in a hard currency hotel easily makes in tips in one morning, social values are profoundly shaken.

The consequence of the Cuban government's dualist strategy of a 'war economy with dollar enclaves' was a split of the economy and society that ended the sharp, but relatively egalitarian fall in the standard of living prior to the legalization of the dollar. Now the ever-sparser provisions available on the ration card contrasted with the opulent offerings in state-run dollar shops. Moreover, since the defense of national sovereignty against U.S. domination is a central source of legitimacy for the revolutionary regime, the legalization of the U.S. dollar also affected its very ideological foundation.

On the macro-economic level, however, the legalization of the dollar provided a desperately needed life vest to the Cuban economy since it opened up a new source of foreign exchange revenue: the dollar remittances from emigrated relatives abroad. An old Spanish saying goes: *Enemigo que huye—puente de plata* (The enemy that flees turns into a bridge of silver). Up until the 1990s, this had not been the case in Cuba. Emigration served as a political escape valve, but not as a source of monetary revenues. But now, with possession of dollars legal—and desperately needed to ameliorate a critical economic situation, emigrated relatives began to send remittances on a massive scale. With figures as high as US$500 to 1,100 million annually, according to different estimates, these remittances probably have become Cuba's principal foreign currency source, ahead of the net earnings of tourism and far surpassing those of the principal export product, sugar.[16] In spite of the sharp political confrontation with large sectors of the emigrated community, the 'export' of emigrants and the 'import' of their remittances to the island effectively have become a key element in Cuba's world market integration of the 1990s (Monreal 1999).

PUSHING FOR REFORM—AND HALTING IT

In Cuba, the crisis of the formal economy led to massive growth in the informal economy, which includes a wide range of activities from black market businesses of every kind to economic dealings along lines of friendship or family. According to an official Cuban study on the issue by González Gutiérrez (1997), between 1989 and 1995 the value of transactions in this 'submerged economy' reached a volume similar to or higher than the value of transactions between the state and the population.[17] This generalization of informal economic relations certainly represented a loss of legitimacy for the socialist system; at the same time, it was one of its survival mechanisms, clearly following a political logic. Where the state economy could no longer sufficiently guarantee provisions, recourse to market mechanisms could take place even without requiring formal market reforms. Since these market activities were not formal and legal, no claims could be derived nor demands voiced from them; no organizing of the producers could take place; and the state always had a free hand to intervene against undesired activities or any growth which it deemed excessive. Precisely this legal insecurity of market relations rendered them chronically dependent on the good will of state authorities. Cuba's 'second economy' is thus far from stepping "from behind the scenes to center stage," as the title of the study by Pérez-López (1995) suggests. On the contrary, the political condition for its existence is precisely that it remains backstage. Nevertheless, the fact that it is becoming a part of everyday life represents a structural transformation within a system which at one time was based on an omnipresent state responsible for practically all the concerns of individuals and society.

In this light, the economic reform discussion of the early 1990s in Cuba turned not so much on the question of the 'introduction' of market mechanisms but rather on the question of what kind of market mechanisms the country should have. Formally and legally recognized mechanisms could develop a self-sustaining economic dynamic; merely tolerated, informal mechanisms would have far less developmental potential but would also be easier to control politically.[18]

Following the legalization of the dollar, three more steps to reform were taken in 1994. First, a range of self-employment professions (*trabajo por cuenta propia*) was allowed, but numerous restrictions were put into place to prevent them from turning into a small scale business sector; as a result, fewer than 200,000 Cubans officially registered in these professions. A second step was the transformation of the large state farms into so-called 'Basic Units of Cooperative Production' UBPC); though this raised high hopes of becoming Cuba's 'third agrarian reform,' the new UBPC cooperatives were granted little autonomy and generally remained firmly subordinated to the state's directives (cf. Burchardt 2000; Pérez Rojas et al. 1996). The third and certainly most far reaching domestic reform was the opening of agricultural markets in the fall of 1994, which greatly increased the supply of goods available for Cuban pesos.

Throughout the early 1990s, when the food situation became increasingly tight, probably no other issue was as prominent within the Communist Party and the mass organizations as the call to re-open the agricultural markets.[19] These calls passed unheard; Fidel Castro repeatedly rejected the idea, arguing that any such markets would lead to the enrichment of the vendors and to the emergence of a bourgeois mentality. Instead, the decision to re-open agricultural markets came as a reaction to other pressures, particularly the open outbreak of the social crisis in the summer of 1994, one year after the legalization of the U.S. dollar. The exit and voice scheme developed by Hirschman (1970) offers a good model for understanding the interaction of emigration and public protest surrounding the issue.

While the emigration flow had slowed to a trickle in the 1980s (Pérez 1999: 20), the profound crisis of the early '90s triggered new emigration, in the peculiar form of the *balseros* (from *balsa* = raft). Since it was very difficult to obtain U.S. visas legally, but the United States received with open arms all Cuban refugees that risked their lives crossing the Florida Straits on boats or improvised rafts, this form of 'exit' greatly increased. It was a violent emigration attempt, the hijacking of boats in the port of Havana to flee the island, that set off social unrest. On August 5, 1994, an open riot against the Castro government occurred when thousands of mostly young Cubans gathered on Havana's coastline avenue, chanting anti-government slogans and breaking the windows of dollar shops and tourist hotels. The 'voice' option, which could not be contained by the narrow channels for articulation within the system, had taken to the streets.

The government had no problems repressing the protests within a few hours and without bloodshed. Nevertheless, the August 5 riots underscored an important development: the depth of the social erosion had become visible. The government reacted by opening up the 'exit' option: It declared the borders open for all those who would leave by sea and with their own means. A mass exodus followed, with more than 30,000 Cubans leaving the country on improvised rafts within a few weeks. Negotiations with the U.S. government ensued and, after reaching an accord on migration, the Castro government declared the borders of Cuba once again closed. Only a week later, Raúl Castro, Vice-President of the country, head of the Armed Forces and brother of the *Comandante en Jefe,* reopened the food markets. Raúl Castro took this step explicitly on the premise that the provision of food was the highest "economic, political and military priority" of the country

(*Granma Internacional,* September 28, 1995). Though a number of restrictions were established, supply and demand in principle define the prices in the markets. "The prices may be high," Raúl Castro explained, "but they will certainly be lower than on the black market" (ibid)—a remarkable argument in a state-socialist country.

Whereas in the case of the GDR in 1989, as Hirschman (1992) analyzed, the interplay of 'voice' and 'exit' options brought about the demise of the socialist state, the Cuban state managed to react differently. It responded to the frustrations that surfaced in the August 5 riot first by repressing them, second by opening an exit option and third it indirectly responded to the social frustrations by implementing a major step in economic reform, the opening of agricultural markets. The agrarian markets indeed proved efficient in improving the food situation, reducing the most imminent social pressure for reform. Moreover, with the inflow of remittances and the growth of the dollarized sectors, the country enjoyed a relative stabilization of the economy by the mid-1990s. The national statistics began to show positive growth rates (see table 7) and the Cuban peso stabilized at a level between 19:1 and 26:1 relative to the U.S. dollar, still low, but far from the informal 130:1 rate in mid-1994.

The successful stabilization did not, however, lead the Cuban government to embark on a more far-reaching economic reform. Very much to the contrary, when the government allowed private small restaurants, the so-called *paladares,* in June 1995, it was not another step in a gradual extension of the reform process. Instead, it already marked the turning point in the reform process. The issue of opening the domestic economy to legal market mechanisms became a matter not of expansion, but of containment, with increased controls and restrictions.

Whereas Cuba's formal political structures had remained largely unchanged by the economic crisis, Cuban society lost much of its earlier homogeneity. In addition, the practical social relevance of the mass organizations had greatly declined, whereas family ties had become more important than ever. The 1990s, too, had witnessed a massive revival of religious communities, particularly with the strengthening of the Catholic Church, the emergence of Protestant groups and the impressive comeback of Cuba's popular religion, the Afro-Cuban Santería. These religious organizations were not only elements in the re-articulation of personal identity: they provided important social networks beyond the official state structures (cf. Kummels 1995). Parallel to the erosion of the socialist mass organizations, the country also experienced a boom of so-called 'non-governmental organizations' (NGOs) in cultural, social and ecological areas. These quickly became important

Table 7 Cuba's Gradual Recovery: Selected Macroeconomic Indicators, 1995–2001

Year	1995	1996	1997	1998	1999	2000	2001
GDP growth rate (in percent)	2.5	7.8	2.5	1.2	6.2	5.6	3.0
Imports (cif, in US$ mill)	2,883	3,569	3,987	4,181	4,349	4,829	4,838.3
Exports (fob, in US$ mill)	1,492	1,866	1,819	1,512	1,496	1,676	1,661.5
Sugar production (in mill tons)	3.3	4.5	4.3	3.4	3.9	4.1	3.5
Cuban peso (extra-official exchange rate to US$)	32:1	19:1	23:1	21:1	20:1	21:1	26:1
Net current transfers (in US$ mill)	646	744	792	813	799	842	850
Tourism (Visitors in millions)	0.742	1.004	1.170	1.416	1.603	1.774	1.775
Tourism/ Gross Revenues (in US$ mill)	1,100	1,333	1,515	1,759	1,901	1,948	1,753

Source: CEPAL 2002, Banco Central de Cuba 2001

counterparts for international aid or solidarity organizations. Though largely directed and founded by loyal Party cadres, they tended to strive for more autonomy than the hitherto existing institutions (cf. Hoffmann 1999).

On the background of this social differentiation and erosion of established structures, the debate about the role of civil society within socialism became an overriding theme in the domestic intellectual reform debate emerging in the early to mid-1990s. A landmark article by Rafael Hernández from the Center for Studies of the Americas (CEA) reclaimed—based on Hegel, Marx, and Gramsci—the term 'civil society' as a useful and necessary instrument for the analysis of Cuban society (Hernández 1994: 31). Hugo Azcuy, also a researcher at the CEA, followed by emphasizing "the necessity of more plural forms of expression of the Cuban society." He emphasized that the civil society concept should "not only be used as an instrument of analysis but also be thought of as a *project*" (Azcuy 1995: 105, italics in the original). A number of publications (e.g. the edited volumes by Dilla 1995 and 1996, and the magazine 'Temas') set out to sketch this reform project along the lines formulated by Azcuy when he set its goal as "the strengthening of Cuban civil society and its necessary autonomy in the context of the revolutionary project, which it reaffirms and of which it is part" (ibid: 108).

It is open to speculation how far this concept of a renewed and pluralized socialism could have gone. For Cuba's political leadership, at this point it already had gone too far. At the Fifth Plenum of the PCC's Central Committee in March 1996, Raúl Castro read a 'Report of the Politburo,' which in martial tones condemned the reform discussion that had emerged on the island, attacking its leading protagonists in the academic centers as a "fifth column of the enemy" who had fallen victim to "ideological subversion" and "divisionism" (R. Castro 1996). The entire 'civil society' debate was decried as a "Trojan horse" of the enemy to subvert Cuba's socialist order (ibid.).[20]

The infamous 'Report of the Politburo' also signalled the halt of the economic reform process. Whereas, on opening the agricultural markets in 1994, Raúl Castro had stated: "If there is food for the people, the risks don't matter,"[21] the government now the reversed its position. It rejected market-oriented reforms explicitly for the reason that any type of opening towards small business or independent cooperatives on the island would endanger the political system. In a follow-up to Raúl Castro's speech, the Director of the PCC's central party academy, Raúl Valdés Vivó (1997: 4), declared: "The creation of seeds of a local bourgeoisie would create a social force which sooner or later would serve the counterrevolution."

The 'Report of the Politburo' is an illustrative example of how internal conflict—in this case, conflict not with opposition groups but with deviating opinions within the Party and the intellectual establishment—became framed as part of the external conflict with the United States and the emigrated Cubans. It is also characteristic of the Cuban situation that the external enemy, the United States, indeed provides ample reason to make this connection plausible.

When the much-awaited 'fall of Fidel Castro' failed to occur after 1989, the United States opted to tighten the embargo policy in 1992 with the so-called 'Torricelli Law.'[22] This law established a number of new sanctions, increased pressure on third countries to discontinue trade relations with the island, and extended the existing embargo restrictions on subsidiaries of U.S. companies in third countries. However, the law also encompassed the so-called 'track two,' a policy of facilitating and increasing contact and communication with the island, with the explicit objective of thus promoting the democratization of the island. In Havana, this 'track two' policy was vehemently attacked as 'ideological subversion.' (We shall see below how this change in U.S. policy greatly affected the question of Internet access and NICT development on the island.)

Four years later, the Torricelli Law was followed by the Helms-Burton Law, which the U.S. Congress passed in 1996.[23] This law extends the reach of U.S. sanctions still further beyond U.S. borders, threatening retaliatory measures against all persons or companies whose business with Cuba involves property expropriated after 1959. For Cuba's economic strategy, the negative effect on potential investment represents a serious problem. In the long-term, even more disastrous politically is the fact that the law—in open violation of international norms—extends the right to file property claims *as U.S. citizens* to all Cuban exiles who were Cuban citizens at the time of expropriation, and who only later, after having emigrated to the United States, acquired U.S. citizenship. This retroactive clause modifies the legal status of one of the thorniest issues in any post-socialist transition scenario. The entire question of post-'59 expropriations and the ensuing restitution claims has passed from being a domestic, Cuban issue to becoming an internationalized one between Cuba and the United States (cf. IRELA 1996, Muse 1996, and Hoffmann 1997).

In addition, the Helms-Burton Law includes an entire section—Title II—which stipulates a long list of conditions a Cuban government would have to fulfil to be accepted by the United States as a "transition government." In this the law states:

> "Sec. 205 (a) For the purposes of this Act, a transition government in
> Cuba is a government that—
> (1) has legalized all political activity;
> (2) has released all political prisoners (. . .);
> (3) has dissolved the present Department of State Security in the
> Cuban Ministry of the Interior, including the Committees for the
> Defense of the Revolution and the Rapid Response Brigades;
> (4) has made public commitments to organizing free and fair elections
> for a new government (. . .);
> (5) has ceased any interference with Radio Martí or Television Martí
> broadcasts (. . .);
> (6) does not include Fidel Castro or Raul Castro; (. . .)
> (Sec. 205, b, 2, D) (. . .) has made public commitments to, and is
> making demonstrable progress in taking appropriate steps to return
> to United States citizens (and entities which are 50 percent or more
> beneficially owned by United States citizens) property taken by the
> Cuban Government from such citizens and entities on or after

January 1, 1959, or to provide equitable compensation (. . .)"
(U.S. Congress 1996).

For anyone in Cuba's political establishment, as reform-minded he might be, this list of conditions dictated by U.S. law does not point to any viable 'transition' perspective of reform with dignity, but rather appears as a submission to the United States and a return to the neo-colonial tutelage, as embodied by the 'Platt amendment' during Cuba's First Republic.

THE RE-ARTICULATION OF CUBAN SOCIALISM: A PRELIMINARY BALANCE

If in the mid-1990s many observers assumed it 'inevitable' that Cuba was heading towards a market economy, it has since become evident that the island did not embark on a process of gradual but steady market reform or 'transition.' Instead, the Cuban government managed to carry out a *sui generis* process of re-articulating Cuban socialism in the new international context. This process differs significantly from that of the other 'socialist survivor' countries, China and Vietnam, which have undertaken a far more profound and comprehensive reform approach. The very term 'market socialism' used by functionaries from Peking and Hanoi has been explicitly rejected by their Cuban counterparts.

Since the 'ideological counteroffensive' launched in 1996, any reform process—in the sense of integrating market mechanisms into the economy or of greater pluralization within the socialist society—has disappeared from the public agenda. Instead, the government has put much emphasis on 'ideological work,' which included a massive turn to ideologically didactic programs in the state media and to state-organized mass mobilizations. These began with the campaign for the repatriation of the refugee child Elián González in 2000 and have continued since. Though imitating the style of the mass demonstrations of the early days of the Revolution, the contrast between them is sharp. Attendance is far from enthusiastic, but rather is seen as a social obligation that keeps one out of trouble; the discourse is highly sterile, and the audience shows an impressive level of disinterest.[24]

In economic affairs, earlier reform steps came to be seen as 'concessions' that were inevitable in the crisis of the early '90s, but which now, wherever possible, should be gradually reversed. After relative stabilization, the direction of the economy did not evolve from 'survival' to 'development,' 'growth' or 'upgrading,' as reform-minded economists had envisaged, but to the 'perfectioning of socialism' (*'perfeccionamiento'*), which essentially translates into the priority of ideological and political considerations above economic criteria.

Nevertheless, the growing hard currency sector remains politically ambiguous. The extension of hard currency shops catering to those Cubans with dollar income—be it through their work in the dollarized sectors, remittances from relatives abroad or other means—contrasts sharply with the official discourse's emphasis on the moral virtues of socialism and the decadence of the Western 'consumerist societies.' In addition, a prominent Cuban academic, Haroldo Dilla, has warned of the emergence of a 'technocratic-entrepreneurial sector,' consisting of the managerial elite associated with the foreign companies, the directors of state companies with world market connections, and successful 'entrepreneurs' operating in Cuba's black or legalized market. In Dilla's view, this sector is gradually eroding the popular roots of the Revolution's political structures.[25]

The economic recuperation in the second half of the 1990s, as remarkable as it is (see table 7), remains precarious. It depended on the rapid growth of tourism and remittances since these represented a highly protected hard currency market for Cuba's domestic production that would not have been competitive on the world market. This phenomenon, called *exportaciones en frontera* (exports within the borders), is a key to understanding the partial recovery of many of the island's typical ISI industries that seemed doomed to extinction. In analyzing this re-vitalization of Cuba's industry through the *exportaciones en frontera* scheme, Cuban economist Pedro Monreal (2000: 12) concludes: "To say it in the clearest way: in the 1990s Cuba has not departed from the industrialization via substitution of imports as the central component of the country's long-term development perspectives."

The software sector only recently became a somewhat more prominent source of hope for a shift towards export orientation. Another high-tech sector, centering on biotechnology and pharmaceuticals, had been designed since the 1980s as a strategic lead sector with export potential (Alvarez 1997; Feinsilver 1992, 1995; Nuhn 2001). Even in the midst of the crisis of the early 1990s the state channeled considerable resources into this sector. However, although a number of medical innovations caught international attention, returns on investment have been slow, since the problems of internationally marketing this type of product proved much greater than expected. Annual export earnings for the sector in 2002 still only hovered around US$30 million, which is certainly modest for a sector with considerable hard currency import needs (Rodríguez Armas 2003). As a result, the biotechnology and pharmaceutics industry probably have been more successful in substituting imports in these fields than in generating export earnings.

In 2001/02, international events demonstrated the vulnerability of Cuba's new world market insertion to external factors. The terrorist attacks on the World Trade Center in New York on September 11, 2001 and the re-sulting recessive international climate affected all major sources of Cuba's hard currency revenues: tourism, remittances, nickel, and tobacco. After September 11, tourism fell sharply, by about 20 percent in the first four months, and remained down by 13.6 percent in the first five months of 2002 (EIU 2002: 22). Similarly, remittances decreased significantly due to the tourism slump and recession in Florida, which affected Cuban emigrants' in-clination or ability to send money to the island. Moreover, in the wake of the September 11 shock, both nickel, which in 2000 surpassed sugar as the country's biggest export, and tobacco, which ranks third in the country's ex-port statistics, experienced steep declines in world market prices.

The impact on the Cuban economy was severe. In 2001 GDP growth fell to 3 percent instead of the projected 5 percent and, in 2002, GDP grew by only 1.0 percent, according to official figures (EIU 2002: 3). The foreign currency crunch became so problematic that the country had to suspend debt service and renegotiate with major creditors, and numerous investment projects were postponed. Prices increased in the dollar shops and, in the peso sector of the economy, shortages began to reappear that were severe enough that the government felt the need to assure the population that no return to the hardships of the *Período Especial* after 1989 was in store.

It can be expected that over time tourist numbers and remittances will return to their pre- September 11 level, but it also seems certain that they will not come close to repeating their spectacular growth rates of the past decade. In fact, as table 7 shows, growth rates in both sectors already had slowed down considerably at the end of the 1990s. Thus, with the tradi-tional leading export, sugar, in decline and with the most important new hard currency earners not likely to maintain their recent, extraordinary dy-namism, the Cuban government will have to seek new forms of world mar-ket insertion. Equally urgent is the need to seek a higher degree of economic coherence that can overcome the economic and social discrepancies and dis-tortions resulting from the prevailing monetary dualism. As long as official salaries in the state economy translate into US$8 to 15 per month, no one can doubt the seriousness of the crisis that continues in the country.

On the whole, the Cuban government can boast a politically successful management of the crisis, maintaining its state-socialist system and the govern-ment led by Fidel Castro against heavy odds. However, the erosion of social support has been severe. Despite the relative economic recovery, the rearticula-tion of Cuban socialism remains precarious and crisis-prone, due as much to its international insertion as domestic conditions, in which a hybrid system suf-fers from highly contradictory economic logics in two parallel sectors.

ROLE AND FUNCTION OF MASS MEDIA IN CUBAN SOCIALISM

Before turning to the role and function of the mass media and the telecommunications regime in Cuban state-socialism, a look back to pre-revolutionary Cuba is necessary. Before 1959, with 1 television set per 25 inhabitants and one newspaper edition per 8 people (Pérez 1995: 296), Cuba ranked first among all Latin American countries in the diffusion of mass media. Cuba was the first Latin American country in which radio was introduced (in 1922), and, together with Brazil and Mexico, it was one of the continent's first countries with TV broadcasting (dating back to 1950). With its modern sectors closely linked to the United States, in 1958 Cuba became the second country in the world—after the United States—to transmit a color TV program (CEDISAC/Prensa Latina 1997).

However, the pattern of media diffusion also reflected the enormous disparities and inequalities of Cuba's dependent capitalist development. TV and newspapers were limited to the urban upper and middle classes, while 42 percent of households didn't even have access to electricity and 24 percent of the population was illiterate. Even the diffusion of the radio was highly unequal, with a full two thirds of the country's 156 private radio stations located in Havana.

Fidel Castro early on learned the importance of the media, above all radio and newspapers, to the political struggles of the Republic. Since the beginning of his engagement in party politics in 1947, Fidel Castro had supported the oppositional leader Eduardo Chibás, whose political activities relied heavily on regular radio programs. In 1951, the same year that Chibás committed suicide, Fidel Castro himself began to use radio and newspapers intensively to publicize denunciations of the corrupt elites and to promote his election campaign for a seat in Congress in 1952 (cf. Bourne 1988: 81, 85–86). While the ruling elites dominated mass media, even after Batista's coup and subsequent suspension of the Constitution in 1952, a number of opposition and radical media continued publication. Fidel Castro himself, before leaving for Mexico in exile in 1955, published widely read attacks on the Batista regime in the magazines *Bohemia* and *La Calle* (ibid: 141f.).

The Revolution caused a fundamental break in the structure and function of the mass media in Cuba. The state took over all private media; independent media—be it radio, TV or print—were not allowed. The state-owned or -dominated media were put into service to secure the legitimacy of the revolutionary power and to promote the construction of the socialist society. Institutionally, the creation of the *Instituto Cubano de Radiodifusión* in 1962 (renamed in *Instituto Cubano de Radio y Televisión,*

ICRT in 1976) consolidated the new media order under an umbrella institution for the nation's radio and TV stations.

In 1976, as part of the 'process of institutionalization,' the state monopoly over mass media also became an explicit part of the country's new socialist Constitution. Article 53 reads:

> "The citizens are granted the liberty of word and press in accordance with the goals of the socialist society. The material conditions for its exercise are given through the fact that press, radio, television, cinema and other means of mass communication are state or social property and under no circumstances can be object of private property" (República de Cuba 1976).

Where the Constitution speaks of "state or social property," it must be noted that the term 'state monopoly' seems adequate, since "social property" refers only to those organizations that are institutionally associated with the state, the party or the armed forces. Typical examples are the daily papers *Juventud Rebelde,* which belongs to the Communist Party's Youth Association, UJC, or *Trabajadores,* edited by the official union confederation, CTC.

Following the Leninist tradition, mass media were regarded as central ideological 'transmission belts' from the party and state to the population at large. As a result, their organizational structures were vertical and highly centralized, and they were subject to far more rigid political control than culture and artistic production. Wilfredo Cancio, who taught communications at Havana University for 15 years, reflected later that "[t]he withholding of information through conscious censorship occurred under the pretext that any pointing to domestic deficiencies would be handing arms to the enemy" (Cancio 1996: 33). The reliance on the Soviet model was obvious and openly acknowledged, for instance in the resolutions of the 4th Congress of the Cuban Journalists' Union (UPEC) in 1981, which called on its members "to study, as to the exercise of criticism in the media of mass communication, the rich and comprehensive experiences of the socialist community, particularly the Soviet Union" (cited in Cancio 1996: 35).

Parallel to the limitations on content and programs, Cuban socialism brought about the massive spread of mass media to all sectors of society. With 6 national, 16 provincial and 36 local stations, state radio covers all parts of the country (CEDISAC/Prensa Latina 1997); since 1980, the two national TV stations, *CubaVisión* and *TeleRebelde,* cover more than 93 percent of the nation's territory (ibid)—in fact, only a few remote villages in mountainous areas remain outside of their reach. Between 1960 and 1980 more than two million black-and-white TV sets from the Soviet Union were imported or—since 1974–assembled in Cuba by a company belonging to the Ministry of Metallurgy and Electronics (SIME). These television sets were often bonuses, distributed through the workplace or unions, for good professional performance or other social merits, and were sold at highly subsidized prices. Due to the government's massive promotion of television at the beginning of the 1990s, two out of every three households on the island had a television set: the density of 23.9 sets per 100 inhabitants is one of the highest in Latin America (ITU 2000: 101).

Thanks to the literacy campaign and the push for a broad educational system, practically the entire population also became a target audience for print media. By 1990, Cuba boasted a total of 733 periodicals (Cancio 1996: 31). Among these, *Granma,* the 'Central Organ of the Central Committee of the Communist Party,' took a lead ideological role. The government also put great priority on the 'international projection' of the Revolution through international editions of newspapers and magazines (such as *Granma Internacional* or *Cuba Internacional*), Cuba's internationally oriented press agency *Prensa Latina* and the international radio station *Radio Habana Cuba.*

MASS MEDIA SINCE THE LATE 1980s: BETWEEN ECONOMIC CRISIS AND IDEOLOGICAL COUNTEROFFENSIVE

When the Cuban government launched the so-called 'process of rectification' in the second half of the 1980s, publicly distancing itself from Soviet models, substantial hopes were raised in Cuba's national media that this could lead to a different model of communication. Even the President of the Cuban Journalists' Union at the time came to describe the dominant practice as "a model which we can call 'officialist,' 'apologetic' or 'unanimist' " (Julio García Luis 1990: 8). The 1989 ban against the Soviet perestroika publications 'Sputnik' and 'Moscow News,' which had turned to a different, more critical type of journalism, soon ended the hopes for a more courageous role of the press, radio, and TV in Cuba. Several of Cuba's journalists who had taken the initial call for a more critical role of the media 'too much' to heart, were sanctioned or fired.

In the early 1990s, not only did the debate about a more critical function of the mass media disappear from the political agenda, but also the profound economic crisis led to a severe contraction of the production of the Cuban media. Two thirds of the 733 print media titles were suspended entirely, the rest suffered strong reductions in volume and frequency. Except for the Party's Central Organ *Granma,* daily papers turned to weekly editions. The state TV channels cut production from a combined 213 to 135 hours a week and radio stations experienced similar cuts (Cancio 1996: 31).

At the same time, the external opening of the economy brought new opportunities for some media serving the tourist industry or the emerging business sectors around the joint-venture companies: a typical example is the Communist Youth organization's *Juventud Rebelde,* which started a weekly business magazine called *Opciones.* Particularly in the area of cultural or academic publications, some external financing became available through development assistance funds, solidarity projects or collaborative editions with foreign institutions. Though to a limited degree, commercial advertisements—generally targeting foreigners or the dollar sectors of society—began to reappear in Cuban publications. Another branch of media that found limited new opportunities was the set of publications for interior circulation in the churches, which is also a legal exception to the state's media monopoly. Though they have very limited possibilities for diffusion and are not allowed to go beyond the realm of ecclesiastical matters, some of these publications, for instance the *Vitral* magazine edited by the catholic diocese of Pinar del Río, in the 1990s became somewhat prominent centers of dissent (cf. www.vitral.org).

As a result of the modest economic recovery beginning in the mid-1990s, the volume of Cuban media production gradually returned towards its pre-'89 levels. The above mentioned 'Report of the Politburo,' which cut off the reform debate in 1996, also made it clear that the government would resist both loosening the state monopoly over the mass media and allowing more pluralism within state media, pointing to "the glasnost which undermined the USSR and other socialist countries [and which] consisted of handing over the mass media, one by one, to the enemies of socialism" (R. Castro 1996: 7). Instead, with the ideological campaigns in the second half of the decade, the government was determined to return the media to its traditional role as an ideological transmission belt. Particularly after the massive campaign for the repatriation of the refugee child Elián González in 2000, state television flooded the airwaves with ideological and didactic programs. Most prominent were the so-called *Mesas Redondas Informativas* ('Informative round-table talks') and hour-long transmissions of the mass demonstrations of the so-called *Tribunas Abiertas* ('Open Tribunes') that often appeared on the two state channels simultaneously.

Television remains the most important mass media format in Cuba and a favorite pastime of its population. To conclude this section, a brief look at the development of television sets since the early 1990s may serve to illustrate the transformations that have occurred. With the dissolution of the Soviet Union, imports of TV equipment came to a virtual standstill and domestic production ceased. Any repairs had to be improvised and it was virtually impossible to obtain a new TV set between 1990 and 1993. This changed with the legalization of the U.S. dollar: the state-run dollar shops now offered color TV's of all sizes from Japan and Korea, though at high prices—making them available only to those Cubans participating success-fully in the new dollarized sectors of the economy or to those receiving suf-ficiently high remittances from abroad.

To counter the resulting frustrations among the excluded population and to signal that the economic recovery is actually 'getting to the house-holds,' the government announced in the fall of 2001 a deal with the People's Republic of China to import one million 'Panda' color TV's in the coming years, payable in hard currency, though via a preferential arrangement (Acosta 2001). The announced method of distributing the sets corresponds to the hybrid character of today's Cuban socialism: a part of the TV sets would go cost-free to schools and other public institutions; another part would go to the state-run dollar shops to be sold for hard currency, just like the Sony and Samsung sets before; and a third part would be distributed along classically socialist lines, with the state and its institutions, rather than money, determining who would have access to these 'privileges.' The govern-ment announced that this last group of TV's would be sold in pesos with in-terest-free bank loans, primarily to "distinguished workers as well as senior citizens" (ibid). What it has not announced is the relative share of each of these segments; it is safe to assume that, with the country's hard currency crunch after September 11, 2001, the volume of imported TV sets distributed for free or at subsidized peso prices might be not as high as initially planned.

U.S. MEDIA POLICY TOWARD CUBA: BROADCASTING FOR POLITICAL CHANGE

While the Cuban government regards mass media as an essential battlefield in the confrontation with the United States and 'the counterrevolution,' the U.S. government and the Cuban exile community have given media a simi-lar high priority in their political conflict with the Castro-led government on the island. Their efforts include attempting to influence U.S. and interna-tional media coverage of Cuban affairs, as well as initiatives to reach out to the population on the island with U.S.-based media.

The latter found its permanent and institutionalized expression in the 'Radio Broadcasting to Cuba Act' passed by U.S. Congress during the Reagan administration in 1983, which was one of the first effective political initiatives of the hard-line Cuban-American National Foundation (CANF). This law created 'Radio Martí,' a radio station financed with U.S. government funds and operated mainly by Cuban exiles. The text of the act explicitly denies any national sovereignty over information, stating

> "that it is the policy of the United States to support the right of the people of Cuba to seek, receive, and impart information and ideas through any media and regardless of frontiers" (U.S. Congress 1983, Sec 2, 1).

The law also clearly states that Radio Martí's mission is not so much to transmit information about the United States (this task is left to the 'Voice of America'),

> "but that there is a need for broadcasts to Cuba which provide news, commentary and other information about events in Cuba and elsewhere to promote the cause of freedom in Cuba" (ibid: Sec 2, 4).

'Radio Martí' began operations in 1985 and the 'Television Broadcasting to Cuba Act' of 1990 followed, establishing 'TV Martí.' Produced in Washington, the radio and TV signals are relayed to Cuba by a balloon over Cudjoe Key in South Florida and via satellite. Administratively, both the radio and TV stations are subordinate to the Office of Cuba Broadcasting (OCB), which is part of the United States Information Agency (USIA); the overseeing body for both is the Advisory Board for Cuba Broadcasting. Jorge Mas Canosa, the leader of the powerful CANF, presided over the Advisory Board from its foundation until his death in 1997, establishing the CANF's dominant influence on the stations' policies and programs. This amalgam of the interests of a particular exile organization and official U.S. government funds and institutions has led to repeated public criticism within the United States of the 'Radio Martí' project. [26]

The Cuban government always has seen Radio and TV Martí as an open political attack and has spent considerable diplomatic effort to counter it politically and technical effort to jam its transmissions to the island. Despite these interferences, Radio Martí reaches the island, though not always and not without problems. The estimates on its audience vary widely. Though exile groups involved in its production suggest that as many as 70 percent of Cubans tune into Radio Martí at least once a week (The Miami Herald, June 22, 1993; cited in Press 1996a: 19), this number is certainly grossly exaggerated. Nevertheless, Cubans are generally familiar with Radio Martí and, though often in indirect ways, its programs do reach a relevant audience on the island. A joke circulating in Cuba may be illustrative: Why is 'Radio Martí' called "Radio Accident"?—Because when people tell other people news they heard on 'Radio Martí,' they will not admit the source, but say: "By accident I heard that . . ."

The same cannot be said of TV Martí. Since its inauguration, its transmissions have been consistently jammed by the Cuban authorities and the number of Cubans on the island who have ever seen any of its programs— never mind a regular audience—is minimal. The only effect the TV station has on the island is of a different nature: the Cuban state has to incur considerable coststo jam the signal, since these operations require well trained specialists and costly technical equipment.

There is yet another effect of Radio and TV Martí that should not be underestimated. As the U.S. government pours eight figure sums annually into the Broadcasting to Cuba program, it provides a formidable financing and employment source for those Cuban exiles that channel these funds or otherwise work in these programs. In fact, an essential part of the Cuban American National Foundation's power base within the exile community is its powerful influence in distributing such funds and positions.

THE CUBAN TELECOMMUNICATIONS REGIME: A STATE MONOPOLY WITH VERY LOW PRIORITY

Before the Revolution, Cuba had the highest telephone density of all Latin American countries. But, as with the high diffusion of print media and television, the telephone system also followed the lines of the country's profound social and spatial disparities: no less than 73 percent of the main line telephones were concentrated in Havana (Press 1998: Table 7) and postal and telegraph offices also were concentrated in the capital and other urban centers. After the Revolution, a process of nationwide extension began and, in 1989, the island had a net of 790 post offices, including at least one in each of Cuba's 169 municipalities (CEDISAC/Prensa Latina 1997).

After 1959, the foreign-owned electricity company and the *Cuban Telephone Company,* a subsidiary of the U.S. company ITT, were among the first companies to be expropriated. The electricity sector received highest priority: it is one of the historic accomplishments of the Cuban Revolution that it connected practically the entire country, including small rural communities, to the national electricity net. In contrast, as in the state-socialist countries of Eastern Europe and the Soviet Union, the telephone system had low priority. As much as the state promoted the one-to-many media of radio, television and newspapers, it showed little interest in a major expansion of the horizontal one-to-one communication of the telephone network. The number of main line telephones increased from 170,000 in 1959 to no more than 349,471 in 1994 (see table 8), giving Cuba the lowest growth rate in main line telephony of virtually all Latin American and Caribbean countries. With a growth in population from 7 to 11 million inhabitants in the same period, telephone density merely rose from 2.44 main line telephones per 100 inhabitants in 1959 to 3.18 telephones per 100 inhabitants 35 years later (ibid).

Costs for telephone calls within Cuba were very low, nearly token amounts. In addition, a significant mitigation of the regional disparities of telephone diffusion took place: if in 1959 almost three quarters of the nation's telephones were concentrated in the capital, less than half were by 1994 (see table 8). This was made possible by keeping telephone expansion in Havana to a minimum, as the network grew from 124,100 main line telephones in 1959 to 156,937 in 1994 (ibid). With the population of the capital more than doubling in this period (from 860,798 to 2,142,100 inhabitants), the per capita rate of telephone diffusion in the capital actually was cut by more than half, from 14.42 to 7.16 per 100 inhabitants (ibid).

The quality of the telephone system as well as the quantity of phones reflected the low priority given to the sector. Up until the early 1990s, an estimated 40 percent of the telephone infrastructure consisted of U.S. technology installed before 1959. Spare parts were nearly unavailable due to the U.S. embargo and the system had limited compatibility with the technologies from the Comecon countries that entered the island from the 1960s onward. Numerous defects, such as a high percentage of incomplete calls, crossed lines and bad tone quality, were the consequence (Press 1996a: 12).

Table 8 Telephone System Indicators in Cuba, 1959 / 1994 / 2000 / 2003 (Data for end of year)

Year	1959	1994	2000	2003
main lines in service	170,000	349,471	488,606	*709,439
percent of these in Havana	73.00	45.00	46.5	*47.00
telephones per 100 inhabitants - entire country	2.44	3.18	4.36	*6.40
- city of Havana	14.42	7.16	10.16	*15.20
- rest of Cuba	0.70	2.10	2.92	*4.20
percent of digitalized lines	0	0	44.25	80.90
public phones in operation	n.d.	5,814	14,620	25,000

*data for November 2003
Source: 1959 and 1994: Press 1998; 2000: MIC 2001b, ITU 2002a; 2003: Fernández Martínez 2003, del Puerto 2004

Because the telephone system was not seen as a major resource for economic or social development, the only sectors where communications infrastructure was given high priority were military and security—and it is telling that, since the early days of the Revolution through 2000, an army general always headed the Ministry for Communications.

Telephones for private, residential users were regarded as a dispensable luxury. Access to a new main line phone was not a service that could be bought: it depended upon political and social criteria and could be won—as a reward for *mérito* (social merits) or in response to the professional needs of the person, for instance in the case of medical doctors who needed them in order to be available during emergencies in their off-hours. (Of course, as for almost anything else there was and continues to be a black market for telephone access, as people invented ways to share one line between multiple parties, for example.) As a consequence of these developments and policies, at the end of the 1980s Cuba's telecommunications infrastructure lagged behind most Latin American countries in the majority of indicators.

The profound economic crisis that hit Cuba in the late 1980s and early 1990s aggravated the problems still further. Spare parts and supplies from the Comecon countries were no longer forthcoming and the hard currency crunch minimized the possibility of compensating through trade with capitalist suppliers. Symptomatic of the problems is the shrinking number of public telephones in service, which fell from 10,003 in 1992 to only 5,814 two years later, according to official data (Press 1996a: 9). A telling indicator for the declining quality of the telephone service is the rate of faults per 100 main lines per year, which in the same period increased from 14.9 to 29.2 (Press 1996a: 9; see also Press 1996b). New investments in the sector virtually ended.

As to international telephone traffic, the United States has been by far the most important country for Cuba ever since, in 1921, AT&T inaugurated the first under-sea cable between Havana and Florida. After the revolution, telephone traffic between both countries became a target of the U.S. embargo: AT&T was allowed to continue telephone operations with the island through the existing connections, but any modernization of these was prohibited. In addition, U.S. laws established that all revenues corresponding to the Cuban share of bilateral telephone traffic could not be paid to the Cuban government, but instead were deposited in an escrow account in the United States.

Over time, the 1921 cable connections became hopelessly inadequate. The U.S. Federal Communications Commission estimated that of 60 million annual call attempts, less than 1 percent were completed (FCC data, cited in

Press 1996a: 3). This situation only changed after the 'Torricelli Law' of 1992 brought a major turnaround in U.S. communications policy to the island.

MODERNIZATION BY JOINT-VENTURE: THE TRANSFORMATION OF THE TELECOMMUNICATIONS REGIME SINCE THE 1990s

While the U.S. Radio and Television Broadcasting to Cuba Acts had promoted a one-way communication—programs are made in the United States to be received by the Cuban population—the 'Torricelli Law' of 1992 opened the way to promoting individual two-way communication between the United States and the island in order "to seek a peaceful transition to democracy" (U.S. Congress 1992, Sec. 1703). While reinforcing U.S. sanctions against Cuba in other areas, this so-called 'track two' of the Torricelli law selectively lifted hitherto existing sanctions on telecommunications and postal connections to the island, stating that:

> "Telecommunications services between the United States and Cuba shall be permitted. Telecommunications facilities are authorized in such quantity and of such quality as may be necessary to provide efficient and adequate telecommunications services between the United States and Cuba. (. . .) The United States Postal Service shall take such actions as are necessary to provide direct mail service to and from Cuba, including, in the absence of common carrier service between the two countries, the use of charter service providers" (U.S. Congress 1992, Sec. 1705, e-f).

Explicitly excluded from this easing of sanctions, however, was any U.S. "investment in the domestic telecommunications network within Cuba" (ibid.).

In 1994, a decision by the U.S. Treasury Department confirmed the Torricelli Law's provisions enabling the resumption of direct telephone traffic between the United States and Cuba, declaring any data and information flow legal except for money transfers (Valdés 1997: 6).[27] Finally, in October 1994, the U.S. Federal Communications Commission approved the agreements a number of U.S. telecom companies had reached with the Cuban government regarding the sharing of revenues from telephone services between both countries. And though the Cuban government vehemently attacked the 'track two' policy of the Torricelli law as 'ideological subversion,' it did not prevent the restoration of direct bilateral telephone connections that resumed on November 25, 1994 (ibid). In fact, with the reestablishment of direct telephone traffic between both countries and the revenue-sharing arrangement, international telecommunications became a substantial hard currency earner for Cuba.

After 1989, the need for Cuba to re-integrate into the globalized world economy and the growth of tourism and other dollarized sectors had brought increasing economic pressures for a modernization of the telecommunications system. At the same time, the steep fall of exports and foreign exchange revenues minimized the state's ability to finance the needed modernization program.

In this situation, one year after the U.S. Congress passed the Torricelli Law, the Cuban government decided to open up the telecommunications company to foreign capital. A joint-venture deal was signed by the Mexican *Domos* group (holding 49 percent of the shares) and the Cuban Ministry for Communications (holding 51 percent). Law number 190 formalized the arrangement on August 17, 1994 and, in early 1995, the country's new telephone company—still a monopoly, though a joint-venture—was officially established under the name of ETECSA (*Empresa de Telecomunicaciones de Cuba, S.A.*).

This option of modernization via joint venture would have been impossible without the Torricelli Law, since this was of crucial importance in making the Cuban market attractive for any foreign investor. Only after the reestablishment of direct telephone connections with the United States did Cuba represent a significant hard currency telecommunications market that offered adequate returns on the high capital investments that were needed for the modernization of the sector. Though the positive function of the Torricelli Law for the modernization of the Cuban telecom sector was politically inconvenient for either side to acknowledge (in consequence, it has received little public or even academic attention), it is of central importance for any political analyses of the transformation of the telecommunications regime and the development of the NICT in Cuba.

The joint-venture agreement with *Domos* was short-lived, but the new approach to telecommunications lived on with a new international partner. In the initial contract, Domos committed itself to a total investment volume of US$1.5 billion in Cuba, but all through 1995 little investment effort was seen. Instead, as early as April 1995, *Domos* sold 25 percent of the ETECSA shares for US$291.3 million to the *Società Finanziaria Telefonica*, fully a subsidiary of the Italian state-owned company STET (also known as Telecom Italia). In the following months, *Domos* sold all of its remaining shares. Since February 1997, STET has owned 29.29 percent of ETECSA shares, the Cuban communications ministry 51 percent and the rest is in the hands of a consortium of Cuban para-statal entities.[28] As a result, the national carrier is effectively a joint-venture between the Cuban state and the Italian company STET.

Since STET began its involvement in Cuba, a major investment process has been initiated. The number of main line telephones more than doubled from 349,471 in 1994 to 709,439 in (Fernández Martínez 2003; see table 8). Given the minimal population growth in this period, the growth in main line telephony translates almost directly into increases in the per capita data. Nevertheless Cuba's telephone density is still one of the lowest in Latin America (see table 3). At the same time, the process of balancing regional disparities in telephone diffusion came to a halt, with Havana accounting for 47.0 percent of all Cuban main line telephones in 2003 (see table 8). In spite of this growth in recent years, the capital still has barely surpassed the pre-1959 rate of telephones per capita.

The joint-venture with STET not only brought an extension of the telephone net, it also initiated the technological switch from analogue to digital lines. In the five years between end of 1994 and end of 1999, a total of 233,700 digital lines were installed; of these, only 100,000 were new connections—the rest were substitutions for existing analogue lines, according to ETECSA's President Marrero (2000). This brought the nationwide share of digitized lines to 44.25 percent at the end of 2000 (MIC 2001b); by the end of 2003, according to official data 80.9 percent of the telephone net were digitized (del Puerto 2004; see table 8). The annual number of faults per 100 main lines dropped from 29.2 in 1994 to 10.0 in 2000 (ITU 2002a) and, for the first time in many years, the joint-venture ETECSA published telephone books that were distributed for free to all subscribers.

The public pay-phone system also was revitalized and extended. While only 5,814 public telephones were left in operation in 1994, their number had recuperated to a total of 25,000 by the end of 2003 (del Puerto 2004), according to official data. Almost all of these are new phones and many require hard currency prepaid telephone cards, though even the authorities recognize that the number of prepaid cards sold for Cuban pesos is far below demand (ibid). This increasing dollarization of a service considered to be a public good has raised considerable frustration in the population, leading to vandalism and creative ways of using phones without paying.[29] As a result, the government has ceased expanding the net of hard currency pay-phones; it has actually removed some and it placed others which are still in operation under closer vigilance (ibid).

Still, the investment and modernization policies of recent years underscore a strategic shift from the traditional state-socialist approach, which gave only low priority to the sector, to a new approach, which sees telecommunications as an important element in the infrastructure necessary for the attainment of national development goals. Given the tremendous backwardness of the sector and the structural hard currency limitations of the Cuban state, the alliance with a transnational company was indispensable for implementing a modernization strategy of such scope.

Parallel to the joint-venture with the Italian STET, the People's Republic of China has become Cuba's principal supplier of telecommunications equipment. The *Gran Kaiman Teleco S.A.*, a joint-venture between the Cuban *Grupo de la Electrónica*—the most important state holding of technology companies—and the Chinese *Great Dragon Information Technologies* group, has played a central role in this relationship. This joint-venture not only arranges the importation of Chinese technology, it also has begun to build assembly factories in Cuba.[30]

MOBILE TELEPHONY: LATE START, SHIFTING PRIORITIES

Mobile telephony made a very late and very slow entry in Cuba. The country's mobile telephone company, *Cubacel,* was founded in December 1991 as a joint-venture with 50 percent foreign capital. Though separated from the national telecom carrier, it too was institutionally subordinated to the Ministry of Communications. After the build-up of the basic infrastructure, it began operations in February 1993, offering services only in Havana and the tourist beach resort of Varadero, a two-hour-drive east of the capital. Its target group was almost exclusively foreigners residing on the island and high-level tourists. In 1996, Fidel Castro still spoke contemptuously of "*esos telefonitos*" ("those little telephones") and presented with pride the fact that the Cuban leaders had not fallen for this type of capitalist vanity:

> "As you see, the leaders of the Revolution do not run around with those *telefonitos;* it is a matter of austerity, which one Minister may emphasize more than another, but there are many foreigners who right away must have their little telephones to walk around with" (Castro 1997: 25).

Nevertheless, in the past years the attitude towards mobile telephony has changed. Leading functionaries or managers of important joint-ventures or state companies may be seen using cellular phones and the extension of the national mobile phone system has become an important part of the sector's development plans. This change in attitude was greatly influenced by the example of socialist China, where mobile telephony has experienced an explosive growth from 3.6 million phones in 1995 to 206.6 million at the end of 2002 (ITU 2003).

In Cuba, the change in attitude was far from mimicking the Chinese approach, and the expansion of mobile telephony remained slow, reaching a mere 17,900 subscribers—0.16 mobile phones per 100 inhabitants—at the end of 2002 (ITU 2003); for the end of 2003, official Cuban data speak of 30,000 mobile phone subscribers (Fernández Martínez 2003). The most important clientele still are foreigners residing on the island, but the percentage of users who are representatives of Cuban institutions and companies is increasing. While *Cubacel,* which utilizes the 'American' TDMA technology, for ten years held a monopoly position in mobile telephony on the island, in 2001 a second company named *C_Com* began to offer mobile phone services based on the 'European' GSM technology; a third company, *MoviTel,* offers trunking and paging services.

With the government's 'ideological offensive' since the late 1990s, the state and joint-venture companies of the dollarized sectors of the economy increasingly had to show tangible contributions to the social development goals of the socialist society beyond the generation of hard currency revenues. A typical example was the (hard currency) donations made by employees of tourist hotels for a nearby kindergarten or hospital. The mobile phone company *Cubacel* reports to its clients on its social contribution as follows:

> "If at night in some neighborhood (. . .) you see comrades, which, in fulfilling their duty in the night watch of the Committees for the Defense of the Revolution (CDR), make use of a mobile phone just like yours, then you should know that this is a service our company offers in cooperation with the CDR. It is part of the integration of *Cubacel* into the everyday life of our country, a contribution to the citizens' security and the fight against crime" (www.cubacel.cu, December 2, 2001).

This example shows the capacity of the socialist state to divert resources from companies in the dollarized sectors for political use even where they are organized as joint-ventures. In addition, it also shows that the state is well aware of the possibilities mobile telephony offers for the organs of public security and, although there are no data available for this, it may safely be assumed that not only the 'amateurs' of the CDR's night watch, but also the professionals from the police and security organs of the Ministry of the Interior have been provided with this new technology.

Despite the high priority given to the telecommunications sector in general, mobile telephony has remained a low priority on the government's agenda all through the 1990s. Only as late as the end of 2000 had use of mobile phones become possible in all provincial capitals (www.cubacel.cu). In comparative perspective, Cuba trails all other countries on the continent in mobile telephony diffusion, with Argentina, Venezuela and Brazil having a per capita diffusion of mobile phones that is 100 times higher, while a nation like Haiti still has a rate of diffusion that is 10 times higher (see table 5). But given the analysis in chapter three, which argued that the extension of main-line telephony is far more important than the spread of mobile telephony for the social diffusion of advanced NICT services, the backwardness in this area certainly can be considered one of the minor problems Cuban develop-ment faces today.

Nevertheless, 2003 saw a remarkable shift in attitudes towards what once were disparagingly labelled *"telefonitos."* At the end of November 2003, ETECSA President Fernández Martínez characterized the "massive diffusion of mobile telephony" as the single most pressing task in Cuban telecommunications (Fernández Martínez 2003). This change is not yet re-flected in present user data, but it already weighs heavily on the sector's in-vestment plans: for 2004–2008 the resources dedicated to mobile telephony are projected to increase greatly, accounting for no less than one third of total investments in telecommunications from 2004 onward (see figure 14 and table 9).

The projected expansion of mobile telephony will be concentrated in the dollarized sectors of the economy and among the personnel of political and social institutions deemed most important by state authorities. For these, mobile phones will come as an addition to their already existing main line telephone connections, not as an alternative to it. As the investment

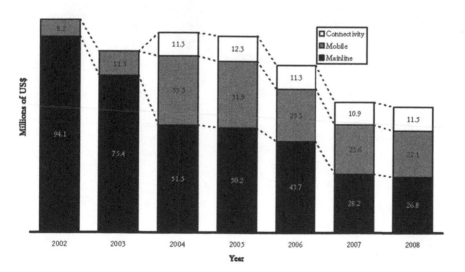

Figure 14 Cuba: Investments in Telecommunications, 2002–2008 (Realized and Projected) (main line telephony / mobile telephony / connectivity)

Source: Fernández Martínez 2003

Table 9 Cuba: Investments in Mobile Telephony as Share of Total
Investments in Telecommunications, 2002–2008 (Realized and Projected)

Year	2002	2003	2004	2005	2006	2007	2008
mobile as a percent of total investments	8.0	13.2	34.7	33.8	31.5	37.7	36.6

Source: Fernández Martínez 2003

plans for the coming years clearly show, the increase in mobile telephony is at the cost of slowing down the extension of the main line network. This turn underscores that the main priority in Cuba's telecommunications policy still is not so much achieving universal service, but rather improving the techno-logical facilities for those that either promise hard currency earnings or that are closely linked to the official structures of the socialist state.

Chapter Six

From the Rejection of the Internet to the 'Informatization of Society': A Political Anatomy of Change

COMPUTER TECHNOLOGIES AND DATA NETWORKS: CUBA IN THE PRE-INTERNET ERA

Before 1959, Cuba was one of the leading Latin American countries in the introduction of electro-mechanical devices for data processing, as it was in mass media and telephone density. It was therefore only natural that, in the 1930s, IBM opened its regional branch office for the Caribbean and Central America in Havana (Altet Casas 2001: 12). After the Revolution, IBM had to abandon the island like any other U.S. American company. But unlike its attitudes toward the telephone system, the revolutionary leadership from the very beginning pinned high hopes on computer technology. At the beginning of 1960, the Cuban government acquired two first generation mainframe computers of Western technology, followed by a second generation British computer in 1963, destined for the National Center for Scientific Research (CNIC)—and marking the beginning of the study of computer science as a professional discipline in Cuba (Casacó 2002).

Only five years after the Revolution, the Ministry of Industries, headed by Che Guevara, established a Department of Automation (mainly focused on the sugar industry) that used a handful of computers imported from Poland (Valdés 1997: 1). Cuba nevertheless continued to import major pieces of its computer technology from capitalist countries. The government imported two second-generation mainframe computers in 1968 for the processing of all data gathered during the population census of 1970, and it later brought in the IRIS series third generation computers from France as well (Casacó 2002).

In 1969 a Center for Digital Research (CID) was founded at the University of Havana as part of the government's efforts to build up a national computer manufacturing capacity. In 1970, the first prototype of a Cuban-made minicomputer, the so-called 'CID-201,' was developed. In the 1970s, the government engineered a closer and more institutionalized alignment with the Soviet Union and the Comecon countries in the area of computer technology, in accordance with the general political climate. In a number of bilateral agreements, the Soviet Union committed itself to supporting the development of a computer industry in Cuba, and the 'CID-201' model gave rise to a whole series of computers designed and produced in Cuba in the following years. When the rise of the micro-computers (the so-called PC's) began in the 1980s, the state-owned COPEXTEL corporation imported Western computers, but also built a large and modern factory on the island to assemble IBM-clones (the 'LTEL 24' computers) (Casacó 2002). In total, the country's computing resources increased from twelve mainframe computers and about 100 minicomputers in 1976 to 28 Soviet-made mainframe computers, more than 200 Cuban-built minicomputers, and almost 4,000 microcomputers in 1987 (Valdés 1997: 1).

On the political and administrative level, in 1976 the *Instituto Nacional de Sistemas Automatizados y Computación* (INSAC) was established as the central institution for the execution of the state's policy in the area of computer technologies (Altet Casas 2001: 13). The 1980 Cuban Communist Party Congress explicitly emphasized the need to promote computerized telecommunications and data-based transmission facilities (Valdés 1997: 1); in 1982, the *Centro Nacional de Intercambio Automatizado de Información,* CENIAI, institutionally subordinated to the National Academy of Sciences, was founded as the country's central institution for all national and international computerized data networks. CENIAI has since resided in a prominent building in Havana, the *Capitolio,* a replica of the Capitol in Washington D.C. that, until the Revolution, had housed the Cuban parliament.

Cuba's first international satellite connection was inaugurated in 1983, giving access to some 50 Soviet databases that were primarily used in economic planning (ibid.). In the following years, Cuba became continuously connected to the IASNet computer network operated by Moscow's All Union Scientific Research Institute for Applied Computerized Systems, VNIIPAS. However, domestic computer networking among Cuban institutions did not start until 1988, when a branch of the Academy of Sciences established a network providing data transmission facilities, e-mail service, and access to Cuban databases. Domestic networking expanded rapidly: as of November 1992, Cuba had nine computer networks serving a reported

1,500 users and, by the summer of 1994, the country had 26 networks in operation, including eleven for scientific use, eight serving the social sciences and libraries, and four educational networks connecting universities, high schools, and technical institutes (ibid.).

A FIRST CONNECTION TO GLOBAL NETWORKS: INTERNET ACCESS VIA E-MAIL

The technological capacity of the Soviet IASNet, based on X.25 protocols, lagged behind the standards of Western networks and the breakdown of the Comecon and the Soviet Union largely eliminated the political and economic value it did have. Yet already in 1988, first talks had begun between academics from the United States and Communist Party officials from Cuba regarding the possibility of connecting the island to the global computer networks (Valdés 1997: 3). Nelson Valdés, a professor of sociology at the University of New Mexico who was born in Cuba and brought to the United States as a child, played a key role in this endeavor. With a long record of solidarity with the Cuban Revolution in his academic writing and political engagement, he could credibly fulfil the role of an 'honest broker' in the eyes of the Cuban authorities. In the words of Valdés (1997: 4):

> "A key reason for success of the negotiations was that security and ideological issues did not surface. Neither the Cubans nor the Americans involved attempted to bring pressure to bear over who should or should not be allowed access to the E-Mail connection. In short, no attempt was made to frame the issue of external connectivity in the context of the clash between the United States and Cuba."

In addition to Valdés and a number of solidarity-minded colleagues from the academic world, one of the most important NGOs of the international networking scene, the Association for Progressive Communication (APC), played a crucial role in implementing Cuba's first Internet connection. Founded in 1989, the APC is a global network of civil society organizations dedicated to empowering and supporting groups and individuals "working for peace, human rights, development and protection of the environment through the strategic use of information and communication technologies" (www.apc.org). In April 1991, negotiations had proceeded to the point that the Cuban Communist Party's Department for Ideology, then headed by Carlos Aldana, gave the green light for the establishment of a so-called UUCP (Unix-to-Unix Communication Protocol) connection from CENIAI to Web Networks, an APC affiliate in Toronto, Canada. In January

1992 this connection began operations: Cuba was, though in rudimentary form, connected to the global computer networks that make up the Internet. The connection via Canada was necessary since until 1994 the U.S. embargo also prohibited any data network communication between the United States and Cuba (Valdés 2002: 58).

In contrast to the Internet Protocol (IP or TCP/IP), which already was standard in Western network connections, a UUCP connection is what can been called a 'store and forward' form of data transmission: it allows for the asynchronous exchange of data through e-mail communications, but it does not support interactive applications such as the search for information on the World Wide Web or Gopher servers. In the practical Cuban experience, the UUCP connection to the global networks was operated in the following way: once every day a long-distance phone call from Toronto to Havana was made and, in about ten minutes, all outgoing e-mails accumulated in the past 24 hours were exchanged between Canada and Cuba. The Canadian APC affiliate then distributed Cuban e-mails around the globe, and CENIAI forwarded incoming international mails to their diverse recipients on the island (Valdés 1997: 4). U.S. and Canadian counterparts and support groups covered all costs of the UUCP connection and the long-distance calls (adding up to about US$1,000 per month).

Obviously, from the Cuban government's point of view, communication through such an asynchronous, centralized, once-a-day mail transfer as provided by the UUCP connection is far more easily controlled than the continuous and multi-level data exchange via an IP connection. For the users, compared to full IP Internet access, this form of access to international networks was certainly limited—and not only from today's perspective. In addition, the capacity of the UUCP connection to Canada became more and more inadequate for the growing traffic and the connection repeatedly suffered interruptions due to technical, financial or administrative reasons. Nevertheless, for Cuba, the UUCP connection was a milestone. For those connected, e-mail was a sensational alternative to traditional international mail, which could take weeks to arrive (and it was not sure if it ever did), and international telephone calls, which were not only expensive, but simply out of reach for the majority of academics, managers or others, even if they worked in the more privileged of the country's official institutions.

Even though a UUCP connection does not allow navigation in the World Wide Web, Cuban users managed to access WWW content through e-mail in various forms, ranging from personal requests and the subscription to mailing lists, to more sophisticated forms such as the sending of e-mails with automated commands to download specific Internet content to the sender's mailbox. An example is provided by LANIC, the Latin America Network Information Center of the University of Texas at Austin (www. lanic.utexas.edu) and one of the leading sources of Internet material on the area. In order to accommodate the often limited technological options for users in Latin America, LANIC for many years in the 1990s operated an e-mail server through which most of LANIC's content could be accessed. As senior LANIC coordinator Carolyn Palaima reported, "a lot of Cuban scholars came this way, so they had e-mail access to many of the texts we put on the web."[1]

The CENIAI staff was well aware of the technological limitations of the e-mail link to Canada. As early as 1991/92, parallel to the inauguration of the UUCP connection, CENIAI began to experiment internally with networks based on the far more potent Internet Protocol (IP). As for know-how and technical equipment, preparations were made for the upgrading of the international networking connection to a full, IP-based Internet access. However, this was not to be a matter of technological capacities: it was an eminently political question to be decided at the highest political level.

FRAMING THE ISSUE IN A POLITICAL CONTEXT: INTERNET AS A THREAT OF IDEOLOGICAL SUBVERSION

The issue of the country's Internet access came to the forefront of Cuban politics with the passing of the Torricelli law by the U.S. Congress in 1992. This shift in U.S. policy had been preceded by a study commissioned by the U.S. Defense Department and carried out by its principal think tank, the RAND Corporation, which outlined the new proactive information and communications policy towards Cuba as it was formulated in the law. This study, authored by González/Ronfeldt (1992), argued that increased communication would facilitate the emergence of civil society structures in opposition to the present regime. Whereas the text of the Torricelli law only mentioned telephone and postal service, the RAND study was unambiguous about the fact that increased communication not only meant phone calls from the uncle in Miami but that it would include the whole range of new digital information and communication technologies transmitted via telephone lines. In the words of the RAND study, it would "build bridges across computer networks (. . .) in the expectation that free information flows should foster pluralist tendencies" (González/Ronfeldt 1992: 71).[2]

The Cuban government condemned this 'track two' strategy, but it did not block the reestablishment of direct telephone communication, nor did it reevaluate the existing UUCP computer network connection to Canada. However, any further steps towards full Internet access were frozen. Valdés (1997: 5) writes:

"Interpreting the bill's telecommunications section as signaling a program of penetration and subversion, the Havana government put the brakes on further networking with the west. The use of E-Mail continued to grow, but movement toward other forms of connectivity (especially those involving real-time interaction) became much more problematic. Whereas establishment of the APC E-Mail link had required only a moderately simple decision by a single party official, the new political climate dictated that the move to a full Internet connection could take place only when the highest authorities in Cuba so decided."

In the following years, a number of political incidents stemming from the use of computer networks by U.S. and exile groups further strengthened those forces opposed to opening up a full Internet connection. One such case was even brought about by the U.S. interest section in Havana, which had obtained an account at one of the Cuban networks, TinoRed, and posted official declarations of the U.S. government on the network's bulletin board (Valdés/Rivera 1999:145). The account immediately was withdrawn, but the political damage was considerable. Another incident was reported by Steve Cisler from the U.S. computer company Apple who, during a stay in Havana in summer 1994,[3] had learned of a case in which anti-government diatribes went out from an e-mail account with the telling name of 'abajofidel@aol.com' (*abajo Fidel* = down with Fidel) to all accounts in the TinoRed network—"a bad way of promoting international connectivity," as Cisler (1994) commented.

In 1993, at the height of the country's economic and social crisis, CENIAI for the first time proposed to the state authorities the adoption of a full IP Internet access. But since international network connectivity had become framed within the context of state security and external confrontation, political considerations impeded any such step. However, UUCP connections still increased and, by 1995, there were four Cuban networks with international UUCP links: CENIAI; the Center for Genetic Engineering and Biotechnology (CIGB); TinoRed, a network linking so-called non-governmental organizations and the Youth Computer Clubs administered by the Communist Party's Youth Organization (see below); and an X.25-based tourism network. International e-mail traffic surpassed 60,000 bytes per month and an estimated 2,600 Cubans were using e-mail (Casacó 2002).

CUBA JOINS THE INTERNET: THE POLITICAL
DECISION-MAKING PROCESS

Though the plan for IP connections was politically blocked, the Cuban networking institutions continued technical preparations for shifting away from UUCP connections (cf. Cisler 1994, Press 1998). The process also advanced on the international side: on January 12, 1995 InterNIC, the U.S. based organization handling the registry of networks connected to the Internet, granted CENIAI a so-called Class B Internet address, in effect giving Cuba permission to join the Internet (Valdés 1997: 6). With this, the question of Cuban access became an essentially domestic issue and, though no public debate on the issue was allowed, divergent positions within the country's political leadership were felt more strongly. Fidel Castro expressed the tenor of the government's attitude at that time on August 5, 1995 at a mass rally on Havana's *Malecón:*

> "They [the United States] speak of 'information highways,' new ways that serve to fortify this economic order, which they want to impose on the world, through propaganda and the manipulation of human mentality (. . .)" (cited in Sánchez Villaverde 1996: 39).

In accordance with this line of thought, strong forces within the Cuban leadership argued that joining the Internet would be ideologically detrimental and a threat for the country's national security. A text representative of this line of thought is by Lieutenant-Colonel Ricardo Sánchez Villaverde, professor for automatization at the *Instituto Tecnológico Militar José Martí,* who wrote:

> "As part of the hegemonic project of the United States [the Internet] represents an invasion *sui generis,* which is not led by marines but by the information which is moved via satellite, fiber optic cables and Hertzian waves" (Sánchez Villaverde 1996: 40).

And when the text continues that "the position of our Party has always been clear: the essence of these phenomena is political" (ibid.), this de facto translates as: whatever the economic, social or academic arguments for joining the Internet may be, they have to be subordinated to the related political and security considerations. In a 1995 study for the Cuban military by the same author, the title immediately makes it clear that new information and communication technologies have to be seen primarily not as a development issue, but in the context of the confrontation with the United States and the Torricelli law: "The Informatization of the Society: A Weapon of War of the 'Track Two'" (Sánchez Villaverde 1995).

Given this official stance, the proponents of bringing in the Internet could not simply focus on potential economic benefits or cost reductions, but had also to show that political and ideological control of the NICT was possible. In analyzing this internal conflict it is essential to understand that the conflict was not divided strictly between civilian 'reformers' and 'hardliners' from the military and security sectors, as is often assumed. Instead, the Ministry for Communications, which was headed by an army general (and which stood to gain in influence, revenues and resources with the expansion of telecommunications and NICT), took a position in favor of connecting to the Internet and argued for the possibility of guaranteeing sufficient technical and political controls against its potentially negative implications (Valdés 1997: 6).

Finally, there was a particular actor that played a pivotal role in influencing the decision-making process of the Cuban leadership. Ramiro Valdés Menéndez had fought alongside Fidel Castro even before 1959 as a *Comandante* of the Rebel Army; he became a member of the first Politburo of the PCC and for many years he was Cuba's Minister of the Interior. After being replaced in this position in the 1980s, Ramiro Valdés became head of the 'Grupo de la Electrónica,' the most important holding of state companies and para-statal 'Sociedades Anónimas' in the electronics and high-technology sector, which was ascribed to the Ministry of Metallurgy and Electronics Industry (SIME). Though he had no formal post in the government, Ramiro Valdés remained a heavy-weight in Cuban politics. He was de facto the *hombre fuerte,* the strong man behind the scenes, for Cuba's electronics and computer development and in the mid-1990s he became perhaps the most influential advocate of joining the Internet, which he considered held strategic importance for the country's future.[4]

No one in Cuba interviewed for this study doubted that the highest political authority, that is Fidel Castro himself, made the ultimate decision about permitting broader access to the Internet. Therefore, certain aspects of his personality, such as his deeply rooted enthusiasm for scientific and technological progress as an intrinsic part of his vision of Cuban socialism,[5] came into play in the final decision. Nevertheless, all indicators suggest that the decision to bring the Internet to Cuba was a complex process for Fidel Castro, who spent months weighing the potentially positive and negative effects, as he perceived them. The specific timing of the decision and the developments in the overall political context therefore hold great importance; three factors shall be underscored.

For one, in 1995/96, when the decision-making process took place, the economic and social crisis had already bottomed out and indicators pointed to a gradual recuperation. The sense of instability and latent social unrest of 1993/94 had significantly decreased and the U.S. government and exile groups had to face the fact that Cuba was not 'the next domino to fall.' With the focus passing from the 'survival of the Revolution' to longer-term perspectives and with a significant increase in hard currency revenues, the new information and communication technologies could come to play a greater role.

A second factor is the experience of socialist China, which had connected to the Internet earlier than Cuba. Since 1987, China operated a UUCP connection to the global computer networks (via the University of Karlsruhe, Germany) and, in 1993, the country had joined the Internet with a full IP connection. Not only did the Chinese experience show that Internet access did not (at least in the short and medium term) represent insurmountable problems for the maintenance of Communist Party rule (cf. Press/Foster/Goodman 1999; Wacker 2000), but the government in Peking also was willing to share with Cuba its technological and administrative know-how for minimizing the NICT's potentially destabilizing effects.

The third factor was the Cuban government's decision in March 1996 to crack down on the intellectual reform debate and to curtail civil society articulations. While the 'track two' strategy of the Torricelli law postulated the fostering of pluralist civil society structures through increased communication, Havana accepted the latter, but only while rigidly rejecting the former.

Though the infamous 'Report of the Politburo' read by Raúl Castro at the 5th plenum of the PCC's Central Committee certainly did not respond only—or even primarily—to the exigencies of the NICT integration, there certainly was more than a temporal coincidence between its attack on pluralist civil society and the government's decision to join the Internet which was publicly announced a month later.[6] In fact, this coincidence reflects the basic pattern of the government's strategy in dealing with the NICT: accept and incorporate the technologies, but minimize their negative effects through authoritarian socio-political mechanisms of control.

Though little noticed, this two-handed strategy was apparent precisely at the Central Committee's 5th plenum. While the 'Report of the Politburo' monopolized public attention, Carlos Lage, the country's chief economic policymaker, made a speech emphasizing, more explicitly than any other high-ranking government official before, the importance of the new information and communication technologies, including the Internet, for Cuba's development:

> "In the area of computers and informatics, an extraordinary transformation is taking place. What a few years ago had been called the 'information highway,' the computer connections, are no longer an idea of the future but they are already a reality which has practical consequences for the economies of our countries. To note just one example of this process that has to do with electronic mail: one telex can cost twelve

dollars [whereas] the same message costs 75 cents in the form of a fax and 3 cents via the Internet. This means that as a result of the spread of the new computer technologies and the advances in informatics on a global level, costs and access to information will become increasingly important (. . .). In spite of our blockaded circumstances, we are in a relatively good position to face the challenges of the global scientific and technological change, due to the educational and scientific work developed by the revolution" (Lage 1996: 9).

Three months later, in June 1996, the government published law-decree number 209 to regulate the administrative competencies and structures for computer networks and Internet services (see more in detail in the following section). The necessary hardware for the international IP connection— bought with the help of US$250,000 financial aid from the United Nations Development Program—was installed in July and CENIAI began the first experimental Internet sessions in August. Finally, on September 9, 1996, CENIAI's Director Jesús Martínez announced Cuba's connection to the Internet in a brief e-mail to the Forum of Networks in Latin America and the Caribbean (Martínez 1996). It is noteworthy that this first 64 kbps IP connection was not from Cuba to Canada, as initially announced, but that the U.S. company Sprint provided the service, linking Cuba directly to the United States. Since this was made possible in the U.S. through the provisions of the Torricelli law, it remains remarkable that, on such a politically sensitive issue, Cuba in fact made use of the opportunity to establish its connection to the worldwide computer networks via the United States.

The official act of Cuba's entrance to the Internet took place on October 11, 1996. The last step was taken in February 1997, when the administration of the Cuban top-level domain '.cu' was transferred from the APC affiliate in Toronto to CENIAI in Havana and the UUCP connection to Canada was suspended.

CHECKS AND BALANCES: DISPUTED COMPETENCIES IN THE ADMINISTRATIVE STRUCTURE OF THE NICT SECTOR

If there were tensions in the Cuban administration over the principal question of whether—and if so, when—Cuba should join the Internet, there also were conflicts about the organizational aspects related to such a move and the competencies the different administrative institutions would have in this area. As early as 1994, in the context of a general reform of the state administration, the army-led Ministry for Communications (MINCOM) had issued a resolution intending to centralize all networking-related activities (Valdés 1997: 4). This initiative did not evoke enthusiasm from other

Ministries and institutions related to the sector and, in the end, it was a complete failure. Instead of the MINCOM taking on responsibilities formerly assigned to the Ministry of Metallurgy and Electronics Industry (SIME), it was the latter (which, we may recall, includes the powerful *Grupo de la Electrónica* led by *Comandante* Ramiro Valdés) that gained influence. The *Instituto Nacional de Sistemas Automatizados y Técnicas de Computación* (INSAC) was dissolved, and in its place the *Dirección Nacional de Informática* (DNI) was created under the administration not of the MINCOM, but of the SIME (Altet Casas 2001: 13).

Two years later, the Council of Ministers' executive committee implemented a major reorganization of the NICT-related administrative functions by passing law-decree number 209 in June 1996.[7] This established a structure very different from the centralization of competencies under the Ministry of Communications, as the MINCOM had proposed. Supreme authority over all questions of regulation, administration, security and development of the Cuban networks and their international connections was assigned not to any single Ministry, but to an Inter-Ministerial Commission (*Comisión Interministerial*), which brought together six different Ministries:

- The Ministry of Science, Technology, and the Environment (CITMA) remains responsible for the administration of the national top-level domain '.cu,' the registration of IP addresses and the regulation and control of all Cuban networks with international connectivity. A number of central institutions of the Cuban NICT sector are assigned to the CITMA: a) CENIAI, the country's central Internet gateway; b) the Cuban Network Information Center, or Cuba-NIC (www.nic.cu) for short, which administers the '.cu' domain and gathers and delivers all Cuban-related information concerning the Internet; c) the *Agencia de Información para el Desarrollo* (AID), dedicated to the development and marketing of software and databases; and d) CITMATEL (*Empresa de Tecnologías de la Información y Servicios Telemáticos Avanzados*), one of Cuba's major companies producing information technology products.
- The Ministry of Communications (MINCOM), which includes the national telephone company ETECSA, is the supreme institution for the entire Cuban telecommunications system and the cable connections on which the nation's computer networks operate.
- The Ministry of Justice (MINJUS) is the institution responsible for providing and adapting the juridical framework to the exigencies brought about by the NICT development.

- The Ministry of the Interior (MININT), according to General Alejandro Ronda, the head of the MININT's *Dirección de Protección*, has the duty "to guide, control and implement the policy of informatics security" (Ronda 2000). This includes as much political-ideological control functions as the technical security of the networks and connections.
- The Ministry of the Revolutionary Armed Forces (MINFAR), similar to the MININT, ensures that the Internet connection and computer networks on the island will not affect the security of the state. Obviously, the precise means available to accomplish this task are not in the public record.
- Finally, the Ministry of Metallurgy and Electronics Industry (SIME) not only includes the 'Grupo de la Electrónica' and other NICT relevant industries, but also has added the newly founded *Dirección Nacional de Informática* (DNI) as the guiding institution (*órgano rector*) for Cuba's informatics and electronics industry. In accordance with this move, the chair for the Inter-Ministerial Commission was given to the SIME. Thus, the new administrative structure confirmed the position of *Comandante* Ramiro Valdés as the *hombre fuerte* of Cuban NICT policy, even though he still did not hold any official public mandate.

Parallel to the government ministries is a new consultative body, the *Comisión Nacional de Informática,* which brings together academics, managers and state and party officials working in the sector. A separate commission exists for questions related to e-commerce, the *Comisión Cubana de Comercio Electrónico* (Altet Casas 2001:13).

In this reorganization of the NICT sector, competencies between the different institutions involved often are not delineated clearly (at least not transparently) and therefore seem to partly overlap. This may not be an unintended 'problem' so much as it follows a characteristic pattern of Cuban politics, according to which parallel administrative structures and informal competencies are developed and sustained as a *sui generis* way of generating 'checks and balances' within the state apparatus.

SIGNALLING CHANGE: THE CREATION OF THE MINISTRY FOR INFORMATICS AND COMMUNICATION (MIC)

The government followed its decision to join the Internet with a turn to a pro-active NICT policy aimed at putting the new information and communication technologies at the service of the country's development while maintaining political and social control. On the basis of law-decree number 209, the Inter-Ministerial Commission formulated the first Cuban NICT development plan a year later under the title "Strategic Guidelines for the Informatization of the Cuban Society" (*Lineamientos Estratégicos para la Informatización de la Sociedad Cubana;* see SIME 1997). This title is remarkable: in 1995, Lieutenant-Colonel Sánchez Villaverde from the army's *Instituto Tecnológico Militar José Martí* had declared that "the informatization of the society [is] a weapon of war of the 'track two'" (Sánchez Villaverde 1995; see above). Yet this same term, informatization, now was reinterpreted as the central slogan of the government's program to put the NICT at the service of the socialist state and society.

In 1998, the Fifth Congress of the Cuban Communist Party ratified the new course, stating in its final economic resolution that "the country must decidedly move on the path of information technology modernization" (PCC 1998:10). In March 1999, a National Information Policy took up the plans sketched in the *Lineamientos Estratégicos* (IDICT 2001). The most widely visible signal of the change in policy however came in 2000: the creation of the Ministry for Informatics and Communication (MIC).

This step came in accordance with the rising influence of the Ministry for Metallurgy and Electronics (SIME) analyzed above. In a conference organized by SIME in June 1999, *Comandante* Ramiro Valdés gave a keynote speech which, in retrospect, can be read as an indirect announcement of the plans for the new Ministry:

> "This end of the century is marked by the technological convergence of industries which until recently were independent from each other. This convergence of electronics, informatics and telecommunications finds its supreme expression in the vertiginous growth of the Internet. (. . .) These so-called information technologies are the core of a multi-dimensional transformation economy and society" (Valdés Menéndez 1999: 5f.).

This emphasis on the convergence of electronics, informatics and telecommunications anticipated what, half a year later, was institutionally fused in the new Ministry for Informatics and Communication—and the government's announcement of the creation of the MIC by the law-decree 204 of January 12, 2000 almost literally repeated Ramiro Valdés' words.[8] In addition, the government argued that, with the creation of the MIC,

> "all these activities which hitherto have been distributed among different institutions, will be integrated into a single ministry in order to promote the harmonic development of the country" (cited in Gil Morell 2000: 5).

However, this was not the case. Instead, the MIC only fused the previous Ministry for Communications with the informatics and electronics departments of the SIME, without addressing the responsibilities of any of the other ministries involved in the Inter-Ministerial Commission. Whereas a continued division of tasks related to the NICT sector seems absolutely logical in the cases of the Ministries of Justice, the Interior, and the Armed Forces, the Ministry of Science, Technology, and the Environment also retained its responsibilities. Many of its previous competencies, such as the fundamental functions of CENIAI or Cuba-NIC,[9] certainly should have become part of a ministry that claimed to integrate all the country's NICT related activities.

In consequence, the establishment of the MIC did not overrule, but rather modified, the previous policy of keeping NICT related competencies spread among different institutions and assigned to different ministries. But there also is no doubt that, if before the SIME was *primus inter pares* in the inter-Ministerial commission, now the MIC clearly has the lead-role in Cuba's NICT policy. As its mission statement formulates:

> "In representation of the Cuban state the MIC regulates, guides, supervises and controls the policy adopted by the state and the government in regard to informatics, telecommunications, networks for information exchange, radio diffusion, postal services, the electronics industry and automatization" (cited in Linares 2001).[10]

In line with this shift in power was the designation of Ignacio González Planas, hitherto Minister of the SIME, to head the new Ministry for Informatics and Communications; behind the scenes, the key position of former Interior Minister Ramiro Valdés was not only confirmed, but strengthened by the move.

Prior to the creation of the Ministry, a series of high-level visits by Cuban officials to the People's Republic of China had taken place and MIC Minister González Planas openly confirmed the model role Peking had played:

> "We oriented ourselves somewhat by the experiences of China, the other country in which the state plays the role it has here [in Cuba], particularly in the regulation of these matters. Two years ago the Chinese created a new ministry, and they transmitted us their experiences and their perspectives on the concentration of these forces (. . .)" (González Planas 2000: 6).

With the creation of the Ministry for Informatics and Communications, the 'Informatization of the Society' program (short: InfoSoc) was further elaborated to include numerous sectoral programs and specific projects covering a wide spectrum of the country's economic, political, social and cultural life

(MIC 2000, 2001a, 2001b). The 'Informatization of the Society' program also outlines numerous e-government initiatives. These include expanded NICT use in the communications internal to the state's administration through the creation of institutional intra-nets. These range from the federal 'Network of the Council of Ministers' to the local level where, for example, in a pilot project in Havana's Playa District, the local *Asamblea del Poder Popular* and the local Communist Party branch have been connected to the domestic Cuban networks (cf. del Puerto 2000, Project C).[11]

The new Ministry for Informatics and Communications became the prime symbol of the importance the government had come to attach to the new information and communication technologies. Beyond this, the fact that, for the first time since the Revolution, the ministry responsible for communications was headed by a civilian—a metallurgic engineer reputed for his administrative-managerial capacities—and not a high ranking military officer also signalled the change in attitude towards the sector. No longer perceived primarily as part of the nation's security apparatus, communications were instead seen as a lead element in the nation's economic development. This change was made visible even at the Ministry's building: the huge neon letters that for many years had proclaimed "*En la guerra como en la paz mantendremos las comunicaciones*" ("In wartime as in peacetime we will maintain communications") were replaced by a sober "*Ministerio de la Informática y las Comunicaciones.*"

INTERNET FOR SOCIALIST DEVELOPMENT: PROMOTING NICT, LIMITING ACCESS

If Cuba's socialist government long had a favorable attitude to the progress of computer technology, it began—as we have seen—to express major reservations when such technology became associated with transnational network connections. As for the use of these, the law-decree 209 of 1996, "Access from the Republic of Cuba to Information Networks of Global Reach," sets out the government's basic approach: Internet access will be employed "in function of the interests of Cuba, giving priority to juridical persons or institutions that are of greater relevance to the life and development of the country" (Gazeta Oficial 1996). In other words, Internet access shall be provided primarily for the official institutions and state companies, not for individual users. All programs for the development and diffusion of the NICT and Internet access on the island stem from this premise, which contains the most celebrated and the most criticized sides of the Cuban approach. For one, some applaud the preference given to social interests and development objectives in the use of a scarce resource, as Internet access is in a Third World country; yet at the same time, critics have denounced the restrictions against independent NICT use by the Cuban citizenry as authoritarian censorship. The organization *Reporters sans Frontières* (2001a), for instance, has given Cuba a prominent place on its list of "enemies of the Internet."

The government openly acknowledges the restrictions on individual access to the Internet. However, the official explanations emphasize that this is not because of the undesired political implications that access to the pluralism of the World Wide Web might have, but that it is due to the technical and financial constraints of the country. One of the first texts on the Internet published by the Party's central organ *Granma* in December 1997, under the emphatic title: "Internet—like the printing press for the Middle Ages!" (Valencia Almeida 1997), frankly admitted the limited diffusion of the service, explaining that "the capacity of the existing connections is not sufficient to additionally connect private users" (ibid: 8). Many others since have followed this line of argument. For instance, in September 2001 Vilma Isabel Altet Casas, a representative of the Cuban Ministry for Informatics and Communications, explained:

> "Cuba does not escape the general situation of Third World countries where costs for Internet access are very high. (. . .) Because of this, there are limitations on the free opening of public Internet services, in order that in this first phase we can guarantee the scientific and academic centers, the health system and other governmental services are not adversely affected by an offer which at the moment surpasses the real technical and infrastructural possibilities of the country" (Altet Casas 2001: 17).[12]

MIC Vice Minister Melchor Gil Morell gave another typical explanation when, at an international press conference, he was asked if the limitations to Internet access are a way of preventing the population from reading anti-government propaganda. The Vice Minister denied this, saying: "We are not afraid of counterrevolutionary information. Counterrevolutionary information is based on lies" (cited in Snow 2001). This position, however, simply does not correspond to the general media policy of the socialist state, which—in accordance with the country's Constitution—insists on a strict state monopoly over mass media precisely in order to keep "counterrevolutionary information" out. If the government spends considerable sums each year to jam the Radio and TV Martí signals from the United States, why would it be indifferent to the same programs entering the country via the World Wide Web?

As for the country's Internet capacity, it certainly is a plausible strategy, from a development perspective, to concentrate scarce resources on those sectors regarded as social and economic priorities. In coming sections of this work, we will look more closely at some of the developmental benefits Cuba has derived from its exclusive strategy. Nevertheless, this official line of argument loses credibility to the degree that it is offered not as a complement, but as a negation of the—rather obvious—fact that the prohibition of individual Internet access might serve a political function. Certainly with politically more neutral products considered luxuries, the Cuban government's attitude is different: the dollar shops offer giant screen TV sets and mountain-bikes, US$1,000 Rolex watches and a lot more that is far beyond the reach of the common Cuban. In this case, the state argues that the high prices serve as a de-facto luxury tax that finances the free health and education system and other social accomplishments. The 'luxury product' Internet, however, receives different treatment: the state Internet service providers are not allowed to provide Internet access for ordinary Cuban citizens, not even at astronomically high dollar prices.

There is, however, a notable social impact of the official discourse emphasizing the limited capacities 'in this first phase' as the impediment to individual Internet access: it creates expectations for the future. To the degree that the technical infrastructure expands, the implicit promise of expanded access becomes increasingly relevant.

THE STATE MEDIA MONOPOLY IN THE DIGITAL AGE: COMPUTER NETWORKS, USER DATA, AND INTERNET SERVICE PROVIDERS

Since joining the Internet, Cuba has gradually expanded its international connectivity. Yet all of Cuba's connections to the global networks still are transmitted via satellite, which implies higher transmission costs and smaller bandwidth capacity than access via sub-marine fiber optic cables, which remain an important objective in the national NICT development plans (Altet Casas 2001: 17). In November 1999, Cuba announced its willingness to connect to the ARCOS 1 trans-Caribbean sub-marine fiber-optic cable and signed the required documents (Valdés 2002: 59) but so far no connection has been established. According to some sources, Cuba is to connect to the ARCOS cable by an offshoot—called ARCOS 2—expected to be complete by 2004 (Ashby 2001). In a presentation in 2000, ETECSA President Rafael Marrero (2000: 7) would not elaborate this option any more than to point out three possible destinations for a Cuban submarine fiber-optic connection: Cancún (Mexico), Nassau (Bahamas), or Miami (United States)—and he followed each with a question mark.

Cuba has repeatedly blamed the United States for blocking any such connection (e.g. del Puerto 2004). However, the Torricelli law in principle has opened the way for a direct fiber-optic connection between the United States and Cuba, as long as the sub-marine cable connection is not considered an "investment in the domestic telecommunications network within Cuba" (U.S. Congress 1992). Given the complete lack of transparency in the negotiations involved, there is no way of independently assessing to what extent the U.S. embargo policy has been the crucial hindrance. However, Cuban authorities seem very hopeful of completing such a connection in the foreseeable future: At the end of 2003, the President of the Cuban telecommunications company ETECSA listed as part of the nation's network development plans "the construction of the domestic part and the accessory infrastructure of the submarine cable in the period 2004–2005" (Fernández Martínez 2003).

Domestically, Cuba today has several dozen wide area networks (WAN—mostly nationwide networks for diverse sectors), and hundreds of local area networks (LAN—mostly intra-institutional or intra-business networks). A considerable part of the domestic networks still is based on the older X.25 technology, but a new IP-based network is being prepared as the 'National Public Network' (*Red Pública Nacional*, RPN) (Lage 2000: 1). The establishment of a high-capacity national 'backbone' is a priority for ETECSA's modernization investments; by January 2004 all of Cuba's provinces except for Pinar del Río in the West of the country had been connected to a fiber-optic network of more than 1,000 kilometers length (del Puerto 2004).

While we have pointed earlier to the deficiencies of the available Internet user data in general (chapter two), in Cuba this problem is still more complicated because the Cuban authorities have followed a restrictive and incoherent information policy in regard to NICT data. On Cuban statistics, a Cuban author has noted that:

> "Paradoxically, in the era of information and technology, researchers in Cuba face immense obstacles while searching for information. Cuba-NIC (www.cubanic.cu) (. . .) not only has the control over the '.cu' domain registration but it also gathers and delivers all Cuban-related information on the Internet. However, such delivery seldom occurs" (Casacó 2002: 295).

Also, the newly created Ministry for Informatics and Communications has not been providing data on a regular base, but has been referring requests for Cuban data to the ITU statistics. This restrictive information policy takes on a seemingly absurd form when even Ministry personnel must use the ITU as data source (see e.g. Altet Casas 2001)—given that the ITU, as an intergovernmental organization, receives such information from precisely the Cuban authorities of the sector. In addition, in Cuba, only a minority of NICT users have full access to the Internet, including the World Wide Web and similar applications; for the majority of Cuban users access is limited to domestic networks—which authorities refer to as the "CUBANET" (Lage 2000: 2)[13] or even "the Cuban Internet" (www.infocom.etecsa.cu)—and / or international e-mail services. Cuban data, however, generally do not reflect these different categories of access in any coherent manner. The compilation of Cuban NICT data therefore is a complicated puzzle with incomplete and often rather uncertain data. Nevertheless, such information can be illustrative. For instance, while for 1997 a number of 7,500 Internet users is given (see table 10), in December 1997 the Party newspaper *Granma* wrote, regarding the diffusion of Internet services on the island: "Today, there are 700 clients in the country, of which 100 have full access to the Internet" (Valencia Almeida 1997: 8). The clients, obviously, refer to institutions and the calculation assumes roughly 10 users for each access point. It is noteworthy that 6 out of 7 of these institutional customers could not access the 'full Internet,' including the World Wide Web. Another data set was given by CENIAI director Jesús Martínez (1999: 1), according to which only one tenth of the 22,000 computers connected to one of the 30 national networks in place at the end of 1988 had access to the World Wide Web. Growth in full access was slow: although by June 2000 the number of computers on the island had long surpassed 100,000, the number of computers with WWW access had increased merely to 3,625 (Grogg 2000)—truly a low figure for a country of 11 million inhabitants.

According to the most recent ITU data, Cuba had 120,000 Internet users, 1,133 host computers and 359,000 personal computers in operation at the end of 2002 (see the following table 10). In January 2004 MIC Minister Ignacio González Planas announced official Cuban data for the end of 2003, according to which the country counted 98,000 "Internet users," 1,100 sites registered under the ".cu" domain (presumably an approximation to the number of host computers in operation), and 270,000 computers, of which 65 percent were said to be "*en Red*" ("connected to the Net")—referring not only to the Internet, but to any one of the existing networks on the island (cited in EFE 2004; see also del Puerto 2004). In addition, the Minister noted that Cuba had 488,000 e-mail accounts (ibid), again not specifying if these were for international or merely domestic use.

These data obviously leave many questions open. Moreover, it is noteworthy that the 2003 data are lower than the official ITU data for 2001 and 2002 (see the following table 10). In either case, with Internet users still barely accounting for one percent of the population, one host computer per 10,000 inhabitants, and less than 4 computers per 100 inhabitants, Cuba continues to have one of the lowest per capita rates on the continent in each of these categories (see table 4).

Table 10 Internet Users, Host Computers, PCs and E-mail Accounts in Cuba, 1997–2003

Year	1997	1998	1999	2000	2001	2002	2003
Internet Users	7,500	25,000	34,800	60,000	120,000	120,000	98,000
Hosts	51	80	169	n.d.	878	1,133	1,100*
PCs	n.d	.n.d.	110,000	n.d.	220,000	359,000	270,000
E-mail accounts	n.d.	25,000	n.d.	35,170	60,000	n.d.	488,000

Source: Internet users, hosts, PCs for 1997–2002: ITU (data for end of year); all data
for end of 2003: press conference by the Cuban Minister for Informatics and Communication in
Havanna, January 16, 2004 (cited in: EFE 2004); the data on
e-mail accounts stem from different sources and refer to different times in the following years: 1998–e
nd of year (Martínez 1999:1); 2000–June (Grogg 2000); 2001–March (MIC 2001b and Snow 2001);
2003–end of year (MIC cited in: EFE 2004).
* the data given refered to "sites registered under the '.cu' domain,' presumably an approximation to
the number of host computers in operation

Cuba currently has five Internet service providers (ISPs). The island's major provider is *Infocom* (since June 2002 also under the name of *Enet*), which is operated by the telephone company ETECSA. The oldest and still very important provider is CENIAI, now formally operating under the name of *CENIAInternet* as part of the company CITMATEL, and institutionally subordinate to the Ministry of Science, Technology, and Environment. CENIAI has been the 'natural' provider for most of the academic and scientific sector. However, it also offers Internet services to all other types of state institutions, companies, or foreigners eligible for access from the island. A smaller provider affiliated with the Ministry of Informatics and Communicationsis *Teledatos-Columbus,* operated by *Copextel,* a large, para-statal *Sociedad Anónima* created in 1985 for the import and export of high-tech products.[14] For cross-border data traffic, *Teledatos-Columbus* depends on the international connections operated exclusively by CENIAI and *Infocom.* In November 2000 an additional license for internet services was granted to *Telefónica Data Cuba S.A.,* or short: *TDATA CUBA.* This ISP, also subordinate to the Ministry of Science, Technology, and Environment, is dedicated specifically to operate for joint-venture and foreign companies as well as diplomatic representations on the island (Acuerdo 3807/2000 of the Council of Ministers, November 16, 2000). Cuba's fifth ISP is the mobile phone company *Cubacel,* which since 2000 offers Internet services for the mobile telephony sector. However, due to the very low numbers of clients for these services, this ISP still has very little practical relevance.

By international comparison, the prices of Cuba's ISPs are very high (see table 11). Though representatives of foreign companies tend to complain about these high prices, the cost generally is negligible for their practical operations in Cuba; it is much more important for them that this type of service is now readily available. For national institutions and state companies, what matters is not so much the price but the administrative decision to grant them access (and if so, of what type); once access is granted, they may be assigned the necessary hard currency resources or they may enter into self-finance schemes (in dollars or in pesos), or make some other arrangement. For the larger part of the nation's companies and institutions, however, access to the Internet still only means access to e-mail functions and domestic networks.

Although ETECSA is a joint-venture company, *Infocom* acts as much as a state ISP as the other three because of the majority share held by the Cuban state. Though there is some separation of their markets, at least in part these ISP appear as competitors on Cuba's hard currency market. Whereas in telephony socialist Cuba maintained the principle of a monopoly company, in the provision of Internet service it gave way to a kind of diversified *de facto* state monopoly. All existing ISPs guarantee compliance with any administrative, economic or technical functions demanded by the government in its efforts to guide the nation's NICT development according to the objectives of the socialist state, and they can adapt the state media monopoly to the exigencies of the Internet.

CONTROLLING INTERNET ACCESS AND USE: ADMINISTRATIVE AND TECHNICAL MECHANISMS

For the government, maintaining a *de facto* state monopoly on the provision of Internet services is a crucial prerequisite to guaranteeing control of

Table 11 Tariffs of Cuban Internet Service Providers, January 2004

Type of Service (all as dial-up modem connections)	CENIAInternet (sector comercial)	Infocom
initial installation fee	US$35–80	US$70
domestic-network-only + international e-mail function 10h per month (excl. tel. fees)	US$25	US$20
full Internet access 24h flat rate per month	US$234	US$260

Source: www.ceniaInternet.cu; www.infocom.etecsa.cu

Internet access and use. Nevertheless, the Cubans' spirit of invention in getting Internet access beyond the official structures is remarkable. Although greatly limited by the extremely low numbers of computers and modems in private households, people have increasingly been using their work-place accounts from home. In addition to this form of private uses of authorized accounts, there also exists an actual black market for Internet access. This is possible through legal accounts, whose legitimate users illegally rent out their passwords and configuration data, typically to be used only in the off-hours of their institution or work-place. A second source of black market access comes from accounts that network administration personnel create illegally to obtain a parallel source of hard currency income. Both of these methods carry considerable risks. In consequence, black market Internet access tends to be hard to find, unreliable and expensive. Snow (2001) gives black market prices for illegal Internet access in the range of $30 to $60 a month.

During a field study in Cuba, the author of the present work also found Cubans that 'organized' their access at less expensive rates. Even the subcontracting of black market Internet access was available: one person who acquired the right to access the Internet between 8:00 p.m. and 8:00 a.m. for US$20 resold half of this time span for US$10. In addition, he made a small business of sending and receiving other people's e-mail at a price of US$1 per message. In any of these cases, prices are very high in relation to the hard currency purchasing power of Cuban salaries. Overall, such informal access to the Internet has been rather limited and, technically, all black market accounts run over one of the state Internet service providers just like the legitimate accounts.

The number of people accessing the Internet in unauthorized ways is a matter of speculation. In December 2000, one estimate suggested that about 4,000 people use their regular work account informally from a computer at home (CNI en Línea; December 29, 2000; www.cnienlinea.com). In January 2004, according to a different estimate 40,000 Cubans had obtained Internet access through the black market or other illicit means (Encuentro en la Red 2004).

The government has not been idly watching these illegitimate forms of internet access spread. An early formal reaction was the decision in 2001 to prohibit any sale of computers or related hardware to individual citizens, even if they had sufficient hard-currency income to buy such items in the state-run dollar shops. An even more hardline move came in January 2004, when the government announced an offensive against the spread of pirate Internet accounts. According to Resolution 180/2003 of the Ministry for

Informatics and Communication, signed by Minister González Planas on December 31, 2003 (but not made public until January 9, 2004), new regulations were decreed that formally prohibit any Internet access through the regular main-line telephone system paid for in Cuban pesos. According to these regulations, access to international networks will be possible only to those using telephone lines paid for in U.S. dollars (largely reserved for joint-venture companies and foreigners resident on the island) and to those Cuban institutions that are explicitly authorized to have Internet access (AP, February 12, 2004; Reporters sans Frontières 2004). The MIC's resolution requires the telecommunications monopoly company, ETECSA, to "use all the technical means necessary which allow detecting and preventing access to services of Internet navigation from telephone lines operating in national currency" (MIC 2004).

These restrictions sparked sharp criticism by explicit opponents of the Castro government as well as by international human rights organizations such as Amnesty International. They also seem to have caused considerable tensions within the Cuban state apparatus: in a highly unusual move, the Cuban government retreated, as ETECSA announced on January 24—precisely the date the new regulations were to come into effect—that the implementation of these measures would be put on hold for an unspecified time (afp, January 24, 2004). Whatever the final outcome of this conflict may be, this incident vividly underscores the difficulties facing the state-socialist regime in its policy of controlling access to the transnational computer networks.

We have seen above that the first—and most important—level of the socialist government's measures to limit the negative impact of the Internet is the diverse set of restrictions put on the access to it. However, on a second level, there also are different mechanisms of controlling those who are authorized to have access. These function on the basis of two pre-conditions: first, that all Internet service providers are state-run and support the technical control functions and second, that Internet access is allowed only in official institutions or public access centers, not in the privacy of personal homes, making available a whole other set of administrative control and sanctioning mechanisms.

Generally all Cubans who receive permission to access the Internet—be it the full Internet or simple international e-mail connections—have to sign a 'self-commitment' contract stating that they will not use this access to seek or receive any content violating national laws, including material of racist, pornographic or 'anti-Cuban' nature. The term 'anti-Cuban' means content critical of or hostile to the political system, but it does not specify how narrowly this might be interpreted in any specific case.

In the institutions or public access centers, the first level of vigilance consists of social control through colleagues, superiors or supervising personnel. In addition, for any institution to be permitted Internet access it must have a person responsible for security matters, which involves as much technical security (e.g. viruses) as guaranteeing proper behavior by the users. Each of these security commissioners regularly receives from the institution's Internet service provider a print-out of the proxy server's logbook, showing which Internet pages were hit by each account.[15] With the scrutiny of this 'hit-parade,' as the jargon in Cuba ironically dubbed it, the responsible superiors have to decide what sanctions to take against those that have exceeded their limits. Authorities may choose from simple admonitions to the withdrawal of the permission to access Internet or even, in grave cases, dismissal from the work-place or, in the most extreme cases, the initiation of a penal process for the violation of national laws. The combination of the fact that users know that they will leave a trail when surfing the Internet plus the threat of sanctions is sufficient to induce a fairly high degree of 'self-discipline' in the practical usage of Internet access at the work-place or in public institutions.

As to e-mail communication, the state monopoly on Internet service providers makes it technically easy to control its content. A different issue is to what degree the state makes use of these capabilities. In a rare statement on the matter, a Cuban government spokesman, Luis Fernández, declared that only e-mails of those people who are "under investigation" for anti-revolutionary activities are monitored (cited in Scheeres 2001). In any case, users in Cuba seem well aware of the potential for control of e-mail traffic. Quite a few of those with full Internet access at their workplace and an official e-mail address at a Cuban state-run provider prefer, for international communication, to use an alternative e-mail address at one of the web-based free-mail services like Hotmail or Yahoo, where the communication is not stored by a Cuban ISP. Even then, users tend to use caution when writing about matters regarded as sensitive.

It is noteworthy that the margin of tolerance in many cases is higher than might be expected from the official political discourse. At the same time, however, exemplary cases illustrate the potential for heavy sanctions. For instance, in one case, an entire research facility in Havana lost Internet privileges after one person, using an account and password of the institution, accessed material that, in an unspecified way, violated the self-commitment signed by all legitimate users.

Another example, however, may underline the change in the attitude to the Internet by Cuban institutions in recent years. In 1999, a student in the literature department at the University of Havana used Internet sources in the bibliography for her senior thesis, provoking a highly negative reaction among the department faculty. Only after lengthy discussion, going up to the faculty's directorship (and probably even beyond that) was the thesis accepted (though the bibliography still was criticized as 'too extensive').[16]

Two years later the situation had changed considerably: the same department now had a room with 20 computers fully hooked up to the Internet and students could not only send e-mails around the globe, they could also navigate the World Wide Web with relatively little formality (although waiting lines for access to one of the 20 computers were frequent). During the author's visit in April 2001, students called up pages like El País or CNN without raising any fears of negative sanctions. Nevertheless, they also seemed well aware that certain websites—e.g. from the island's dissident groups—were off limits, and it certainly would be considered risky to read, for instance, Amnesty International's human rights reports on Cuba on one of the 20 monitors (not to mention printing it at the one central printer, administered by the supervisor).

The possibilities for access have doubtlessly improved for these students, and for many others in similar institutions across the country. It is a very different question, though, if their perceptions of the matter have improved similarly. Whereas earlier, the possibilities of the Internet were little known and their absence therefore not felt, it seems that once the wealth of information available on the Internet is practically experienced, the limitations on its use are felt all the more onerously. Especially in the case of the Internet, where websites of all sorts are within the reach of one mouse-click, it seems that the individual user feels the imposed (or self-imposed) restrictions more painfully than in the case of other media, where it would require considerable extra effort (special antennas, smuggling literature etc.) to access information sources beyond the state's media monopoly. In addition, it may be expected that the new restrictions on Internet use announced in January 2004, although officially only targeting unauthorized access by individual users through residential main line telephones, will also lead to a less relaxed atmosphere in officially authorized Internet access centers.

A different type of control mechanism is involved if Cuba censors the Internet by blocking pages regarded as politically intolerable, as has been reported repeatedly. For instance, Pablo Alfonso of Miami's major Spanish language paper *El Nuevo Herald* reported in 2000, after interviewing a Cuban networking expert that had defected, that:

> "Some websites cannot be directly read from Cuba since the regime blocks the access to them. For instance this is what is happening to many

websites of the opposition and the independent journalists which are
hosted outside of the island. To access them from Cuba one must follow
a link to them from some other website. A Cuban technical expert explained to 'El Nuevo Herald': 'All Internet traffic passes through a central computer, and this allows the blocking of any specific URL or
Internet address (the IP number) and even of specific text content
through previously defined key words, though the latter is rather complicated' " (Alfonso 2000).

Hernández Castillo (1999: 156) gives a similar description, listing the homepage of the dissident journalists' website CubaNet (www.CubaNet.org) and
the site of the *El Nuevo Herald* as two websites at least temporarily blocked.

There is little doubt that these technical censorship possibilities exist.
As Press writes (1998: 20):

> "Security and control over access and content are still high priorities,
> and are relatively easy to assure with the centralization of international
> connectivity at CENIAI, Singapore-style Web proxies, email filtering,
> and firewalls. . . ."

Although since then more than one provider has taken up operations
and CENIAI and ETECSA both have direct international IP access, centralization remains sufficiently high to make these filter and control functions
possible. In his statement in February 2001, cited above, government
spokesman Luis Fernández admitted that the Cuban authorities filter out
'subversive' Web pages just as they jam counterrevolutionary radio or TV
transmissions from the United States: "We need to do the same with the
Internet because we can't have things that undermine Cuban society" (cited
in Scheeres 2001).

In an internal document, General Ronda, the man in the Cuban
Interior Ministry responsible for computer network security, lists four principal protection measures: authentification (e.g. the use of passwords), control of access (implemented by technical or administrative methods, as
analyzed above), encryption (guaranteeing the confidentiality of transmitted
information) and firewalls (permitting censorship of particular Internet content). However, Ronda immediately indicates that, in the case of firewalls,
"their implementation is complex" (Ronda 2000, slide 7). In fact, during
both of the author's field stays in May 2000 and April 2001, the actual execution of any blocking of Internet websites could not be confirmed. At different times computers at different locations were tested by calling up
politically sensitive websites without any problems.[17] It seems likely that
this kind of technical censorship of the World Wide Web's content, which

would have to be either highly selective or else would represent a Sisyphean task, is currently not implemented by the Cuban authorities. The decisive control mechanism does not focus on filtering the web's content, but on limiting and controlling the access to it.

TELECENTERS, CUBAN STYLE? AN ASSESSMENT OF DIFFERENT TYPES OF ACCESS CENTERS

We have discussed above the diffusion of Internet access (and its limits) in Cuba through official institutions and state or joint-venture companies. More limited have been the initiatives to provide NICT services beyond the work-place and to the population at large. Different initiatives in this direction shall be analyzed in this section, which considers to what extent they represent a Cuban version of 'telecenters' in other Third World countries: public access centers for the use of the NICT by those large sectors of society without residential access to them.

A first 'Internet café' was inaugurated in Cuba in Havana's *Capitolio* building, which also houses Cuba's central Internet gateway, CENIAI. An hour of use on any of eight terminals costs US$ 5; each printed page is US$0.20. Not only are the prices prohibitively high for the general Cuban population, but use of these services is explicitly restricted to foreigners. Even though it can be observed that Cubans manage to organize access to the web or to their mail-box by accompanying foreigners (and this is widely tolerated), this type of 'Internet café' certainly is not a public access center, since it formally excludes the Cuban public.[18] The Internet services offered by a number of major tourist hotels, for instance the 5-star *Meliá Cohiba* in Havana, present a similar situation, as they are considered installations off limits to the general Cuban population.

Internet access rooms, which have been established in a number of organizations and institutions in recent years, make up a different case. The Cuban chamber of commerce has opened a computer room like an Internet café for business people: besides foreigners, this group also includes Cuban managers of joint-venture businesses or state companies, but it is not for public access in general. Here, too, tariffs are as high as US$6 per hour (cf. www.camaracuba.cubaweb.cu). In November 2000, the Cuban National Union of Writers and Artists (UNEAC) also opened its own cybercafé, called *El Aleph,* in its central offices in Havana. Here, five computers with full Internet access are available free of charge to all registered members of the association. The Journalists' Union (UPEC) and other associations and institutions have announced similar intentions. Although these intra-organizational access centers may considerably extend the possibilities for Internet use to their specific clientele, they too are members-only services and do not service the public at large.

In the Latin American Association of Telecenters (www.tele-centros.org), a different type of institution, the 'Youth Computer and Electronics Clubs' (*Los Joven Club de Computación y Electrónica*, JCCE), represent Cuba. These Clubs, founded in 1987, are organized as an activity of the youth organization of the Cuban Communist Party, the *Unión de Jóvenes Comunistas* (UJC). A typical local *Joven Club* center consists of a room with eight computers and double the number of seats, as well as a small electronics lab. The mission of these clubs is to introduce young Cubans, in an organized and educational way, to the new technologies of computers, informatics and electronics. Since their foundation, 250,000 people have participated in classes and training programs, according to the organization's data (see www.jcce.org.cu). The Youth Computer Clubs also maintain a national computer network, TinoRed, which, beyond servicing the *Joven Club* and their members, also serves as the network used by the Communist Party's youth organization.

In the early 1990s, the Youth Computer Clubs suffered strongly from the economic crisis, as their technical material became obsolete and many centers were practically put out of service. In the second half of the decade, however, a major initiative for their rehabilitation got underway. The number of clubs increased to 300 nationwide and they were equipped with modern hardware—financed directly from the 'office of the President,' that is, directly by Fidel Castro and not the *Unión de Jóvenes Comunistas* or the Ministry for Informatics and Communication. By now, most Youth Computer Clubs also have been provided with either limited or full Internet connectivity.

But even where the *Joven Club* have Internet connectivity, they don't precisely fit the characterization of telecenters as public Internet access centers. Although Boas (1998) characterizes the Youth Computer Clubs as "walk-in centers for relatively unstructured computer use as well as a number of computer classes," this perception is quite misleading. It certainly is at odds with the organization's own view of itself, as it prides itself precisely on providing relatively structured approaches to computer use, with a clear emphasis on learning computer skills, not on personal use. Nor are they exactly 'walk-in' centers, either. When asked by a Cuban journalist how an interested Cuban youngster could approach the Youth Computer Clubs, the national Chairman of the Joven Club, Pedro Martínez Piñon, replied:

> "There are different ways to come to the *Joven Club,* but the most adequate one is to access it through the 'Organization of Pioneers José Martí' or through the UJC. As a pioneer, you speak with your guide who

in turn contacts the comrades of the local *Joven Club*. It is not advisable that children come to the *Joven Club* by themselves, since experience has shown a lack of an adequate follow-up to the activity. Therefore we put it as a condition that children come with their teacher, their pioneer guide or some other responsible individual" (in Márquez González 1999: 4).

The author's observations during his field stay in April 2001 confirmed the rather formal and not so 'walk-in' character of the Youth Computer Clubs. For instance, in the *Joven Club* of Havana's Plaza de la Revolución district, renovated and equipped with state-of-the-art equipment, the director explained explicitly that the offer of access is limited to computer classes. For these, due to the high demand, a prospective student needs a letter of recommendation from his school or employing institution. Only in exceptional cases, for example if someone needs it for work or studies, is individual use of computers possible. Writing or checking e-mail is not an activity supported by the Club, and only the computer in the office of the director has access to the Internet.[19]

At the same time, experiences during the author's field stay also showed that the character of the Youth Computer Clubs in regard to public accessibility and Internet uses can vary considerably, and also that some Clubs are more informal than the one described above. Nevertheless, since the *Joven Club* belong to the Communist Party's youth organization, certain social 'filters,' though they may vary in form and degree, almost inevitably apply.

It is true that the Youth Computer Clubs include quite strong "pedagogical components of a different nature than those employed in the standard telecenter model used throughout Latin America and the world" (Rushton 2000). But it is also true that they don't fulfil elementary functions typically performed by community telecenters in the Third World. To walk in, write an e-mail to an aunt in Miami or search the web for any particular item of personal interest for a modest fee, then to say good-bye and leave, definitely is not an acceptable use of the Youth Computer Clubs in Cuba.

The Cuban authorities are well aware of the fact that no access centers with such services are available to the population at large. To answer this shortcoming, the plans for the 'Informatization of Society' include a 'New Postal Service' project (*Proyecto Nuevos Correos*), which foresees the integration of e-mail services and network access with the national postal offices (del Puerto 2000). The Minister for Informatics and Communications, Ignacio González Planas, sketched out the plan as follows:

"We have 1,044 post offices in the country; probably there is no other national administrative institution with a comparable presence in the entire territory. (. . .) The postal service can perfectly well establish an

e-mail system on the basis of the national networks so that a teacher in
Guantánamo who wants to contact someone in Havana or New York
can do so (. . .) from his local post office in Guantánamo, where he may
have his electronic mailbox just in the same way as he now may have his
post office box" (González Planas 2000: 6).

It was not until the first of these re-modeled post offices opened in
2001 that for the first time, Cuban citizens had a legitimate form of NICT
access for private purposes. As we noted earlier, there are considerable differences in what access to the NICT can mean and it is true that the 'New
Postal Services' project provides access only in limited form. So far it includes domestic and international e-mail communication, but normally only
access to the domestic Cuban networks and not international networks. The
plans for the 'New Postal Services' outlined by MIC Vice Minister Carlos
Martínez Trujillo in September 2000 explicitly formulated the target of providing access to what he called "the Cuban *Intra*net" (Martínez Trujillo
2000; italics added), that is the closed national networks. However, in line
with prior official discourse, MIC Minister González Planas extended this
shorter-term goal also to include, eventually, access to the Internet "at a later
moment when we dispose of sufficient bandwidth and connections"
(González Planas 2000: 6).

By early 2004, e-mail services and domestic network access were integrated into 11 post offices in Havana and eight more in the rest of the country (www.correosdecuba.cu/correostuisla.asp). Payment occurs through a
system of prepaid cards which provide an e-mail address in a web-based e-
mail system operated by *Correos de Cuba* and its provider *Columbus*. If
(and only if) paid for in U.S. dollars, at a price of US$4.50 for three hours,
these services will also facilitate access to international e-mail functions.
Official announcements have been ambiguous regarding whether this might
also include the 'full Internet' (including the World Wide Web); according to
reports in January 2004, in the post offices' '*salas de navegación*' only domestic Cuban networks were accessible, even for users paying in U.S. dollars
(Boadle 2004). Though the prepaid card system suggests anonymity, this is
not so. Any first-time user must register his name and ID number and, in return, he receives that login and password for anytime he uses the service.
These identifications are non-transferable and the client is held responsible
for the privacy and confidentiality of his login and password, and for all the
uses made with them. The contract each first-time user must sign with
Correos de Cuba includes a "commitment to comply with the Security req-
uisites and the limitations of use of the service" stated in the regulatory
framework (Correos de Cuba 2002).[20]

This extension of some type of e-mail and networking services to the population at large certainly comes late by any standards, and it still is a rather slow process—and Cuba's hard currency crunch since 2001 certainly will not help to accelerate it. The integration of NICT services into the postal offices could, if gradually expanded, represent an important step in the diffusion of NICT use to larger sectors of the population. However, the restrictions announced in January 2004 do not bode well for the future. Here too, though, the introduction of a larger public to the domestic Cuban networks noticeably increases the pressure for more access to the Internet, as can be seen on *Correos de Cuba*'s website: *"Por qué no hay Internet?"* ("Why is there no [access to] Internet?") is among the top five 'frequently asked questions' (www.correosdecuba.cu/correostuisla.asp). The answer given is a remarkable combination of the argument of limited technical resources and the need for censorship: "The use of the Internet shall be introduced after the network has been expanded and parameters for informatics security have been established, which make impossible hacker attacks as well as the access to pages that are not in accordance with our regulations" (ibid.).

MODERNIZING THE SOCIAL ACCOMPLISHMENTS: NICT INTEGRATION INTO THE HEALTH AND EDUCATION SECTORS

The health and the education sectors, two of the central social accomplishments of the Cuban Revolution, suffered greatly in the economic crisis after 1989. Technical equipment in large part became obsolete. When the economy started to recover in the second half of the decade, in both sectors an important technological modernization, marked by the incorporation of the NICT, was necessary for the systems to regain their high standards. The current 'Informatization of Society' programs single out health and education as high priority areas for NICT use.

In the health system, the core of the NICT integration is the *Infomed* network, operated by the National Information Center for Medical Sciences of the Cuban Public Health Ministry (MINSAP). Its beginnings are linked directly to an emergency situation in the most critical moments of the economic crisis in the early 1990s. Largely due to deficient nutrition, at the end of 1992 Cuba witnessed the outbreak of a rare eye disease, the so-called optical neuritis, which in a short time infected no fewer than 50,000 Cubans. In this situation, ad-hoc aid to the tune of US$250,000 came from the United Nations Development Program (UNDP), with support from the World Health Organization (WHO) and the Pan-American Health

Organization (PAHO). This aid made possible the development and rapid deployment of a nationwide network which provided e-mail communication and a bulletin board system (BBS) for data gathering, epidemics control and the exchange of medical experiences and knowledge in the treatment of the affected persons. Pedro Urra González, today's director of *Infomed*, and Jardines Méndez, Vice Minister for public health, write in retrospect: "Perhaps this is the only case in history where the introduction of telematics in the health sector had its origins in a nationwide epidemic disease in the midst of crisis" (Urra González/Jardines Méndez 2000: 40).

The *Infomed* network that emerged from these emergency structures is today a broad network, based on modern IP technology and connecting all major medical institutions of the country, either through dedicated lines or dial-up access. All its services are free of charge. *Infomed* offers a comprehensive portal (www.sld.cu) that interlocks Cuba's medical institutions and includes a daily information bulletin, announcements, and numerous databases. A virtual library connects to the nation's medical libraries and even provides a considerable amount of literature online, including the full text version of all of Cuba's 27 medical journals. *Infomed* also includes a 'Virtual Medical University' (www.uvirtual.sld.cu), which offers numerous classes and workshops for the specialization of Cuba's medical students, doctors and other health sector personnel.[21]

However, the extensive services offered by Infomed are a substitute for, not an addition to, what the WWW could offer to those working in Cuba's health sector. In a number of areas of medical interest, *Infomed* does go beyond the national network and provides access to international sites such as 'Medline,' the world's largest medical database and part of the National Library of Medicine in the United States, and to foreign medical journals. Nevertheless, though recognizing the merits of the *Infomed* network, critics within Cuba's health sector insist that the screening done by *Infomed* inevitably leaves out valuable resources available on the World Wide Web, and that results could be better still if, in addition to *Infomed*, the Cuban medical staff also were given freer access to full Internet.[22]

Although this criticism certainly is not unfounded, *Infomed* is an extraordinarily successful effort of NICT integration into Cuban society and it has received considerable international recognition.[23] Most importantly, *Infomed* seems well adapted to the practical needs of the Cuban health sector. Particularly given the limited printed resources of any kind available in the country, *Infomed* has greatly enhanced the possibilities for hospitals, universities and medical doctors to access medical information from national and international sources. Nevertheless, important bottlenecks in the use of the NICT persist. Their diffusion to provincial medical centers and local medical doctors' offices and health posts is still insufficient, data transmission speed is often slow and diffusion of access is relatively low even within the major medical institutions, where generally only a part of the staff has adequate access to connected computers. Therefore, extending the reach of the *Infomed* network (and corresponding NICT equipment and capacities) is a central goal of the health related parts of the 'Informatization of Society' program (cf. del Puerto 2000).

Despite these problems, the NICT integration in Cuba's health sector remains remarkable, especially considering the difficult economic situation and the fact that Cuban health system still is effectively accessible for the entire population. This modernization process particularly stands out in an international situation in which, in the context of neo-liberal reforms, the public health sector of most Third World countries is experiencing severe budget cuts and often painful losses of quality and quantity in services.

Like the health sector, education also has become a priority area of NICT integration for the Cuban government. Although the education system has not yet recovered its pre-'89 levels in several indicators, the 'Informatization of Society' program puts a great deal of emphasis on the introduction of the new information and communication technologies into the universities and other schools of the island. Besides the acquisition of the necessary hardware and the further diffusion of connections, the plans include a strong training component for teaching staff.

In a number of public speeches in 2001, Fidel Castro personally and emphatically endorsed the high priority given to the NICT integration into Cuban schools:

> "We are currently working to distribute 20,000 computers to secondary schools and to higher education centers. The goal is the massification of computer knowledge, its teaching from pre-school to the final university class" (Castro 2001).

In 2001, almost all Cuban universities had local intra-nets and connections to the domestic networks, and most also had access to the Internet. Low data transmission speeds of 64 or 128 Kbps, however, remained a common technical limitation (Altet Casas 2001:19). According to Lage (2000: 4), the Cuban universities operated a total of 5,000 computers, of which 3,000 were connected to some kind of network. Knowledge of basic computer applications has been integrated into the curricula of all faculties (Altet Casas 2001:19) and, in many institutions, computer rooms have been set up in which students can type their class papers, use e-mail services, or surf the domestic networks or even the World Wide Web. Though secondary and higher education command higher priority, a program for the introduction of computers into elementary schools was started as well. In *Ciudad Libertad,* a large Havana school complex housed in former army barracks of the Batista regime, a major model project is under way.

For a socially inclusive diffusion of NICT knowledge and usage capabilities, it certainly is a fundamental asset that the NICT are being integrated with a universal education system in Cuba, which has relatively even quality between social sectors and regional areas. However, NICT related educational efforts do suffer from the weak points of the system as well. For one, there are still many basic material needs in schools. Perhaps more important, though, is the discrepancy between the peso salaries paid to teachers and the prices, exorbitantly high by comparison, on the peso or dollar markets: this situation has led to motivational problems for the teaching staff and a substantial 'brain-drain' from education to the dollarized sectors of the economy, such as tourism. To fill these gaps after 40 years of Cuban socialism and its great pride in its education achievements, the government had to resort to an emergency program (called *maestros emergentes*) in 2001, in which high school graduates received a one year crash-course on teaching. Moreover, as in any other plan involving major hard currency investments, the government's ambitious plans for NICT integration in the education system will likely be scaled down significantly or postponed by the recent downturn in Cuba's foreign exchange earnings.

Cuba's first 'Virtual University' (www.cursosenlinea.cu), inaugurated in October 2000, merits a special note. Operated by the state company *Citmatel,* it is not part of the system under the Ministry of Education; this has allowed it to break with the sacred principle that all education be free. For an 'Introduction to Power Point,' for instance, the Universidad Virtual charges US$120 for "diplomats, joint-venture or foreign companies and tourist enterprises" (www.cursosenlinea.cu); yet even for participants from "Cuban companies," the class costs a fee of 188 Cuban pesos plus US$12.

Quite in contrast to the 'Virtual Medical University,' which is well integrated into the country's health sector and is working with good results, the *'Universidad Virtual,'* though it presents itself as part of the 'massification of culture' (Iglesias 2000), seems instead to be designed as a hard currency earner, directed at foreigners and foreign companies resident on the island. The NICT certainly offer considerable possibilities for distance learning, but the relatively low diffusion of network access and the almost nonexistent access from private homes make it unreasonable in Cuba to reach out to the public through net-based education.

DODGING WASHINGTON'S SANCTIONS? E-COMMERCE AS A MEANS TO REACH OUT TO THE U.S. MARKET

E-commerce raised particularly high hopes in Cuba, since it promised overcoming many of the island's problems in marketing its products abroad. Cuba's e-commerce activities, however, did not originate with Cuban products, but with the use of the Internet for money transfers from emigrated Cubans to their relatives on the island. In 1995, well before the country connected to the Internet, the Cuban government set up a website (www.cubaweb.cu) hosted in Canada and operated as a joint-venture between a Canadian company and the Cuban state company *Grupo de la Electrónica del Turismo* (GET). Politically, this site served as a first quasi-official World Wide Web presentation of socialist Cuba; economically, it was at its heart a service for remittances via the Internet (called 'Quickcash') that was integrated into a homepage.

As remittances have become a central hard currency earner, today money transfers still constitute one of the most important commercial Internet applications for Cuba. While sanctions impose limits on how much money Cuban exiles may send to their family or friends on the island, enforcement of these provisions against traditional ways of sending money always has been difficult. It is no less difficult in the case of remittances via a Canada-based Internet site, since enforcement efforts almost inevitably would imply a systematic infringement on the privacy rights of U.S. citizens. E-commerce in tangible products proved to be far more difficult for Cuba, however. The state music company, Artex, made a first attempt in 1996 by establishing a specific e-commerce portal to market compact discs and tapes of Cuban musicians. This pioneering experience encountered numerous problems, and its poor results had a sobering effect. The Cuban economic journal *El Economista* de Cuba writes in retrospect:

> "Everything went very slowly due to the limits of communication connections, but also due to strict customs regulations. Another weak point was the shipping and delivery of the products we sold (. . .) The experience showed that there was still much work to do to transform the internal conditions in Cuba in a way that allows us to integrate into this kind of trade, which requires technological infrastructure, financial capacities, excellent business logistics, and a supportive legal framework" (Bermellon/García 2000).

In the wake of the problems shown by this early experience, a National E-commerce Commission (*Comisión Nacional para el Comercio Electrónico*) was created to bring together different Ministries to cooperate on essentially three tasks. First was the establishment of the necessary technical and logistic organization. Second was the knowledge transfer and training of the personnel involved; this form of education not only applied to the technical staff responsible for the enterprise's web presence, but also the employees in the buying and selling departments of those state companies seeking to engage in some kind of domestic B2B commerce. The third task was the adaptation of laws and norms to the exigencies of e-commerce, involving regulations regarding security, privacy, and authentification.[24] A 'digital signature' law was passed and the national Chamber of Commerce was chosen to operate an authorized registry for this type of digital certification, established under the name of *Arcámara* (see www.camaracuba.com).

A first study that came from the National E-commerce Commission (see Bermellon/García 2000) had recommended, for the time being, limiting e-commerce to two areas. The first one was business-to-business (B2B) commerce dealing with domestic trade between Cuban state enterprises; implementation, however, took some time, and a pilot project elaborated by the Cuban state company *Softel* (and limited to a number of participating companies associated with the Ministry of Metallurgy and Electronics) was presented at the *Informática 2000* fair in Havana. The second area the study recommended was business-to-consumer (B2C) trade directed at international consumers and focused on immaterial goods and services, such as software or databases, that would need little supportive infrastructure, since they could be delivered directly via the Internet (Bermellon/García 2000).

Quite a lot in attitudes towards e-commerce has changed in recent years. A key for making it more viable was the government's decision to at least partly drop its earlier reservations and organize outward-oriented B2C e-commerce in the form of joint-ventures with foreign companies. This change allowed cooperation with a Canadian company that in 1997 produced a first system for credit card billing, indispensable to resolving the payment problems in e-commerce (cf. www.icc-cuba.com). A central focus for this e-commerce is services in the tourist sector: now, numerous hotels, travel offers and car rentals can be booked via Internet. Another joint-venture with mostly Canadian companies offers a wide range of tangible goods, particularly items such as music CDs, books (see www.discuba.com) and Cuban cigars (www.cuban-cigar-shop.com). Not only are such websites hosted 'off-shore' in Canada, but products also are stored and shipped from Canada, which turns over most of the logistics and marketing problems to the foreign joint-venture partners.

Since e-commerce potentially enables Cuba to dodge the U.S. trade sanctions and to thus reach out to the forbidden U.S. market, it is no wonder that Canada, due to its geographic proximity, has become Cuba's prime partner for e-commerce operations. In fact, Cuban cigars, ordered from one of the Cuban-Canadian joint-venture websites to an address in the United States, normally are delivered without problems. Officially, of course, the U.S. embargo applies to buying online just as much as to any other form of importing Cuban goods, and U.S. authorities claim to have confiscated some shipments.[25] For obvious reasons, there are no data available on the volume of these transactions. However, Cuban sources estimate that the total volume of Cuban online sales still is rather limited, while stressing that growth has been significant in recent years and expectations for the future run high.[26]

A different type of e-commerce activities are those that serve as remittances in kind, nurturing what has been called Cuba's 'exports within borders': the delivery of dollar shop products to Cubans on the island, but ordered and paid for by their relatives in the United States or other countries. For this purpose, the state hard currency retail store *Tiendas Universo* (www.cuba-shop.com) and other websites, such as 'www.cubagift-store.com,' put catalogues of goods on the web from which the relatives can select products for their relatives. The embargo policy imposed by Washington prohibits any U.S. citizens from shopping on these sites (and it does not make any difference if the websites are hosted outside of the country or if they use a '.com' address, instead of the Cuban national domain '.cu.'), but enforcement is understandably difficult. As to the volume of these transactions, again for obvious reasons, no data are available. However, experiences in Miami and in Cuba have led the author to assume that, in general, emigrated Cubans prefer to send money. Since e-commerce transactions like these depend on a relatively high level of trust, the animosity and distrust for Cuban authorities that are dominant in the emigrant community seem to play quite strongly against a more dynamic development of this business.

A particular problem arises from a similar payment-from-abroad arrangement offered by *Opticas Miramar,* the eye-glasses and opticians' services branch of the Cuban tourist company *Cubanacán.* On the island, the provision of eye-glasses to the population has become quite problematic and waiting-lines are long. In this situation, the website—obviously directed at readers abroad—suggests a remedy by announcing:

> "*Opticas Miramar* is glad to put at your disposition a new option by which you can facilitate optical services for your friends and family in Cuba. You can pay online (. . .) and the beneficiary will receive the desired consulting services and articles in Cuba" (www.optica-miramar.com).

Whereas *Opticas Miramar* initially served only tourists and other foreigners living on the island, such exclusivity has been dropped. This, however, touches upon a politically highly sensitive point. Since the free health service is considered a central accomplishment of the Revolution and plays an important role in its legitimacy, the government had always rejected the creation of any type of parallel hard currency market for medical services as a 'two class' system incompatible with Cuba's socialist order. Though limited to one specific and non-vital service, the e-commerce offer by *Opticas Miramar* represents the first open break with this principle.

PRODUCING NICT: HIGH HOPES FOR THE SOFTWARE SECTOR

The crisis of the early 1990s virtually halted the long-standing efforts to produce or assemble computers in Cuba. Even with the gradual recuperation of the island's economy in the second half of the decade, computer hardware production has not reappeared in the government's development plans; any more ambitious hardware production plans would entail high capital investments that conflict with the country's severe hard currency limitations.

Historically, software had been considered an accessory to computer production and use, rather than an economic activity in its own right. In the 1990s, this attitude greatly changed. Initially, software development was entirely oriented towards domestic needs, such as security, administration, science, biotechnology, and medical services. But the so-called 'exports within borders' and particularly the rapidly growing tourism sector played a central role in the rearticulation of the sector in the 1990s. Institutionally, most of the software companies created in the course of the 1990s are associated with the 'Information Technology Group' of the Ministry for Informatics and Communications, while a few others are spread out among the rest of the Ministries.

Table 12 not only shows the high growth rates of Cuba's software sector since the mid-1990s, it also illustrates the enormous weight of the 'exports within borders' in its development. If, between 1996 and 2001, the software production for the domestic economy increased seven-fold and for exports ten-fold, then growth of the sales to the dollarized sectors of the Cuban economy was almost twenty-fold. In 2001, out of a total of US$46.8 million in hard currency sales, actual exports beyond the country's borders account for less than 20 percent.

The growth rate experienced by the software industry in the 1990s and the high number of Cuban NICT professionals with solid technical expertise

Table 12 Growth of Cuba's Software Sector, 1996–2001

	Year					
Currency/Market	1996	1997	1998	1999	2000	2001
Cuban Pesos/ Domestic Economy	8,642,300	16,439,900	21,955,500	25,502,300	n.d.	61,599,300
US$/'Exports within Borders'	1,970,400	3,211,800	4,484,600	8,554,700	11,190,900	37,744,790
US$/Exports fob	931,600	1,512,400	3,878,900	2,788,400	2,909,600	9,023,830
Total US$sales	2,902,000	4,724,200	8,363,500	11,343,100	14,100,500	46,768,620

Note: These data include only those firms belonging to the MIC, and not those subordinated to other Ministries.
Source: Ministry for Informatics and Communications, as cited by Casacó (2002: 295)

have nurtured the hope to turn software production into a strategic sector for national development. Some entrepreneurs even began to speak of Cuba as a "Silicon Island" which could "become the India of the Caribbean"[27] for offshore software development. Fidel Castro most prominently highlighted the software optimism in late 2002, when he announced the conversion of the large former Soviet/Russian military surveillance post near Havana into a top-flight computer science academy and a state-run 'cluster' for software development (afp, August 23, 2002).

These developments merit cautious evaluation. Although Cuba certainly has strong assets in its notable software producing capacity and highly skilled yet low-cost labor force, there are also important drawbacks. In contrast to official optimism, a sober analysis by a Cuban author (Casacó 2002: 292) also points to the underdeveloped state of the telecommunications and NICT infrastructure, marked by limited access to Internet services, which still pales compared to internationally competing software developing centers, despite recent efforts. In addition, Casacó questions the economic wisdom of maintaining low salaries for the highly skilled Cuban NICT experts (ibid: 296).

When it comes to exports, the Cuban software sector faces problems similar to those experienced by other ambitious high-tech sectors, such as the country's pharmaceutics and biotechnology projects. Even where the production process is successfully mastered, product marketing represents a major obstacle for Cuban industry. In addition, full subordination to the dominant international norms and regulatory institutions is required for many markets, from patent laws to intellectual property rights, and Cuba has failed to respect some of these in the past, some still today. For instance,

Microsoft's Windows and Office programs are standard applications in Cuba, but they are imported in cost-free pirated versions, an open violation of international copyright norms, which in Cuba is legitimated as a form of retaliation against the damage inflicted by the U.S. economic sanctions.

Doubtlessly, the U.S. trade embargo represents a central obstacle for Cuban high-tech exports. Even though Cuba seems quite successful in acquiring U.S. products through third countries, similar mechanisms to circumvent the embargo legislation are not as easily available for its own export production. Therefore, although free production zones have been set up and well qualified labor tends to be abundant and inexpensive, the country's attractiveness for foreign companies in this sector is minimal as long as any Cuban-made production is excluded from the world's most important market.

The country's software companies certainly can find niches in particular export markets, particularly in the other Spanish language countries of Latin America, and in certain production segments, such as in medicine, where the country can build on its specific expertise. This indeed can provide for growth in the coming years. But given the restrictions imposed by the international marketing structures plus the effect of the U.S. sanctions, it will be very difficult to expand beyond market niches and become a large-scale producer of software services and applications for the world market. Euphoric visions that Cuba "is preparing to become a key Internet communications hub" (Ashby 2001) and "a prime location for IT industries" (ibid) seem unrealistic without a full integration of Cuba into the world economy. Such integration, however, would presuppose a normalization of relations with the United States—a condition dependent on a fundamental political change either in U.S. Cuba policy or in the political system on the island.

THE NICT AS CROSS-BORDER MEDIA: AN INTERNATIONAL PLATFORM FOR THE GOVERNMENT—AND ITS FOES

For the Cuban government, an important aspect of the Internet is the possibility it holds for the international projection of Cuban socialism and its institutions. In fact, even before Cuba joined the Internet, the government had begun to use the World Wide Web as a global communication medium. As mentioned above, in 1995 the first official Cuban Internet portal (www.cubaweb.cu) was opened as a Canada-based website. Even today, *CubaWeb* is considered to be the most widely used official Cuban portal, claiming to be among the 5 percent of websites most frequently visited worldwide (Valencia Almeida 1997: 8). Beyond facilities for money transfer, which were its core economic activity from the very beginning, the *CubaWeb* site offers up-to-date Cuban news and government announcements, as well as numerous links to the official institutions of the island and to e-commerce offers of all sorts.

Cuba Web also operates Cuba's main server outside of the island, with so-called 'mirrors' for the great majority of Cuba's WWW sites. For instance, anybody from outside the island who calls up the website of the Cuban Communist Party's newspaper *Granma* will use an address ending with Cuba's national domain name '.cu,' but will access the *Cuba Web* server in Canada, which is continuously fed an updated copy of the *Granma* site from its home server in Havana. Due to this function, the *Cuba Web* representative, not inclined to false modesty, remarked at Havana's Informática 2000 fair: "We *are* the Internet in Cuba!"[28] Since Cuba still does not connect to the Internet via fiber-optic cables, these mirrors have the great advantage of reducing traffic on the island's costly international connections. In addition, the Canadian server has a high-bandwidth connection to the U.S. networks and therefore provides a much faster and more reliable data transfer to users outside of Cuba.

By early 2001, more than 200 Cuban websites existed and that number was growing rapidly. Of these, 175 had 'mirrors' off the island. They included more than 16,000 pages and, according to official data, registered about 10 million hits per month (MIC 2001b). Besides *Cuba Web,* a number of other official Cuban portals with similar functions have been designed (e.g. www.cuba.cu). Practically all Cuban newspapers have an online edition and Cuban TV is live-streamed via the Internet. The government has its own website (www.cubagob.cu) as do the Communist Party (www.pcc.cu) and most other major public institutions of the country. All of Fidel Castro's speeches are transmitted worldwide via the Internet, as are other important government declarations. Special websites launch for political issues and official campaigns, whether they are in memory of Che Guevara (www.che.islagrande.cu), document numerous activities and accusations of Cuba against the U.S. economic embargo (www.cubavsbloqueo.cu) or defend the five Cubans that are being tried for espionage in the United States (www.antiterroristas.cu).

The Cuban Internet presence showed particular ability to attract attention in the Elián case. According to the Communist Party's newspaper, its Internet edition reached a record of over one million hits in one week at the height of the struggle between the Cuban exile community, the Cuban government and the U.S. administration over the refugee child (*Granma International,* April 1, 2000). The extraordinary international attention paid to the Cuban websites in this episode supposedly made a strong impression on Fidel Castro, which in turn led to an increased push by the government to promote Cuba's Internet presence (*El Nuevo Herald* 2001b).

The call for an 'ideological offensive' in the second half of the 1990s is reflected in the country's outward presentation on the Internet, as the previously cited text by Lieutenant-Colonel Sánchez Villaverde underscores. After citing Fidel Castro's call to win the public opinion inside and outside of the country's borders, the author writes:

> "In this sense important use can be made of the possibilities the new information technologies, particularly the INTERNET, hold. New circumstances oblige us to search for new forms of fight. Also in this field we must confront the enemy who already has placed manipulated information about our reality in this and in other computer networks. Doubtlessly, there is an element of equilibration that we have to benefit from, since the dependence on material circumstances is much lower than in the case of paper for printed magazines or of the technical requirements for the production and diffusion of TV programs. There [in the Internet] we can tell our truths, which are solid, and we can transmit the rationality, the justice, and the dignity of our socialism, which are unquestionable" (Sánchez Villaverde 1996: 49).

However, as the author noted, not only the government but also its critics and foes will make use of the Internet as cross-border media. In the World Wide Web, the sovereignty of the nation-state is limited to the websites under the country's top level domain, and the Cuban state's media monopoly also is limited to sites with a '.cu'-ending. However, hundreds of other websites are dedicated to Cuban affairs and opposition forces and Cuban exiles are responsible for many of them.

Probably the most important development of recent years, however, has been the emergence of an active scene of independent—that is, oppositional—journalists who in the Internet found a worldwide platform for their texts. Although the first 'Association of Independent Journalists in Cuba' was founded as early as in 1988,[29] initially they were limited to sporadic appearances in U.S. or exile Cuban media. It was not until the spread of the Internet that Cuba's dissident journalists gained a continuous forum for their articles and publications on their own websites. They have received considerable international attention, and a reported dozen such oppositional press agencies together employ a hundred odd journalists (Reporters sans Frontières 2001b). Internationally, the most famous of these agencies is 'Cuba Press,' which is directed by Raúl Rivero who, until his public break with the political system in 1989, had been a prize-winning poet and officially honored journalist.

Given the state monopoly on Internet service providers in Cuba, oppositional journalists cannot put their articles directly onto the Internet. Even though the authors live in Cuba, supporters outside of the country, typically in Miami's Cuban exile community, inevitably have to technically administer oppositional websites. Since the dissident journalists also are not allowed e-mail access, they extraordinarily illustrate how the new information and communication technologies depend on the old ones: dissidents must use the 'plain old telephone' to transmit their texts vocally to Cuban exiles living in the United States, who then transcribe the texts and post them on the Internet. Pablo Alfonso, of Miami's *El Nuevo Herald,* even says:

> "More than from the Internet, the independent press in Cuba has profited from the restoration of the telephone communication between the United States and Cuba made possible by the Torricelli Law. Essentially, what has broken up the information blockade is the communication via telephone. This is the basis for everything else."[30]

The leading opposition website, which publishes the articles of various dissident press agencies on the island, is *CubaNet* (www.cubanet.org), which is administered by exile Cubans in Miami. The origin of *CubaNet* goes back to the Cuba newsgroup within Usenet, an early computer network based on the UUCP protocol. [31]

In these Usenet newsgroups, one of the later founders of *CubaNet* began to circulate selected articles on Cuba that had appeared in the U.S. press.[32] The establishment of *CubaNet* as an independent Internet activity from there required the confluence of three different factors: expansion of the audience through the popularization of the World Wide Web; the continued existence in Cuba of dissident journalists, which the government decided to tolerate up to a certain point; and funding for such a non-profit project, which would consume considerable sums each year in salaries, equipment, office space, and very high telephone bills. The last condition was met when, from 1995, *CubaNet* began to receive substantial annual funding from U.S. institutions such as the National Endowment for Democracy (NED) and the United States Agency for International Development (US-AID).[33] Given this background, the oppositional Internet journalism cannot escape the internationalized context of Cuba's confrontation with the United States, and the Cuban government attacks it as an 'anti-Cuban' activity at the service of the imperialist power.

Given the restrictions on Internet access on the island, the possibilities of Cuba's dissident Internet journalists reaching a substantial audience in Cuba itself are minimal. *CubaNet* Vice President, Rosa Berre, admits: "This is our great problem: that the natural audience does not have access to the work of the independent journalists. And in any mid-term perspective we do not have a solution to this problem."[34] Criticizing the Cuban government's policy of facilitating Internet access only through official institutions, *CubaNet* President José Alberto Hernández Castillo writes: *"CubaNet,* on the contrary, sees the individual subscriber, without the controlling ties of government, as the catalyst for these needed changes" (Hernández Castillo 1999: 155). However, this "individual subscriber, without the controlling ties of government" does not exist. Though their websites are hardly read by the Cuban population at large, the Internet journalists have tapped into prominent international supporters—such as the France-based NGO in defense of journalists' rights, 'Reporters sans Frontières' (www.rsf.fr)—and they have received numerous press awards and human rights prizes. This extraordinary international response in turn has also given them a relatively prominent role within the still small opposition groups on the island. The wave of repression in 2003, when the government condemned Raúl Rivero and 70 other dissident journalists to long prison sentences, underscored and in fact magnified the political relevance they have attained.

On another level, as well, the relevance of opposition journalism for the country's future should not be underestimated. *CubaNet* Vice Director Berre explicitly speaks of the dissident journalists as "the nucleus of a post-Castro press in Cuba." This attitude signals a significantly different approach than in the regime changes in the state-socialist countries of Eastern Europe and the Soviet Union. These were so unexpected that no oppositional structures had time to prepare themselves. In consequence, the new media structures emerged largely from within the established media and retained many of their journalists. (An exceptional case has been the GDR, where in most cases West German journalists rapidly took over the command posts in the newspapers and the radio and TV stations.) In Cuba, even though the dissident journalists at present are not read on the island, they are establishing working routines and international support structures, which—if one day a political opening should come—will find them in an extraordinarily strong position to challenge the official institutions.

Other oppositional groups on the island have followed the example of the dissident journalists and are using the Internet as an international platform of articulation. Most of the established dissident organizations by now have their own homepage on the Web; these include the *Corriente Socialista Democrática Cubana* (www.corriente.org) and the *Movimiento Cristiano de Liberación,* headed by Osvaldo Payá, who recently has been awarded the European Union's Sakharov Prize for Freedom (www.mclpaya.org). Beyond this, the Internet has also given rise to the emergence of numerous other dissident websites announcing 'independent libraries,' 'independent trade unions,' 'independent agricultural cooperatives,' 'independent economists' and other associations, which represent virtual 'counter-institutions in waiting-position.' It is often difficult to tell to what degree these 'cyber-dissidents' really represent existing structures of any relevance. Nevertheless, they are waiting for their day to come, when they—with the expected support from exile groups, the United States and other international actors—will claim to be the genuine institutions of Cuba's society, ethically superior to those who have 'collaborated' with the official institutions of the socialist regime.

Political actors from outside Cuba, including the exile community, are using the Internet to transmit their own messages. The most ambitious project is that of a daily Cuban Internet newspaper, launched in December 2000 in Madrid and, until his death in 2002, directed by the prominent Cuban writer Jesús Díaz. This project, the *Encuentro en la Red* ('Encounter on the Web'; see www.cubaencuentro.com), came to light as a sort of daily sister product of the cultural print magazine *Encuentro de la Cultura Cubana* that, under Jesús Díaz, had become the leading intellectual forum of the Cuban emigrant community. The official Cuban authorities have fiercely condemned both, the cultural magazine as much as the daily Internet paper. Although *Encuentro en la Red* makes it possible for people without access to the World Wide Web to receive its content in an e-mail version, the lack of individual private access and the political pressures against receiving such material effectively limit the number of its more or less frequent readers on the island to, at best, a few hundred. Nevertheless, in intellectual circles on the island the print magazine, smuggled onto the island, and the Internet newspaper have become two of the most well known intellectual and political activities of the Cuban exile community. Similarly to *CubaNet,* the *Encuentro de la Cultura Cubana* and *Encuentro en la Red* also depend on substantial financial backing from governmental and non-governmental funds from the United States, Spain and other countries. Also similar is the Cuban government's vehement condemnation of this support, which it sees as proof that *Encuentro* is "a political operation of the U.S. government, as it has been financed by the National Endowment for Democracy, (. . .) a smoke-screen of the CIA."[35]

It should be noted that also *Radio* and *TV Martí,* the two stations directly integrated into the U.S. government's international media system, have come to use the Internet. They do not only publish written news, but they also facilitate the download of audio and video files of their radio and TV programs from their website (www.martinoticias.com). As the failure of *TV Martí* to reach any considerable audience on the island has not led to any reduction or re-orientation of the costly production, it will not worry significantly its operators that their Internet initiative shares the same problem. The website of the most powerful organization of the Cuban exile community, the Cuban-American National Foundation (www.canfnet.org), avoids these problems in its own way: the entirely bilingual site, supporting the organization's lobby work for a staunch anti-Castro U.S. policy, is directed not at the population on the island, but at an audience in the United States.

NICT IN CUBA: KEY TO DEVELOPMENT OR THREAT OF SUBVERSION? A PRELIMINARY BALANCE

The new information and communication technologies have often been described as a 'democratizing force' with subversive potential towards non-pluralist political systems. However, it is now more than a decade since Cuba established its first connection to the Internet and there still are no signs that "the Internet will topple Fidel Castro." At the same time, Cuban state-socialism indeed does face a serious dilemma that comes fully into focus only when political and economic perspectives are linked. On the one hand, the cross-border character of the Internet does potentially undermine political control over media and communication, while on the other, economic pressures and developmental considerations make NICT use increasingly indispensable.

U.S. policy has been particularly important to this dilemma. With the passing of the Torricelli Law in 1992, the U.S. government has turned to a strategy of fostering communications with the island in order to erode the political system. This put international telephone and NICT communications explicitly in the context of the Cold War-type confrontation between the two nations.

However, U.S. hopes that "liberty will spread by cell phone and computer modem" (Clinton 2000) have been disappointed. Political change is never brought about by information and communication (of whatever sort) alone—it always needs articulation, organization, and, in the end, public action by concrete social actors. As Hirschkop (1996: 97) wrote: "One cannot buy democracy off a shelf, or download it from a website." Therefore, it has been most significant that, precisely at the moment of Cuba's joining the Internet, the government launched a frontal attack against those on the island that had called, however moderately, for greater societal autonomy and pluralism. While the United States had meant to foster oppositional civil society articulations, the Cuban government responded by closing political space even for those that argued for a renewed understanding of civil society as a project conceived within the socialist order and Revolutionary project.

The resulting strategy combined a renewed hard-line emphasis on ideology in domestic politics with a gradual extension of NICT use under state control. So far, this has proven a viable path for Cuba's state-socialist government. Internet access has been made possible only through official institutions, which allows for strong administrative and social control, and the *de facto* state monopoly on Internet service providers facilitates a wide array of technical control mechanisms. Moreover, where access to the NICT has been expanded to a broader public, in many cases it has been limited to e-mail functions and domestic networks. Access to 'full Internet,' including the World Wide Web, is considered a far more serious threat to the state's monopoly on mass media and therefore is restrained severely. Oppositional groups and dissident journalists accordingly have found the World Wide Web to be an efficient platform to reach a wide international audience, but not the national public.

At the same time, the government has managed to initiate a notable modernization process of the island's largely obsolete telecommunications infrastructure, based on the transformation of the state monopoly telecom operator into a joint-venture with Italian capital. Moreover, the NICT strategy launched in the second half of the 1990s, which concentrates the available resources on official institutions, has brought remarkable results in some sectors. The 300 Youth Computer Clubs nationwide certainly are a notable educational approach to introducing the young generation to the new technologies, and many public health systems in Third World countries may envy Cuba's *Infomed* network. As a result, whereas state-socialist Cuba for some is a violator of human rights due to its access restrictions, others hail its exemplary developmental use of the NICT.

From a developmental point of view, good arguments certainly can be made for prioritizing institutional and public Internet access over individual and residential use in a Third World country. However, the restrictions and controls on Internet access that can be observed in Cuba clearly stem from a political logic, grounded in the state's monopoly over the country's mass media as established in the Constitution's Article 53. It should be noted that the 'Report of the Politburo' of April 1996 that, parallel to Cuba's joining the Internet, launched the campaign against reformist tendencies on the island, explicitly underscored the importance of a tight preservation of the state media monopoly by pointing to the negative example of the *glasnost* policy that, in Raúl Castro's reading, brought about the collapse of state-socialism in the Soviet Union by giving up control over the mass media.

The new information and communication technologies do not lead to political change in any direct way, but they do confront Cuban socialism with a whole set of complex challenges. Carlos Lage, the Politburo member in charge of economic affairs, articulated the government's attitude toward the NICT, according to which Cuba seeks "to make use of their extraordinary benefits while forestalling their negative effects" (Lage 2000: 7). This, however, does not answer the question what is to be done if these interests are in conflict—and precisely this conflict is the rule, not the exception, in the use of the Internet in a state-socialist country like Cuba. The question if the NICT are a key to development or a threat of ideological subversion for Cuba must be answered: they are both.

Chapter Seven

The Politics of the Internet in Third World Development: Conclusions in Comparative Perspective

The new information and communication technologies (NICT) have become a prime focus of attention in Third World development. The analysis presented in this work certainly has confirmed our original hypothesis: technology is not neutral. The NICT neither spread around the globe in a uniform way, nor do they have universal political and developmental implications, nor is their diffusion a mere function of economic wealth. Politics matters.

The dominant discourse tends to address the developmental issues associated with the NICT under the concept of the 'digital divide,' focusing on the quantitative disparities between North and South in the diffusion and use of the Internet and other NICT. These inequalities indeed are staggering. Moreover, although some observers proclaim a gradual closing of the gap, this is so only superficially. The number of NICT users in Third World countries has in fact been growing rapidly in recent years, but a closer look at the data shows the qualitative differences in NICT use between the 'information rich' and the 'information poor' have been growing even stronger.

Whereas the concept of the 'digital divide' tends to imply 'digital solutions,' the present study has shown that the questions raised by the NICT must be understood in the broader context of Third World development. The lack of access to the NICT in the countries of the South is not an isolated issue, but it comes as part and parcel of—to paraphrase the term—the global 'development divide,' reproducing its sharp economic and social cleavages. Technocentric promises, according to which technological advances automatically will translate into social progress, are as false in the case of the NICT as they are for any other technology.

For Third World countries, access to the NICT is a necessary, but never a sufficient condition for development. The essential question is not how to maximize the numbers of computers or Internet users in any particular country, but how the new information and communication technologies can help foster economic, social and political development goals. The central problems raised by the NICT are not technical, but social, economic, and political.

The technocentric mentality itself relies on a much-used term, 'information technology revolution,' that tends to ignore the evolutionary character of the development of the NICT. Instead, these have emerged as a result of the convergence of computer technology, telecommunications and traditional media. The 'revolutionary' approach also tends to overlook how intrinsically the digitally-based new information and communication technologies interact with the 'old ICT.' In fact, the main line telephone system continues to be the key form of infrastructure for Internet access and other NICT applications. Moreover, it is a central conclusion of this work that the political implications of the NICT have to be analyzed in the context of what we have introduced as the 'telecommunications regime.' This concept is defined as the specific configuration of economic, social and political actors, the regulatory framework, and the patterns of diffusion and use in the telecommunications sector in any given country or region. If the telephone system provides the essential technical infrastructure for the NICT, the telecommunications regime provides the basis for their organization and operation.

In the past two decades, with the emergence and public breakthrough of the core technologies of the NICT, telecommunications regimes worldwide have undergone fundamental transformations. Before this, the industrialized capitalist countries of the North had established the dominant paradigm, which we have termed the 'Fordist telecommunications regime.' Centered on a public monopoly telecommunications operator, it typically followed a socially inclusive strategy of network expansion. This strategy committed the state to the goal of 'universal service,' that is the provision of a main line telephone for every household that desires to have one. Based on cross-subsidization schemes, high tariffs for long-distance calls—which mostly affected business and institutional customers—facilitated low tariffs for domestic calls, the service most widely used by the population at large.

In the Third World, the ISI strategy, the dominant development model in the 1950s and 60s, led to a deformed version of the Fordist industrialization model because it failed to generate the relatively high levels of mass income and mass consumption characteristic of Fordism in Europe and North

America. This asymmetric experience translated into a telecommunications regime that shared the organizational structure of a state-owned monopoly telecommunications operator, but did not provide for a similar policy of socially inclusive network extension. Instead, whereas mass media received much attention, telecommunications were largely seen as of minor importance in Third World development, up to the early 1980s. The often considerable earnings of the state monopolies subsidized other activities or helped balance the state's deficit, rather than be invested in the improvement of the telephone system. The state-socialist development model, though it differed in many ways from the ISI strategy of capitalist countries, took a similar approach, strongly favoring the development of vertical, one-to-many mass media over the horizontal, one-to-one communication of the telephone.

With the decline of Keynesianism in the industrialized countries and the demise of the ISI strategy in the Third World in the wake of the 1980s debt crisis, the emerging neo-liberal paradigm put the dismantling of the state telecommunications monopoly high on the political agenda. In the Third World, the Pinochet dictatorship in Chile spearheaded this transformation of the telecommunications regime; many other countries in Latin America followed. However, in contrast to much official rhetoric, the drive for the privatization of telecommunications was motivated less by specific policy convictions of the national governments and more by their desperate needs, in light of the pressing foreign debt burden, to obtain hard currency revenues and by strong pressures from the world financial institutions, which made new loans dependent upon economic liberalization.

These privatization strategies had important political implications. Telecommunications passed from being seen as a public service, provided by the state to its citizens, to being a commercial good, principally subject to the market dynamics and profit incentives characteristic of a capitalist economy. However, in a political democracy, access to information and communication are goods unlike any other: they must be seen as essential elements of civic rights and meaningful citizenship. Therefore, for the analysis of 'the politics of the Internet' as undertaken in this study, an equitable social diffusion of access and use of the NICT is not just one among many desirable development goals, but a central political concern, since it touches on the very substance of democracy.

Accepting, in principle, their responsibility in this matter, states accompanied privatization initiatives with new regulatory systems. Even in the world's most liberalized telecommunications regimes, the state has maintained some form of commitment to compensate for the socially discriminatory dynamics of pure market mechanisms; the typical example of a case requiring state involvement is the integration of rural and remote areas, where the extension of the national networks would be costly and unprofitable. However, since market forces—which are inevitably oriented towards those sectors of society with the highest effective demand: businesses, the high-income population, urban centers—take the lead in driving telecommunications development, existing subsidization mechanisms tend to be grossly inadequate for achieving socially balanced patterns of diffusion and use.

In the Third World, even in those cases where the expansion of telecommunications had come to be seen as a relevant developmental goal, the Northern model of 'universal service' as residential, individual access seemed hopelessly unrealistic. Instead, the postulated goal was 'universal access,' which means the provision of public telephone service at affordable cost within walking distance of every citizen. For the NICT based on computer networks, the Northern model of individual access is even further out of reach, since the economic and social barriers to access are considerably higher. In addition to the high hardware, software and operation costs, the NICT not only presuppose connection to the telephone net and to stable electricity supply, but also a substantial technical knowledge for which literacy is only the most elementary precondition. In most countries of the Third World, a large part of the population is lacking at least one, and often many of these conditions.

To promote 'universal access' to the NICT in Third World countries, therefore, the key is not technological innovation, as some enthusiastically endorse, but the creation of social structures, such as the diffusion of collectively used, low-cost public access centers. From a developmental perspective, however, access is only the first step. To initiate a process of meaningful use and social appropriation of the NICT by broad sectors of society, the co-operation of numerous social actors, specific educational measures and social community work is required in order to embed the NICT in an adequate social and institutional context.

The empirical country studies undertaken in this work have underscored the great importance the NICT sector has acquired for Third World countries. Moreover, both studies have shown how strongly the specific political regime, the socio-economic development model and the specific telecommunications regime affect the introduction, diffusion and use of the NICT. The new information and communication technologies indeed pose very different challenges in a country like state-socialist Cuba, with its state monopoly on mass media, versus a country like Costa Rica, with its stable political democracy and its 'mixed model' of a socially relatively inclusive capitalist development.

At the same time, although the two countries in the study had been the Third World showcase models of the competing ideological camps during the Cold War, they show a number of remarkable parallels. Even though they were conceived within opposing ideological frameworks, both Costa Rica and Cuba in the past decades have followed a development strategy that was atypically inclusive for Third World countries. As a result, both have exceptionally high educational standards that have generated a highly skilled labor force. This relative wealth of human capital in both countries, created by the development path chosen in the past, is seen as a key asset for both nations' development prospects, particularly in light of the growing importance of the NICT.

In both countries, the telecommunications regime stands in marked contrast to the dominant paradigm of liberalization, privatization and deregulation. In the Costa Rican case, it is notable that in the past, Costa Rica already had deviated from the typical patterns of the telecommunications regime of peripheral Fordism in that it promoted the expansion of the telecommunications network as an essential part of national development. Whereas elsewhere telephone access was long regarded as a luxury item for the urban elites, Costa Rica very early on established a telecommunications regime aimed at geographic and social inclusion. This effort encompassed a continuous push for the nationwide extension of the network, very low tariffs for domestic use and programs establishing public call offices in rural towns and villages. As a result, the country has reached one of the highest and most socially balanced rates of telephone diffusion in the Third World.

When Costa Rica experienced a serious foreign debt crisis in the early 1980s, it was in precisely this moment that the country's role as an ideological counter-model against the revolutionary movements in the neighboring Central American countries became more important than ever. This particular external circumstance generated sufficient financial and political support from the United States to allow a far more partial and gradual version of the liberalizing 'structural adjustment programs' than in almost any other Latin American nation. Nevertheless, international and domestic pressure to dismantle the state telecom monopoly grew. By the 1990s, Costa Rica had gradually entered into what we have described as a 'pre-liberalization telecommunications regime': a state monopoly on the defensive, awaiting liberalization and having its revenues diverted to cover the public deficit. In response to the monopoly's problems, the political elite began to show an attitude of 'interested neglect,' since the resulting deficiencies would only strengthen the argument for liberalization.

In the context of Costa Rica's weakened 'pre-liberalization telecommunications regime,' it was not the state telecommunications company, but academic pioneers from the country's public universities who became the vanguard in establishing Internet access and a non-profit network providing e-mail and Internet services. The inertia of the state monopoly in the introduction of the NICT became a prime argument for those who called for its liberalization. However, when both of Costa Rica's dominant parties joined forces in 2000 to pass a law for the comprehensive liberalization of the country's electricity and telecommunications sector, the initiative provoked the most massive public protests the country had seen in the last five decades.

As the empirical study has shown, this broad public defense of the state's telecom monopoly came as a direct consequence of the atypically inclusive telecommunications regime of the past. With a similarly inclusive model of electricity provision, the *Instituto Costarricense de Electricidad* (ICE), which operates both electricity and telecommunications, had been a core institution of Costa Rica's economic and social development model. Looking beyond the precise issue at stake, it is apparent that these protests therefore signalled a conflict over the underlying development paradigm], with the population turning out to defend the 'old,' state-interventionist Costa Rican development model and oppose its erosion through a liberalization process. In the course of the conflict, the dismantling of the state electricity and telecom monopoly had become a central and symbolic issue in the liberalization debate. The political elite feared that a protracted conflict would lead to a further and more fundamental loss of legitimacy, and the bill was fully withdrawn.

As a result, Costa Rica today presents the rare case of a Third World telecommunications regime that is centered on a state monopoly in mainline and mobile telephony as well as Internet services, and which continues a socially inclusive pattern of network extension. Moreover, the state monopoly emerged greatly strengthened by the conflict over the liberalization law, since the political costs of re-launching such a liberalization initiative are considered prohibitively high.

After the failure of the liberalization law, the state company did obtain sufficient political backing and financial resources to pursue an active strategy of NICT modernization and expansion. By the end of 2002, the percentage of the population that used the Internet had reached 19 percent, more than double the Latin American average. The country study also has shown that, for main line telephony as well as mobile phones and Internet services, the tariffs charged by Costa Rica's state monopoly are far below the average of the continent's countries with liberalized telecommunications regimes.

Whereas the mainstream discourse often presents it as a universal truth that liberalization and privatization will automatically lead to decreasing consumer prices, the Costa Rican case is an impressive example to the contrary.

If both countries, Cuba as well as Costa Rica, pride themselves on their socially inclusive development model, a central question is to what extent this translates into a broad social diffusion of access and use of the NICT. Both countries have almost entirely literate populations and near-universal school enrolment. Both have cost-free public school systems, which have generally good qualitative standards in comparison to other Third World countries; both also have started substantial programs to integrate NICT applications into the educational curricula. As far as education goes, then, both countries start with exceptionally good conditions for universalizing the knowledge and skills necessary to use the NICT and for promoting a socially inclusive diffusion of the NICT.

Yet as to the material conditions for access, even in Costa Rica, with its high telephone density and the highest rate of computers per capita of all Latin American countries, the majority of the population is in no condition to replicate the Northern model of individual residential use. To facilitate NICT access beyond the upper and middle classes, public access centers are indispensable. Since privately operated Internet cafés are largely limited to the urban centers and tourist areas, the Costa Rican state launched a remarkable initiative to promote universal NICT access not only as a material development goal, but moreover as part of an up-to-date concept of citizenship.

In a triangular cooperation among the government, the state-owned telecom monopoly and a private foreign company, a citizens' website and freemail program was established, automatically generating a personal e-mail address for everyone holding a Costa Rican I.D. card. The corresponding e-mail addresses—'*name*@costarricense.com'—symbolically underscored the message that NICT access is part of 'being Costa Rican,' an element of citizenship. In addition, in every municipality terminals were established where every citizen could use the Internet free of charge, for a limited time-period daily.

This experience, however well-intentioned, painfully revealed the pitfalls of any initiative not sufficiently embedded in an adequate social and economic context. The municipal localities did not have the necessary means for maintaining the equipment or providing technical assistance to the users. Many terminals therefore fell victim to the 'rusty tractor syndrome' soon after inauguration. When the foreign company involved in the project had to declare insolvency, the state and the state-owned telecommunications company were left to manage the initiative alone; it can only be hoped that some of the earlier mistakes will be corrected.

This study has additionally shown that not only this state-led program, but also prominent initiatives from non-government organizations fall far short of providing an adequate social integration of the new technologies. A particularly drastic example of the failure of a technocentric approach is an NGO-based project launched by Costa Rica's former President, José María Figueres. It sheds a remarkable light on the preferences and interests prevailing in the international development establishment that, despite the practical failure of this high-tech project without social vision, Figueres has been appointed as the top United Nations Representative for Information and Communication Technologies.

Turning to the Cuban case, although the issue of a socially balanced diffusion of NICT access involves many of the same questions outlined above, the empirical case study has shown that political considerations of a different kind dominate the discussion. The differences result from the particular characteristics of the NICT, which, due to their decentralized and transnational structure, enable actors of all sorts to circumvent many of the 'filters' established in traditional mass media.

Compared with radio, television or print media, relatively few economic resources are required to establish websites or mailing lists that can reach out to a worldwide audience. Many social movements, which before had only a very limited presence in national and international media, have come to use the NICT not only for their internal communication, but also as an important platform for the worldwide diffusion of their positions. The most prominent example of this phenomenon has been that of Mexico's Zapatista uprising in the mid-1990s, in which Internet use became crucial for a highly communicative guerrilla movement to shore up world opinion in its support and to become an articulate and influential actor in the national public sphere.

In addition to this aspect of counter-balancing established economic and media power structures, the inherently border-transgressing characteristics of the Internet are centrally important because they strongly challenge the notion of national sovereignty in regard to media. In principle, this applies to all countries, since everywhere national laws establish the fundamentals of the regulatory framework for media. Even established democracies may battle with certain contents that may be banned from the national public sphere by law but which enter freely via the Internet. However, the political problems stemming from the cross-border character of the Internet are greater the more the media content is regulated and restricted by national norms. Accordingly, political systems with a non-pluralist or largely restricted media regime based on a concept of nation-state media sovereignty face greater challenges from the Internet. Though print media always has been smuggled across national lines, the problem is of a qualitatively different dimension in the case of the intrinsically global medium of the Internet. Here, it takes almost no effort to reach out across national borders—but once connected, it is a major undertaking to separate desirable from undesirable information.

While the particular political regime and its media policies are essential variables for the political implications of the transnational character of the Internet, major communities of emigrants, living outside the nation's borders, also play an important role. Where these communities maintain close social connections to their country of origin, demand for transnational communication will be high—in fact, the existence of emigrant communities often plays a catalyzing role for the spread of NICT use among the popular sectors in the Third World. In addition, the more the emigrants' social and political interests are oriented towards their country of origin, especially if they oppose the prevailing political structures, as in the case of exile communities, the more they tend to push to use the NICT as a political medium to influence public opinion worldwide and, to the degree this is possible, in the country itself. The same is true for other external actors with interests in exerting influence on the domestic situation of another country; this includes foreign governments as much as international non-governmental organizations dedicated, for instance, to the protection of human rights.

In Costa Rica, all these factors have minimal effect. The liberal political system has rather few restrictions on media content and, moreover, Costa Rica is an atypical Third World country in that it does not have any significant group of emigrants living in other countries, and virtually none of those would define themselves as exiles. For Cuba, by contrast, these issues are at the very core of the 'politics of the Internet' that this work analyzes. The state monopoly on mass media, enshrined in the country's Constitution, is a central pillar of the political system. A large Cuban community lives in the United States and, though it is debating its identity as 'exiles' or 'immigrants,' it maintains strong ties to the population on the island; it also demonstrates great interest in Cuban politics and its overwhelming majority strongly opposes the present political system. In addition, the U.S. government maintains an uncompromising Cold War style policy against the island, including a strong emphasis on the use of media to erode Communist rule.

Underlying Cuba's state monopoly on mass media is a vehement defense of 'national media sovereignty.' For many years, the government has spent considerable resources to jam the broadcasting of U.S. sponsored radio and TV stations from South Florida. Moreover, in the Cuban leadership's interpretation of the causes of the demise of state-socialism in the Soviet Union the loosening of media control through Gorbachev's *glasnost* policy plays a crucial role. For the island's socialist government, this reading of events has reinforced the insistence on tight state control over the media.

In this context, any expansion of transnational media is a highly politicized issue, whether it occurs through 'plain old telephone' communication or access to the NICT based in computer networks. Yet while the government is wary of any such opening as potentially undermining political control over media and communication, economic pressures and more long-term developmental considerations make an active promotion of telecommunications and NICT increasingly indispensable. Facing this dilemma, the Cuban government struggled to adopt a policy that would actively foster NICT development while maintaining a maximum of state control.

The strategy adopted by the Cuban government represents a substantial transformation of its telecommunications regime. After the Revolution of 1959, though the government in principle pursued an economically inclusive development model, telecommunications was an insignificant concern, as was typical for the telecommunications regime of Soviet-style state-socialism. The formerly private, U.S. owned telephone company became a state-owned monopoly, but the revolutionary state's priority went to the one-to-many mass media, not the horizontal one-to-one communication of the telephone system. In the three decades following the Revolution, the expansion of the network held such a low priority that Cuba, which in 1959 had the highest telephone density of all Latin American countries, had one of the region's lowest telephone-per-capita ratios by the beginning of the 1990s.

This pattern changed only after 1989, when Cuba was forced to re-integrate into the globalized world economy. The emerging dollarized and world market-oriented sectors in the island's economy, which were crucial for Cuba's economic reactivation during the profound crisis of the early 1990s, brought significant pressure to bear on the push for a modernization of the largely obsolete telecommunications system. Beyond this, Cuba focused the long-term developmental hopes of the island on economic sectors that build on the country's high investments in education and science; in these sectors, it became increasingly unthinkable to meet world standards without adequate NICT access. However, given the steep fall in Cuba's foreign exchange revenues after 1989, the needed large-scale investments for any such modernization only became viable after the conversion of the state telecom monopoly into a joint-venture with foreign capital.

Though in this work we have analyzed the political implications of the NICT in specific national country studies, we have also stressed the need to do so while considering international context. The transformation of the Cuban telecommunications regime is a case in point. The transformation of the state monopoly into a joint-venture company with foreign capital became a viable strategy only due to two external factors: the existence of a large emigrant community with high communication needs to the island; and a major change in U.S. Cuba policy, which occurred in 1992 when telecommunications were exempted from the trade embargo, allowing the reestablishment of direct telephone connections between the two countries. Whereas Cuba's depressed domestic market would have given few incentives for any foreign investor, these external factors opened up a major source of hard currency revenue for the island's telecommunications operator, in the form of internationally shared revenues from long-distance calls. Thus, a political decision by the U.S. government, aimed at eroding regime support through increased communications between the Cubans on the island and their relatives in the United States, has been the cornerstone of the modernization of Cuba's telecommunications since the mid-1990s.

Another important international aspect for Cuba's changing approach to telecommunications and NICT was the example of the main 'socialist survivor state,' the People's Republic of China. Since the late 1970's, China had adopted an alternative version of a state-socialist telecommunications regime, in which the dynamic expansion of telecommunications was defined as a strategic development priority. This policy has resulted in one of the highest teledensity growth rates in the world. While this became the most important international model for Cuba, the island's government followed much more timidly. Expansion of the telephone network has focused on official institutions and the dollarized sectors of the economy, whereas residential diffusion has been very slow. Private telephone access still is considered neither a public service nor a commercial good, but rather a privilege for citizens deemed worthy by the state's official institutions, on the basis of 'social merits.' As a result, Cuba's rate of telephone main lines per capita still, in 2003, remains amongst the lowest in Latin America.

The transformation of Cuba's telecommunications regime created the essential structures on which the country's NICT policy was built. The joint-venture telecom monopoly, in which the Cuban state holds a majority of shares, proved to be the quintessential answer to the dual needs for political control and for major modernization investments. Through the first half of the 1990s, however, the Internet was publicly rejected as a propaganda weapon of U.S. imperialism and the nation's access to it was limited to a rudimentary, once-a-day-e-mail connection to Canada. The present study has analyzed the lengthy and protracted political decision-making process within the government, which led to Cuba joining the Internet by the end of 1996, later than any other country in Latin America.

For the political analysis, it is crucial to see that this step, when it was finally taken, came in the midst of a general turn to a hard-line ideological offensive in domestic politics. This effectively cut off those reformist tendencies within the country that had called for the acceptance of more pluralist and autonomous forms of expressions in Cuban society. The government in Havana opted to accept first, increased international telephone traffic, and second, Internet connection. However, while the United States pursued a policy of promoting increased communications to foster the emergence of oppositional social actors, the Cuban government closed the spaces for the articulation of dissent, even if voiced in most moderate form and from within the state-socialist project.

In addition to this overall political context, three factors were key political conditions in Cuba's turn to a pro-active NICT policy: first, the concentration of NICT resources in the official institutions, while individual use was restricted; second, a differentiated approach towards different types of NICT; and third, where access to cross-border NICT communication was facilitated, the implementation of a combination of administrative, social and technical control mechanisms to minimize politically undesireable uses. For the last to be fully possible, it is essential that the country's telecommunications regime maintain a *de facto* state monopoly over all Internet service providers, which permits diverse technical mechanisms of registration, filtering or blocking of communication or content.

In line with the general political-economic conception of state-socialism and similar to the patterns of telephone expansion, the Cuban state's commitment to universal NICT access is overruled by an explicit prioritization in the allocation of the limited NICT resources. As a consequence, these resources are channeled towards the state-associated institutions deemed important for the nation's economic and social development, to the detriment of individual use. In this work we have argued that in a situation of severe economic constraints, typical of Third World countries, a certain channeling of available resources to key sectors for development is indeed an important policy task. However, in the Cuban case this prioritization goes so far as to greatly curtail the citizens' rights to information, which risks producing severe negative social repercussions that ultimately may erode economic efficiency as much as political support.

Though the Cuban prioritization policy has shown remarkable results in the NICT integration in public service sectors, such as medicine and education, its downside has been staunch restrictions on individual use.

Individual citizens cannot become customers of the state Internet service providers, and any sale of computers or modems to private persons has been explicitly banned, even for those who can afford their high dollar prices. The existing intra-institutional access possibilities, conceived as professional or educative functions, are limited to the employees or affiliates of the particular institution. It was not until 2001 that, for the first time, private use by 'normal' Cuban citizens was given a legitimate and 'organic' form, when the government equipped a number of the country's post offices with public NICT access centers.

The government's approach, however, distinguishes between those NICT applications that facilitate cross-border communication and those that do not. Closed domestic networks, particularly if designed for specific social sectors, do not pose any major ideological problem and it was these that, within the narrow limits of the financial conditions of the early 1990s, were actively developed even well before the country joined the Internet.

The government has a far more reserved attitude towards NICT involving transnational communication. However, here too another distinction applies. Following the turn in the country's telephone policy, the government began to adopt a relatively tolerant attitude towards international e-mail use, which (as outside mass mailings are banned by the state-controlled Internet service providers) is much closer to the one-to-one communication of the telephone than to traditional mass media. In contrast, the cross-border many-to-many medium of the World Wide Web, which includes platforms of virtually all major traditional mass media from capitalist countries, is seen as a serious threat to the state's media monopoly. While there has been some opening to 'full Internet' access in official institutions, this has been in such a limited form as to prevent the World Wide Web from becoming anything close to a mass medium of communication and information on the island. The NICT access provided to the population at large through the conversion of the post offices still is limited to international e-mail functions and closed domestic networks. In its social impact, this policy of providing Internet access in greatly restricted form seems to result in the paradoxical situation that, rather than satisfying existing demand, it creates demand for greater access. This occurs because it raises awareness of the potential of the NICT while making people all the more aware of the limitations imposed on their use.

While in Costa Rica the NICT have tended to reflect existing power and media structures rather than challenge them, in the Cuban case 'cyberspace' has become a field of conflicts, which have been effectively banned from the island's public sphere. In addition to the government, actors from the exile community and domestic dissident groups have come to use the World Wide Web as a prime medium to bring their views and positions to a worldwide audience. Since the state monopoly on Internet service providers prevents these groups from getting direct Internet access, they have become an extraordinary example of how the new information and communication technologies interact with the old: the dissidents use the 'old technologies' of telephone or even written letters to transmit their texts to the United States, where they are transcribed and posted on the Internet by supporters from the Cuban emigrant community. For Cuba's small domestic opposition groups, the World Wide Web has become an important forum through which they have gained much international visibility and support. Nevertheless, under current conditions, the Internet has no way of reaching a broad nationwide audience on the island.

In their development strategies, both Cuba and Costa Rica not only seek to integrate the NICT into their economy and society they also perceive the production of NICT related goods and services as a strategic industry. In Costa Rica this approach has little tradition. In the context of the ISI-led development strategy of the past, the country's domestic market, even if enlarged by the Common Central American Market, would be far too small to think of the production of complex technological goods. New possibilities opened with the turn to an export-led development model and the gradual liberalization of the economy. By the mid-1990s, Costa Rica adopted a strategy explicitly focused on the promotion of high-technology clusters, to be built around foreign direct investments. These efforts culminated in Intel setting up a huge microprocessor plant, one of the most coveted foreign direct investments in Latin America in recent years, in free production zone near the Costa Rican capital. By 2000, this single plant accounted for no less than 40 percent of the country's exports. Any notion of 'technological autonomy,' which had been a central concern for Third World industrialization in the past, has given way to the quest for successful integration into the global production chains of leading transnational companies.

Although the impact of the Intel investment on Costa Rica's macroeconomic statistics is enormous, it has led to a dualization of the economy, rather than a coherent development model for the country as a whole. Aside from the notable impact on the labor market, where the chip-producing site has created relatively well-paid employment for a considerable quantity of highly skilled workers, the linkages between the Intel plant and the rest of the economy are low. If this is typical of industries in free production zones in Third World countries in general, the sophisticated production process plus the company's global production schemes in the Intel case make it particularly difficult for any domestic company to become a supplier, outside of mostly low-tech services. For the ISI industries, integration into the world market was the problem; now the equally problematic challenge is the integration of world market industries into domestic economies.

Despite some efforts to the contrary, the Intel plant maintains an essentially enclave character. In its shadow, however, a small but dynamic software sector of largely domestic companies has emerged. Although its production and export volume is very much lower, this sector is far more densely integrated with the national economy and includes a far higher nationally-added value in its products and services. However, recent international reports celebrating Costa Rica as the Third World's second 'rising star' in NICT related exports, after India, represent an unwarranted optimism. Costa Rica's emergent software sector is a very recent phenomenon: its volume still is small in respect to the national economy and its mid- and long-term perspectives are far from assured.

The software sector as much as the Intel plant certainly can be an important element in a strategy of industrial upgrading that builds on a highly educated labor force. However, given the sparse linkages of 'Intel island' to the national economy and the limitations of the software sector, these sectors are still far from becoming the motor of Costa Rica's economy.

Where the dominant ideology promotes the transformation of state and economy along neo-liberal lines as a *conditio sine qua non* for successful development, the Costa Rican case demonstrates an important lesson. The attraction of the Intel investment, as well as the emergence of one of the most dynamic software sectors in Latin America, have taken place in one of the least liberalized economies of the continent. Moreover, contrary to prevailing theory, the continued existence of a telecommunications regime centered on a state monopoly has not been a serious hindrance to either development. In fact, the Costa Rican example strongly underscores the economic value of a socially inclusive development model that can provide for a good and broad education system, reliable public institutions and stable social and political conditions.

In contrast to the Costa Rican experience, considerable efforts were made in Cuba since the 1960s not only to promote the use of computer technology, but also to develop computer manufacturing as an essentially import-substituting industry. After the crisis of the 1990s led to virtual collapse of this industry, Cuba's efforts in NICT production shifted from hardware to software. Though in the beginning this was entirely oriented towards domestic needs, the growing dynamism of this sector was largely due to so-called 'exports within borders,' that is production for the emerging tourism industry and other dollarized sectors of the Cuban economy.

The continuous growth of software production in recent years has nurtured high hopes that it will turn into a strategic industry for Cuba's development. A prominent manifestation of this optimism came in late 2002, when the former Soviet/Russian military surveillance post near Havana was converted into an ambitious computer science academy, designed to become the core of a large, state-run software cluster. If this sector is to become competitive in a rapidly changing world market, it will depend on positive linkages to other sectors of Cuba's economy and society. The promotion of such a fertile domestic breeding-ground for the sector will almost inevitably entail pressures to promote NICT diffusion and to open up further the possibilities for Internet access in the country.

As of today, although the Cuban software sector certainly has developed notable production capacity, it has not yet passed the test of the world market to become an export-led industry. The protected market of the 'exports within borders' still accounts for more than 80 percent of its dollar sales. Moreover, in many respects, Cuba's software sector is likely to face problems similar to those experienced by the country's ambitious pharmaceutical and biotechnology industries, which have painfully shown the formidable barriers in the international marketing of high-tech products, even when the technological production process itself has been successfully mastered. These obstacles range from oligopolized market structures to the demands for compliance with international norms, patent laws and intellectual property rights, which Cuba, not being a member of the WTO, partly continues to ignore.

On top of these problems, the U.S. trade embargo represents an additional and very severe hindrance to the development prospects for Cuban NICT production. Whereas Cuba seems to have few insurmountable problems in acquiring U.S. computer technology through third countries, circumventing the U.S. embargo in the other direction is far less feasible. As technically viable and inexpensive as the Cuban NICT products may be, as long as they are is banned from the world's principal and standard-setting market, prospects for exports will be largely limited to niches in the world market.

In regard to the supposed 'democratizing effect' of the Internet on nonpluralist political regimes, the analysis of the Cuban case has shown that the Internet certainly is no unstoppable 'virus of freedom,' infecting the world with to a pluralist order of North American design. The social and political effects of the NICT, as with other technologies, always depend on the specific context in which they are used. A political regime like the Cuban one, but also those of China, Singapore, Saudi Arabia and many others that reject Western style media pluralism, have found viable strategies for incorporating the NICT into their development plans in ways that do not threaten the domestic political or social order in any immediate manner.

At the same time, however, the Cuban case has shown equally clearly that the transnational network-based information and communication technologies do represent a major long-term dilemma for countries that follow a nation-state based concept of media sovereignty. Undoubtedly, broad and open Internet access would undermine the state's media monopoly. Restricting access, however, implies economic and social costs, which increase as the transnational network-based NICT gain importance as a means of social communication and as a productive force for economic development. Cuba's NICT policy is profoundly marked by these competing logics. So far, despite the notable turn to a pro-active strategy in recent years, it is fair to say that the government has preferred to err on the side of caution. As to the future, it seems a safe prognosis that, for state-socialist Cuba, the integration of the NICT remains a highly ambivalent exercise.

The present work has centered upon two in-depth empirical country studies, which have inserted the analysis of 'the politics of the Internet' into the broader context of the political economy of Third World development. Its findings will hopefully advance a field of social science research that has been largely dominated by studies focusing on the industrialized countries, using technocentric approaches or taking a bird's eye perspective on global processes.

In this study, the choice of the nation-state as unit of analysis has allowed a focus on the key challenges stemming from the contradiction between the inherently transnational character of the network-based NICT and the essentially national character of political organization and regulation. The selection of contrasting regime types has opened up a comparative perspective that has shown in exemplary form the different challenges that the NICT pose in different political contexts. The analysis of the evolutionary character of the NICT and their persistent nexus with the 'old' technologies from which they have emerged has permitted historic contextualization. This has shown far more precisely the continuities, transformations and fractures involved in the adaptive process than the widely used concept of the 'information revolution' suggests. The concept of the 'telecommunications regime' has introduced a highly useful methodological framework for analyzing the organizational structures of NICT diffusion, use and regulation in the context of the specific political and economic development path pursued by any given country, as much as in the context of the global economic and political transformations in which the country is inserted. The understanding of communication and information not merely as tradable goods, but also as essential parts of civic and social rights has provided a key to show the intrinsic link between the promotion of an equitable social diffusion of NICT access and the quest for a participatory public sphere, substantial citizenship and social justice.

The new information and communication technologies that have inspired this work are a set of technologies that certainly have not yet exhausted their potential, either in their development or in their diffusion. New innovations will spark new questions and problems. However, the most crucial issues raised by the NICT are not 'technical' in nature, but refer to core questions of democracy and development. It is the hope of the author to have contributed to a better understanding of these implications.

Notes

NOTES TO CHAPTER ONE

1. Hamelink (2002a). See also Hamelink (2002b), and the address of Pimienta (2002) to the same 'PrepCom 1' meeting in July 2002, arguing that the so-called information society "holds new challenges and opportunities, and for this it needs new rights such as the right to communicate."

    Note that all through this work, for the sake of readability, Internet sources are cited in briefest possible format, without the preceding 'http://'; this is given only when the web address does not start with 'www.' Except for .pdf files, for citation from Internet sources no page numbers can be given, but they are easily identifiable through the browser's search functions. Wherever Spanish or German language sources—be it Internet or print—are cited, the translations into English are the author's. The same is true for the interviews cited, which were all conducted in Spanish (except for Henkel, which was conducted in German).

2. The "handling of information" includes different processes, namely the capturing, storage, processing, transmission and display of information, and Hamelink (1997: 8) provides a breakdown of products under each of the following headings: capturing technologies: keyboards, mice, trackballs, bar code readers, image scanners etc.; storage technologies: floppy disks, hard disks, RAM disks, optical disks etc.; processing technologies: the systems and applications software required for the performance of digital ICTs; transmission technologies: digital broadcasting, local area networks (LANs), wide area networks (WANs, one of them being the Internet), modems, transmission media such as fiber optics, mobile phones and digital transmission for mobile space communications (satellites); display technologies: computer screens, printers, digital television sets etc.

3. The increasing amount of products and services available in digitized form is a central topic in the fashionable discourse on the 'new economy'; for an academic perspective in this line, see for instance Phujola (1998), who—under the heading "Dematerialization of Economic Activities"—writes: "The truly revolutionary aspect of modern information technology is believed to be the possibility it offers to unbundle information from its physical carrier (. . .) This means that the economics of information can be

separated from the economics of things" (Pohjola 1998: 2); see also Quah (1999) who speaks of the "weightless economy." However, catchy expressions such as 'dematerialization' and 'weightlessness' are misleading, since the production as much as the use of any digitized product depends on a wide array of material inputs from the 'economy of things,' ranging from the human labor necessary for its elaboration to the computer monitor to display the digitized file, from the satellite in orbit to the silicon needed for the computer chips, etc. etc. Therefore, we insist on analyzing 'the economics of information' and 'the economics of things' not separately, but as two closely interrelated considerations.

4. The following passages largely draw on the overviews of the historic development of the NICT in ISOC (2000), Hamelink (1995), Castells (2000: 38–54), Zakon (2002), Everard (2000: 11–24), and Miles (2001).

5. In this work we will generally rely on the data given by the ITU, since these are not only the most authoritative data, but also since the ITU is the only institution providing comprehensive data for all nations of the world by (at least more or less) common methodological standards.

6. The internal discrepancies are also strong in the industrialized countries. Ever since, in the United States, the National Telecommunications & Informatio n Administration (NTIA 1995) published its first survey under the title "Falling through the Net," this and following analyses have shown a strong positive correlation between Internet usage and a) income; b) educational levels; c) ethnic background, with Blacks and Hispanics lagging behind Whites; d) urban versus rural residence; and, e) though declining in recent years, along the gender division (NTIA 1995, UNDP 2001: 57, InfoDev 2000: 17, Bimber 2000).

7. In order to keep a sense of the proportions, it is essential to recall some numbers concerning transmission speed. Whereas dial-up access through modems commonly is about 33.6 Kbps or a maximum of 56 Kbps, 'broadband' generally refers to transmission speeds exceeding 256 Kbps, giving users far more velocity in 'surfing' the web, and the possibility to download larger files and to consume digital goods like music or videos. DSL (Digital Subscriber Line), which has especially in Europe become the current broadband standard, is enabling data transfers from 192 Kbps up to 2.3 Mbps, depending on the specific end-user needs (Hilbert 2001: 21; see also ITU 2002a: 29ff.).

8. Again, this is mirrored in the debate on the digital disparities within advanced industrial countries. For instance, in 2000, the annual survey conducted by the U.S. National Telecommunications & Information Administration was given a more optimistic touch: the title "Falling through the Net" was amended to an undeniably contradictory "Falling through the Net: Toward Digital Inclusion" (NTIA 2000). Following a similar line, a number of authors have argued that the "digital divide is a temporary condition" (Hutter 2001), characteristic only of the initial phase of the diffusion of a new technology. In a rebuttal to Hutter, Kubicek (2001) underscores that the persistently differential use of the NICT in developed countries such as the United States or Germany is no simple 'time-lag,' but a function of the

underlying social and educational inequalities, which will not be overcome by time and patience but only by active initiatives by society and state.

9. Hargittai (2002) suggests five dimensions along which qualitative differences in NICT can be focused: technical means (the availability of software, hardware, connectivity quality); autonomy of use (location of access, freedom to use the medium for one's preferred activities); use patterns (types of uses of the NCT, the purposes for which the technology is employed); social support networks (availability of others one can turn to for assistance with use or to share the results of use); and skill (an individual's ability to use the technologies effectively).

10. The authors form part of what has become a prime forum of debate over NICT use for development in Latin America, the virtual community 'MISTICA' (acronym for: Metodología e Impacto Social de las Tecnologías de la Información y de la Comunicación en América—www.funredes.org/mistica). For a brief presentation of the project see Barnola/Pimienta (2001); central programatic documents are Pimienta (2000a) and MISTICA (2002).

11. The work of Prebisch (1949) was groundbreaking; a comprehensive elaboration of the theoretical concepts is found in CEPAL (1969); see also CEPAL (1998), Alexander (1990), and, for a political economy of the ISI process, Hirschman (1971).

12. Indeed, some of the 'Asian tigers,' such as South Korea, also started out from an ISI approach. However, with the passing of time, South Korea's strategy turned to export-led development based on industrial exports. By analogy, this has been termed 'industrialization via substitution of exports' (e.g. Martinussen 1997: 83), that is, the substitution of traditional agrarian and mineral exports with processed primary products, semi-manufactures and industrial exports.

13. Originally in Spanish '*estado desarrollista*' (e.g. Cardoso/Faletto 1988: 123ff); the international literature contains both the English translation, 'developmental state,' as well as the English-Spanish mix '*desarrollista* state'; generally, the latter is used when referring to Latin American cases and the former in regard to non-Latin American countries. As a noun, the term '*desarrollismo*' is used more or less as a synonym for '*cepalismo*' or 'ISI strategy.'

14. The authors point out that the political nation-building argument was of greatest important even for Friedrich List in his classic plea for infant tariffs in 19th century Germany. In List's conception, external tariffs should be the economic foundation for a united German nation-state, which would emancipate itself from the small and competing states that hitherto made up its territory. From the "protective system depend the existence, the independence and the future of German nationhood" (List 1982: 426).

15. In much of the development debate, the perspective is from the standpoint of an individual country. However, it must be added that, since Fordism is not only based on mass production and mass consumption but also on mass energy consumption, the project of late industrialization following Fordist patterns cannot be pursued by all countries at the same time, due to the global ecological limits (Altvater 1991: 288).

16. There is a significant difference in the role of the state in the Latin American and the South East Asian development paths. The 'Asian tiger' countries were in no way examples of laissez-faire liberalism, as the states there were actively involved in economic development planning, including five-year-plans and protectionist measures. However, the states maintained sufficient autonomy to avoid a similar rentist behaviour by companies (cf. Amsden 1989).

17. In an open reversal of the developmental efforts of earlier decades, deindustrialization was accompanied by a reconcentration on agrarian export and natural resources. For an overview of the fallen share of the industrial sector in the Latin American economies, see Benavente et al. (1996); for the prominent case of the deindustialization process of Brazil's industrial center around São Paulo, see Singer (1996).

18. This perspective was sharply criticized by many of the so-called *'dependencia'* authors, who were far more critical towards the capitalist 'role model' than the CEPAL.

19. 'Free software' does not mean that it necessarily is free of charge, but that its source code is freely accessible and that it can be modified and elaborated at will, as long as all new products stemming from it are made public under the same free licensing arrangement. Though often used synonymously with 'free software,' the term 'open source' is broader and also includes licensing models with restrictions on modifications and further uses of the program (Winzerling 2002: 37f.). See also www.opensource.org.

20. In spite of this resolution in their favor, Great Britain did not rejoin the organization until 1997, while the United States did not return to UNESCO until October 2002. For a documentary history of the struggles over the New World Information and Communication Order, see UNESCO (1988) and Nordenstreng/Manet/Kleinwächter (1986). For a detailed analysis of the conflict around UNESCO and the NWICO see Preston/Herman/Schiller (1989); for a comparison of UNESCO communication policy of the 1970s and after 1989 see Rohn (2002); for a critical analysis of the international media power structures, arguing the need for the NWICO see Smith (1980), Becker (1985), and Hamelink (1995).

21. The term "telecommunications" encompasses the telephone system as well as other forms of data transmission such as telegraph, telex or fax. It does not include the postal services, although organizationally these have often been linked together within a single post and telephone company, following the model of most European countries.

22. For the contrasting case of the East Asian 'tiger states,' see Freeman (1996), who points to the extraordinarily high rate of diffusion of information and communication technologies in these countries. In 1990 Taiwan had 30.6 main lines per 100 inhabitants, and South Korea 28.32–figures in the range of a country like Spain (30.40) and markedly higher than Portugal (20.07) (Prößdorf 1997: 197). Kraemer/Dedrick (1996) conducted a study of 12 countries of the Asia-Pacific region finding a positive correlation between investments in ICT and growth in both GDP and productivity. However, Freeman (1996) as well as Kraemer/Dedrick (1996) move onto shaky ground when they argue that the high rates of ICT investment led to GDP

and productivity growth. Instead, the positive correlation may be interpreted not as the cause, but rather as the result of economic development—or may at least be a result of the interdependence of these factors.

23. This practice continued well into the 1980s. In India, for instance, from 1979 to 1983 no less than 50 percent of telecom profits were passed on to the state. In Brazil, the net flow of funds from telecommunications to the treasury was about US$600 million in 1984; in Mexico, it was about US$350 million in 1987. Both amounts were equivalent to 30 to 40 percent of operating revenues (Saunders et al. 1994: 34).

24. Since the 1970s, the process of technological innovation has sharply increased the capital intensity of the formerly rather labour intensive telecommunications sector. This was the result not only of increased automation in the production and operating processes, but also of skyrocketing investments in research and development activities, which have greatly exceeded the R&D percentage of most other industries (Hamelink 1995: 53–54).

25. One of the most prominent examples of social diffusion via public call offices was the experience of low-income neighborhoods of metropolitan Lima, where more than 100 community telephone centers were set up in the 1980s. Each center served, on average, more than 600 customers, who could not only make outgoing calls but could also receive messages left for them at the center, and could even be listed in the Lima telephone directory under their local center's number (Saunders et al. 1994: 24). This experience was taken up in the 1990s when the Peruvian Science Net (RCP) pioneered collective Internet access centers (*'cabinas públicas'*), which successfully reached out to the poor sectors of society (Herzog 2002).

26. For instance, Castells' influential information age trilogy (Castells 1996/2000, 1997, 1998) is highly ambivalent in this respect. For one, he explicitly states that "of course, technology does not determine society" (Castells 1996/2000: 5) and that he believes "in the chances of meaningful social action, and transformative politics" (ibid: 4); nevertheless, his analysis time and again diffuses social forces and economic interests, treating them as depersonalized, quasi-natural structures. Castells concludes in his first volume: While capitalism still rules, capitalists are randomly incarnated, and the capitalist classes are restricted to specific areas of the world where they prosper as appendices to a mighty whirlwind which manifests its will by spread points and futures options ratings in the global flashes of computer screens" (Castells 1996/2000: 505). With an author as eloquent as Castells, the choice of metaphor is telling: a "mighty whirlwind," an irresistible force of nature. For a lucid critique of the ambivalencies of Castells' work, see Marcuse (2002).

27. Among the notable exceptions are Skidmore (1993) and Fox (1988). However, for the little attention generally given to media or communications, the work by Linz/Stepan (1996)—probably the single most influential book of the U.S. political science debate on democratic transition and consolidation in South America and Post-Communist Europe—is illustrative: although the authors write, in regard to democratization, that "for practitioners and theorists alike, diffusion effects have obviously gained in salience in the modern world owing to the revolution in communications" (ibid: 76), in their

book this "revolution in communication" does not merit more than a three-line note on the influence of television in the upheavals of Eastern Europe.

28. A major cross-country opinion poll in 2000 showed 60 percent of Latin Americans either 'not very satisfied' or 'not at all satisfied' with the way democracy works in their country (Lagos 2000: 142; see also www.latinobarometro.org).

29. We use the term state-socialism to describe the Soviet model, since in fact not 'society,' as alleged by the term 'socialism,' but rather the state (together with the party and military apparatus) became the organizational axis of the system. This also takes into account that undogmatic socialists defended the term 'socialism' against its use by the regimes in Eastern Europe and the Soviet Union, arguing that these countries did not represent what socialism should be, or had to be, to deserve its name (e.g. Müller-Plantenberg 1990). However, for the sake of readability in this text we will not insist on the term 'state-socialism' each and every time, which, especially in the case of the country study on Cuba, would also lead to complications when, in the case of quotes or paraphrases of official sources or other authors, the term 'socialism' has to be used.

30. It is worth recalling that the October Revolution inherited the czarist empire which, beyond the at least partially industrialized Russia, Ukraine and Belorussia, included large territories in Asia that by any standards would fall into what later was called the 'Third World.' Though in a subordinate position, these territories were integrated into the Soviet drive for modernization, industrialization and secularization, dramatically changing the social and economic structures in less than one generation. One of the most remarkable accounts of this process was provided by the contemporary German journalist Egon Erwin Kisch (1935) in his book "Changing Asia."

31. However, in its foreign policy, the USSR opted largely for a rather pragmatic, if not skeptical view on radical Third World revolutionary movements—as long as they had not yet come to power. The Latin American Communist parties aligned with Moscow generally supported a broad popular, national, or anti-imperialist front with bourgeois forces, whereas the armed struggle of guerilla organizations tended to be regarded as 'adventurism.' This resulted in severe clashes, with the most prominent coming over Che Guevara's intent to create a guerilla focus in Bolivia in the mid-1960s when, in a historic conversation, Mario Monje, the secretary-general of the pro-Moscow Communist Party of Bolivia, refused his party's support for Che's guerilla plans (see Guevara 1981: 29f.; Blasier 1987: 75–102).

32. Whereas official Soviet data had shown growth at a two-digit rate annually for the time between 1928 and 1960, these figures have been critically revised in recent years. Still, even by these critical accounts the Soviet economy showed considerable annual growth rates for a long time, averaging 3.2 percent in the period 1928–40, 7.2 percent in 1950–60, 4.4 percent in 1960–65, 4.1 percent in 1965–70 and 3.2 percent in 1970–75 (data by G.I.Khanin, cited in Castells 1998: 9).

33. The following section draws largely on the account by Castells (1998: 26–37), which in turn is largely based on a review of a number of Russian and international studies on the subject.

34. A recent exposition in the Berlin Communications Museum (March 22 to September 1, 2002) on the practice of telephone and postal control in the GDR had the telling title "An open secret." For telecommunications development in the GDR in general, see Günther/Uhlig (1992).

35. Slightly divergent figures are presented in Berlage/Schöring (1992: 3f), which also includes an explanation of the measurement problems in regard to Eastern European telecommunications statistics.

36. In Poland, for instance, more than 7,500 villages had no access to telephone at all (Müller 1991: 55) and the same was true for several thousand villages in Hungary (Prößdorf 1997: 197; Hunya 1995: 5). Where telephone lines did reached rural areas, the technology often was obsolete and access severely limited; a large percentage of rural areas was connected through manual switches, so that there was no telephone service outside the opening hours of the post offices; in the daytime, the congestion of the few available lines often made placing a telephone call a matter of hours (Prößdorf 1997: 195–198). The one major exception in Eastern European telecommunications was Bulgaria, which between 1973 and 1985 tripled its number of main lines to reach a density of 17 per 100 inhabitants (cf. Berlage/Schönring 1992:5; Saunders et al 1994: 317).

37. It is worth recalling that these restrictions from the United States persisted even after the political upheavals of 1989. Only since July 1990 have the Eastern European countries been allowed to import fiber optic cables and digital exchanges of the pre-1984 generation; more modern digital equipment, local area networks of more than 20 Mbps and ISDN-technology remained on the embargo list even then and were eliminated only later (Müller 1991: 58f).

38. The Sixth Congress of the Vietnamese Communist Party in 1986 initiated a process of economic reform, called Doi Moi, promising a "comprehensive economic renovation which aims to eliminate the subsidy-based, bureaucratic, centrally-planned mechanism, develop a socialist-oriented multi-sector market economy under State control" (cited in Lulei 1997). Internationally, the opening of the economy was accompanied by a process of reconciliation with the United States, which in 1994 led to the United States lifting its economic embargo against Vietnam; in 1995 both countries reopened their diplomatic relations.

NOTES TO CHAPTER TWO

1. For the full text of the National Information Infrastructure initiative, see U.S. Department of Commerce (1993); for an overview see Kalil (1997) and UNIDO (1998). Central academic volumes accompanying the NII initiative were published by the Harvard Information Infrastructure Project, amongst them Kahin (1992), Kahin/Keller (1995), Kahin/Wilson (1996), and Kahin/Nesson (1997); for a further discussion of the 'information highway' see Kleinsteuber (1996) and Kubicek/Dutton (1997).

2. This design of the 'universal service' commitment was much narrower than what public interest groups had advocated in the course of the law-making process; under the pressures of corporate interests, it in fact stepped back on

quite a number of important provisions that civil society activists had managed to include in preceding legislation, including a broader definition of the state's 'universal service' commitment to encompass not only telephone service, but also the access to data and broadband networks (cf. Drake 1997, Blau 1997).

3. The agreement is not a single document but a compilation of the commitments made by individual members (which vary in their precise definitions). All current telecommunication commitments can be found under: www.wto.org/english/ tratop_e/serv_e/telecom_e/telecom_e.htm

4. In most cases, the owners of intellectual property today are no longer the individual authors or musicians, but transnational corporations that market these media products (Hamelink 1999: 10f); see also the discussion of the WIPO in Ó Siochrú/Girard (2002: 85–98), particularly on its relation to the WTO and TRIPS agreement (ibid: 90–93).

5. Of great influence were the conferences held by the World Bank in 1987 in Kuala Lumpur (published as Wellenius et al. 1989) and in 1991 in Washington D.C. (see Wellenius/Stern 1991); the World Bank's long-time chief telecommunications advisor has summarized the main elements of telecommunications restructuring as promoted by the institution in Wellenius (1993). See also IMF (2001).

6. While by 2000 more than two-thirds of the Latin American countries had partially or fully privatized their telecommunications companies, in Africa the percentage was 28 percent, in the Arab states 33 percent, and in Asia and Oceania 46 percent (ITU 2000: 1f.). For a comprehensive overview, see the corresponding database at the ITU website (www.itu.org).

7. Again, the Argentinian case is illustrative. "The most significant element of privatization was the method chosen," as Schvarzer (2000: 13) writes: "If the government wanted to create an atmosphere of greater discipline for public companies, it should have prepared some competitive market conditions that would have obliged new owners to contemplate deep-rooted reform of their methods and technology. However, if it wanted to favour the creditors, so they could exchange the sums due to them for shares, certain market privileges had to be maintained that would ensure the profitability of those companies. (. . .) Consequently, it was no accident that at the very beginning of its term of office in 1989, Menem's government launched the sale of two key public corporations—those providing telephone service and air transport—on special conditions. The strategy was carried out to obtain respite from the demands of the banks (. . .) [To increase the company's market value] one of the first measures of the privatization process was to raise charges for telecommunications (. . .) Moreover, the government guaranteed the monopoly for 10 years to give the new owners an outlook of certainty (. . .). On these conditions, it ensured an appreciable profit rate, as subsequently recorded in the balance sheets of the privatized company during the 1990s."

8. Typically, this view is presented by the earlier cited World Bank book: "The continuum along which state operating entities are separated from governments begins with the reorganization of government departments into state enterprises; state enterprises are then transformed into state-owned companies, which are transferred, in turn, to mixed state/private ownership and then from

mixed to totally private ownership. (. . .) All reforms are moving in this direction, although only some have completely privatized state operations" (Saunders et al. 1994: 310).

9. In Latin America, the major exception has been the Caribbean, where exclusive concessions have been granted to a single mobile telephone operator, generally the fixed-line incumbent (ITU 2000: 9).

10. The aggressive push to drive out RCP included often more than questionable methods: in a long fight, RCP more than once had to call upon the telecommunication regulator to intervene against Telefónica de Peru's abuse of its monopoly position (see Herzog 2002: 204–209, 225–229). At the end of 2003, the RCP, with a market share of 15 percent, remained one of the country's three largest ISPs (www.rcp.net.pe).

11. As to the technological standards of second generation mobile phones, Third World countries in Asia and Africa have predominantly adopted the GSM protocol, as used in Europe and the Asian OECD countries (and which serves roughly two thirds of cellular phones worldwide). In Latin America, in contrast, mobile telephony is dominated by the standards favored by the United States (TDMA and CDMA); while still in a minority position, the GSM protocol has been gaining ground and is now used in more than a dozen networks on the continent (ITU 2000: 11f, ITU 2002c: 7f, Lamprecht 2002).

12. In Latin America, there is a continent-wide telecenter network called 'Somos@Telecentros,' which promotes the development of community Internet access centers in the region and organizes—through a mailing-list, conferences and other means—the continued exchange of experiences between the existing centers (see www.tele-centros.org). For an analysis of the telecenter experience in Latin America, see also: Gómez/Hunt/Lamoureux (1999).

13. For a general statement on privacy rights, see the chapter five of the 'Internet Rights Charter' drawn up by the Association for Progressive Communicaton (APC 2001); for an overview over the infringements on privacy rights through the use of NICT in Latin America, see Gregorio (2002); NICT-related violation of privacy rights in Argentina and Peru has been analyzed in Herzog (2002: 131–136 and 290–295).

14. While focussing on the positive cross-fertilization effects of the networking structures between different firms and institutions, the dominant literature on the Silicon Valley success story—e.g. Rogers/Larsen (1984) Saxenian (1994), also Castells (2000: 62–66)—tends to downplay or neglect other factors at the base of the 'new economy,' such as the incorporation of a lowly paid, largely immigrant and female labor force with precarious employment conditions, as Lüthje (2001) has pointed out.

15. The role model character of the Californian 'Silicon Valley' is underscored by the widespread, imitative word creations for high-tech (or wanna-be-high-tech) regions: in September 2002, no fewer than 103 regions were counted with informal names like 'Silicon Beach,' 'Silicon Desert,' 'India's Silicon Valley' etc. etc. (www.tbtf.com/ siliconia.html).

16. For the conceptualization of the cluster approach and its political implementation in Latin America, see also Altenburg/Meyer-Stamer (1999), and Bortagaray/Tiffin (2000: 5–9), who, after debating definitions of the term

and related concepts such as 'industrial districts,' 'industrial parks' etc., present a categorization of five different types of clusters: a) dependent or truncated, b) industrial, c) innovative industrial, d) proto innovation, and finally, e) mature innovation cluster (the last, according to their analysis, is still not to be found on the continent).

17. For instance, the 2001 UNDP Human Development Report has popularized the term 'hubs of technological innovations' as essentially synonymous with the term 'cluster,' listing the world's 46 most important high tech centers (UNDP 2001: 42). The empirical basis for this classification, however, was thin—a rather informal survey by 'Wired' magazine that certainly would not have met academic standards.

18. It is worth clarifying that this communication was not as direct as many assume. The popularized image of Subcomandante Marcos, sitting in a remote place in the jungle and linking his notebook computer directly to the World Wide Web, is simply false. In the Lacandonian rainforest, the Zapatistas had as little Internet access as any indigenous community living there, and the Zapatista communiques and Marcos' letters had to be saved to a diskette and taken, often by foot or donkey, to a modem-connected computer in one of the larger towns of the region, where sympathizers would put them on-line or forward them to their destinations. Hence, the Zapatistas are precisely an example of the necessary combination of old technologies—in this case human couriers—with the new ICT. Similarly indirect is their presence on the Internet: the Zapatista website, www.ezln.org, was set up by a sympathetic U.S. student and it is maintained at a server in San Francisco, not in Mexico (Schulz 1998: 603f).

19. Adapting the term, Castells speaks of the Zapatistas as "the first informational guerrilla movement." (Castells 1997:72). An analysis by the U.S. Department of Defense think tank, the RAND Corporation, examined the Zapatistas as the role model of an emerging "Social Netwar" (Ronfeldt et al. 1998), which will be copied by radical groups around the globe. It then outlined what should be the strategy of "counternetwar" to be developed and implemented by the U.S. government (ibid: 17f. and 128–130).

NOTES TO CHAPTER THREE

1. For a more detailed discussion of the social and economic model established in 1948, see Rovira Mas (1988), Sojo (1984) and Mesa-Lago (2000: 397–534). The classical study on Costa Rica's agrarian development is Seligson (1984), with p. 158ff. particularly focusing on the agrarian policy after 1948 and the establishment of the CNP.

2. The CACM share of Central America's total exports declined from the 1970 peak of 26.1 percent to 20.3 percent in 1979 (Bulmer-Thomas 1987: 208). In the late 1970s Honduras left the Common Market, and the Sandinista revolution in Nicaragua accelerated the political polarization between (and within) the countries of the region.

3. On the crisis and economic policies of the 1980s, see Fürst (1990), Rovira Mas (1989), Minkner (1990), Lizano (1987), Reuben Soto (1988), and Villasuso (1992).

4. As to the US-AID program, the Costa Rican share had dropped to a total of US$181 million between 1973 and 1981, before surging again, on the back of the Sandinista Revolution and the strong insurgency movement in El Salvador, to no less than US$1,764 billion between 1982 and 1995, the bulk of which was endorsed in the 1980s (Fox/Monge 1999: 43ff and 70).

5. The real social expenditures index (1980=100) had declined to 73.3 in 1982, but parallel to the beginning of the structural adjustment program, it increased significantly, reaching 125.3 in 1986 (Mesa-Lago 2000: 496). In the 1990s, financial aid from the United States was strongly reduced.

6. After Oscar Arias was awarded the Nobel Peace Prize in 1987, the Costa Rican model regained high international prominence. The Prize was given for Arias' initiative in the Central American peace process, but it was considered also to show a general appreciation for Costa Rica's civilian, peaceful and democratic system, which so sharply contrasted with the conflicts in the neighboring countries (cf. Arias 1987).

7. Cf. *La Nación*, September 11, 2000: "Desilusión y esperanza en ticos," September 12, 2000: "Fuerte apoyo a la democracia . . . pero crítica a los partidos y poderes del Estado," and September 14, 2000: "Dura crítica a partidos." A study by Seligson (2002) underscores the long-term forces leading to the erosion of support for Costa Rica's bi-party system over the last two decades.

8. This latter program has been the subject of one of the most prominent early empirical studies on telephone use in the Third World: the World Bank-sponsored survey underscored the great practical importance of telephone access for people in rural areas and showed benefits increasing with the distance from the country's urban centers (see appendix C in Saunders et al. 1994: 379–412).

9. Author's interview with Ricardo Monge in San José, April 27, 2001.

10. This section largely draws on Téramond (1995) (which also includes a detailed technical account of the process); see also Krennerich (2000:47ff.) and Sáenz/Galeano (1996).

11. Founding members of CRNet were the *Universidad de Costa Rica* (UCR), the *Instituto Tecnológico de Costa Rica* (ITCR), the *Universidad Estatal a Distancia* (UNED), the *Consejo Nacional de Investigaciones Científicas y Tecnológicas* (CONICIT), the *Fundación Omar Dengo* (FOD), the *Instituto Nacional de Biodiversidad* (INBIO), the *Instituto Centroamericano de Administración de Empresas* (INCAE), the *Instituto Interamericano de Cooperación para la Agricultura* (IICA), the *Escuela de Agricultura de la Región Tropical Húmeda* (EARTH), the *Centro Agronómico Tropical de Investigación y Enseñanza* (CATIE), the *Universidad Nacional* (UNA) and the country's Legislative Assembly, the *Asamblea Legislativa de la República* (cf. www.crnet.cr.).

12. Administratively, the domain name system established a second level order for different sectors, applying, in modified form, the international division of top level domains like .com, .edu, or .org to the national level. These second level suffixes come before the national '.cr' ending; the following second level domain names were established: .ed = educational sector (excluding higher education); .ac = academic sector; .co = commercial sector; .fi = financial

sector; .go = government; .or = organizations, especially NGOs and other civil society associations; .sa = health (Spanish: *salud*) sector. The typical three-step combination of Costa Rican Internet addresses results in the individual name plus the first and second level domain name—for instance for the Ministry of Science and Technology (MICIT): www.micit.go.cr.

13. The RACSA data are from an interview with the company's manager, Carlos Moreno, conducted by Michael Krennerich, January 1, 2000; for the data on the CRNet, see Téramond (2000:86). It is important to remember that the number of accounts is not the number of users, since users share accounts. RACSA itself calculates a multiplication factor of five, suggesting 150,000 users, a figure in line with ITU data (ITU 2000: 10).

14. Vanessa Castro, in an interview in Fundación Acceso (2000: 71); the countries she had been invited to were Germany, France, Chile and Argentina.

15. The following sections draw on numerous interviews, as well as informal conversations led by the author during his field stay in Costa Rica in April 2001; they also draw on documents and statements of diverse institutions and social groups engaged in the conflict, in part obtained directly from these entities, in part found in existing archives of the Fundación Acceso (San José), the Costa Rican 'Revista sobre Movimientos Sociales' and others; on a systematic analysis of the country's leading newspaper, La Nación, for 1998, 1999 and 2000, and the press bulletins of the Legislative Assembly (www.racsa.co.cr/asamblea/); on a comprehensive chronology of events as well as interviews with leading actors in Fundación Acceso (2000); and, finally, on secondary literature, of which the accounts of the failed liberalization initiative given by Monge (2000) and Krennerich (2000) merit particular mention.

16. The term *Combo* alludes to the package deals offered by fast-food stores at discount prices ("a hamburger, french fries and a soft-drink for only . . .") ; when it was first coined in regard to the combined ICE law bills it had an ironic touch to it, but it soon became the generalized term for the bill whose official name is 'Ley para el mejoramiento de servicios públicos de electricidad y telecomunicaciones y de la participación del Estado'; for the full text, see www.asamblea.go.cr/boletines/Proyectos/ 13873r9.doc .

17. Besides the three delegates of the left-wing *Fuerza Democrática*, this was one representative from each of the *Partido Acción Laborista Alajuelense*, the *Partido Renovación Costarricense*, the *Partido Integración Nacional*, and the *Movimiento Libertario*.

18. In this, the government followed an estimate of the *Coalición Costarricense de Iniciativas de Desarrollo* (CINDE), which spoke of the creation of 6,587 new jobs in the sector within three years of liberalization (MIDEPLAN 2000: 3).

19. Pointing to a supposed manipulation of minors, President Miguel Angel Rodríguez tried to downplay the dimension of the anti-Combo demonstration by saying: "The protests are by the same people that have gone to many demonstrations under this and the preceding governments. The only difference is that they have found an additional element: they have taken the kids out of school and put them on the streets, and that is what makes up the mass of the protests" (La Nación, March 26, 2000).

20. In retrospective, Ricardo Monge, who had a lead-role in the draft of the telecommunications part of the bill, gave much of the responsibility for the failed telecommunications reform to its combination with the other two law projects, concerning electricity and the ICE, where the government blundered in 'selling' it to the public: "The government never transmitted convincingly that its aim was not to sell or destroy the ICE" (Monge in an interview with the author, April 27, 2001 in San José).

21. Interview with the author, April 28, 2001 in San José.

22. Interview with the author, April 27, 2001 in San José.

23. Interview with the author, April 30, 2001 in San José.

NOTES TO CHAPTER FOUR

1. For this and the following paragraphs, see ICE (1997c), www.ice.go.cr, Pratt (2000) and http://icetel.ice.go.cr/esp/proyectos/procable.html .

2. In 2002, Costa Rica was also connected to a second submarine fiber optic cable called 'ARCOS' (Americas Region Caribbean Optical-ring System), a ring cable stretching 8,600 kilometers and connecting the United States and 17 countries in the Caribbean or with coastlines on the Caribbean. In addition, connection to a third submarine cable reaching out to the U.S. West Coast and across the Pacific Ocean was projected for late 2002. However, the operating company 'Global Crossing' had to declare insolvency in the same year, leading to much uncertainty about the Costa Rican investment in the cable (La Nación, February 1, 2002; see also La República, June 26, 2001). But even without this project, between May 1998 and May 2002, Costa Rica's installed capacity in submarine cables leaped from 10 megabits per second to 954 megabits per second, an increase of 9,440 percent.

3. Iván Rojas, the Ministry's director for information technology, acknowledged in an interview with the author on April 23, 2001 that this meant the relegation of other areas of science and technology to secondary priority, as academics in the affected fields had criticized.

4. For a comprehensive description of the project, see MICIT 2001 and the Science Ministry's homepage under www.micit.go.cr. Again, in the practical implementation of the project, sponsorship by transnational companies has been central; for the pilot project of the Advanced Internet Network, its technological core pieces (the routers) were donated by CISCO. These donations are not without self-interest, as they set technological standards which greatly help the donating companies to obtain future orders.

5. The Advanced Internet's pilot project that began in April 2001 connected 700 companies and institutions in the capital region to the new broadband network. A year later, the ambitious expansion project got underway aiming for DSL access for 84,000 clients nationwide—if reached without much delay, certainly a high percentage by international standards.

6. Cf. Saunders et el (1994: 397).The author recalls his own experience in the provincial town of Puerto Viejo de Sarapiqui, where he spent three months on a field study in 1989; at that time, it was a small town at the agrarian frontier whose only public phone operated in a small, wooden, make-shift shop on the

main road. When visiting Puerto Viejo again in 2001, the town had grown in size and wealth, and in the store windows household electronics had displaced the *campesinos'* machetes. On the main road, a few blocks from where the telephone stand once was, a brand-new Internet café had been set up, with a satellite connection providing parallel Internet access at eight computers.

7. Information provided by Hernando Pantigoso, the leading advisor for technology to Costa Rican President Miguel Angel Rodríguez (1998–2002) and the man responsible for the government's Digital Agenda program; interview by the author on April 30, 2001 in San José. The predominant position on the web of the country's leading newspaper is significant for the fact that, for Costa Rica, the Internet tends to underscore rather than challenge existing media structures.

8. This public-private partnership program, which channeled many of the country's social programs, has come under sharp criticism for using up to 50 percent of its funds on the staff's salaries and consulting fees (Gutiérrez Araya 2002).

9. This is reflected even in the Foundation's unusual website address 'www.entebbe.org': 'Entebbe' is the name of the residential zone in San José in which Figueres lives. (The LINCOS program has its own website under www.lincos.net.)

10. Proyecto de Ley - Ley de Derecho de Acceso a Internet, Expediente No. 14.029, introduced by Alvaro Trejos Fonseca (PUSC), August 23, 2000.

11. See e.g. the 'Plan to save the ICE' elaborated by the 'social sector' representatives in the 'mixed commission' established after the failure of the *Combo* law (La Nación, September 12, 2000).

12. This and the following analysis of the CAFTA agreement largely draw on the press coverage in *La Nación* (www.nacion.com), the Semanario Universidad (www.semanario.ucr.ac.cr), the *Tico Times* (www.ticotimes. net) and the *New York Times;* the document of the CAFTA agreement signed between Costa Rica and the United States has been published by the Costa Rican Ministry of Foreign Commerce on its website: www.comex.go.cr.

13. Author's calculation on the basis of the data provided by the Central Bank (for total of exports from free zones) and Procomer (for Intel exports).

14. The study by Larraín et al. (2000) is the most comprehensive analysis of the Intel investment in Costa Rica; one of its authors, Andrés Rodríguez-Clare, is the son of Miguel Angel Rodríguez, Costa Rica's President from 1998–2002. In this period he also served as coordinator of the Presidential advisory staff and it can be assumed that he had preferential access to the Intel executives as well as to the company's data. On Costa Rica's attraction of the Intel investment, see also the World Bank study by Spar (1998) and the account by Nelson (1999).

15. Author's interview with Ricardo Monge, April 23, 2001 in San José.

NOTES TO CHAPTER FIVE

1. A panorama of history, politics, economy and society of Cuba is presented in Hoffmann (2000).

2. Author's interview with Hans-Olaf Henkel, October 7, 1999 in Berlin.

3. The most prominent programmatic statement is Fidel Castro's court room defense after the Moncada attack of 1953, which became widely publicized under the title "History will absolve me" (e.g. Castro 1993b). In this, Castro decries Batista's putsch of 1952 and declares the necessity of profound social reform. He legitimates his revolutionary call to arms with quotes from John Milton, John Locke, Jean-Jacques Rousseau, Thomas Paine, the 1776 Declaration of Independence of the United States of America, and the Human Rights declaration of the French Revolution of 1789 (ibid: 104–106), but there is not a single reference to Marx or any other of the 'classic' authors of socialist or Communist inclination.

4. The 'revolutionary mass organizations' are: the Committees for the Defense of the Revolution (CDR), which are supposed to be the watchful eyes and ears of the Revolution in every block and every village; the Union of Cuban Workers (CTC); the Federation of Cuban Women (FMC); the National Small Peasants' Association (ANAP); the Federation of University Students (FEU); the Federation of Secondary School Students (FEEM); and finally, for the youngest, the 'Organization of Pioneers José Martí.'

5. In 1969 Hans Magnus Enzensberger concluded his remarkable essay on the Cuban Communist Party by stating: "Only one thing the Party definitely is not: the political power. Political power in Cuba lies exclusively in the hands of a tiny number of persons around Fidel, for whom no Party discipline is relevant but only their personal loyalty towards the *Comandante en Jefe.* (. . .) Fidel needs the Party but he doesn't like it. He finds it annoying. He hardly goes to its meetings. He cannot do without its apparatus, but he fears it becoming a mill-stone around his neck. With great endurance he runs away from the vanguard he calls upon. Never will it catch up with him" (Enzensberger 1969: 215). Three decades later, Rafael Rojas (1997: 25) notes essentially the same dynamic when he writes: "It is obvious that the Communist Party of Cuba is not a Gramscian or a Leninist institution. Fidel Castro's power is not delegated through institutions, but through persons."

6. See the exchange of letters between Kennedy and Khrushchev, as documented in Chang/Kornbluh (1992: 223–227).

7. In the 1960s, different economic approaches were debated, most prominently reflected in the so-called 'planning debate,' involving Che Guevara on the one side and foreign trade minister Alberto Mora and the French economist Charles Bettelheim on the other—see Bettelsheim et al. (1969), for example. Whereas Mora and Bettelheim argued for the necessity of introducing the "economic calculus" method with some financial autonomy for the production units and material incentives for the workers, Guevara argued for a budgetary system of centralized planning and the insistence on moral incentives to foster revolutionary *conciencia* (consciousness). Fidel Castro tended to maintain an ambivalent position above the disputing forces, pushing the balance at times in favor of one side, at times in favor of the other.

8. According to an estimate by Mesa-Lago (2000: 378), the differential between the 'just prices' of Soviet-Cuban trade and world market prices amounted to almost US$40 billion between 1960 and 1990. In addition,

loans equivalent to more than US$25 billion were disbursed (ibid), and these to a great extent were not paid back. Moreover, Cuba had considerable hard currency earnings from 're-exporting' 2–3 million barrels of Soviet oil annually in the late 1970s and 80s. According to a Western estimate (ibid), Soviet military aid amounted to US$13.4 billion between 1960 and 1985. It should be noted that all these figures are only very rough estimates, since the conversion of calculations based on Soviet transfer roubles into U.S. dollars leaves ample room for different interpretations.

9. The lack of tolerance also affected other areas. For instance, people who publicly defended their religious faith—be they Catholics, Protestants or followers of the syncretistic Afro-Cuban *Santería*—generally faced considerable social pressure and discrimination. Another prominent example was the aggressive discrimination against homosexuals, an aspect which was brought into public debate in Cuba in the mid-1990s by Tomás Gutiérrez Alea's award-winning movie 'Strawberry and Chocolate.'

10. The name itself indicates a significant change in the community: the transformation from 'Cubans' to 'Cuban-Americans.' In other words, while still strongly attached to their country of origin, the Cuban community debates between its identity as 'exiles,' which implies a rather temporary status, or as an immigrant group, as the term 'Cuban-Americans' implies. For the change in the identity of the Cubans in the United States, see Angeles Torres (1988), Rieff (1995), Forteux (1999) and Portes/Stepick (1993).

11. Fidel Castro used this expression for the first time in the inauguration of the IX International Fair in Havana, November 3, 1991, according to Alonso (1992: 164).

12. Since this touched upon a fundamental part of the Revolution's social accomplishments, at first the government denied categorically that the outbreak of the disease could have been a consequence of deficient nutrition (at one point even putting forward the thesis that the epidemic disease might have been deliberately introduced as a sabotage act from abroad). However, a comprehensive study of an expert commission of the Pan-American Health Organization confirmed the causal impact of the lack of proteins and vitamin B; now this explanation has also been accepted by the Cuban authorities (cf. UN Department of Humanitarian Affairs 1993, and 'die tageszeitung' [Berlin], January 22, 1996: 13).

13. For a comprehensive assessment, elaborated in cooperation with official Cuban sources, of Cuba's economy and economic policy in the first half of the 1990s (including a statistical as well as a legal annex), see CEPAL (1997). See also Dirmoser/Estay (1997) and Hoffmann (1995 and 2001).

14. Author's interview with José Luis Rodríguez, the country's current Economy Minster, published in 'die tageszeitung' (Berlin), November 11, 1993.

15. The amount of money in circulation increased by 47 percent between 1989 and 1991 while consumer sales simultaneously fell 30 percent, according to Cuban estimates (Carranza 1992: 139).

16. In the first year after the legalization of U.S. dollars, the balance of payments for 1994 showed an entry of US$574.8 million under the heading "current transfers" which, as is expressly explained, is "mainly due to the income from

donations and remittances" (Banco Nacional de Cuba 1995: 20f). For 2001, the Cuban Central Bank showed "current transfers" of US$812.9 million, which essentially are remittances plus informal tourism earnings (Banco Central de Cuba 2001).The estimate of US$1,100 million is from CEPAL for 1996 (CEPAL 1997, Cuadro A.1). For a discussion of different estimates and their methodology, see Monreal (1999: 74). It should be noted that family remittances from Cuban-Americans in the United States are exempt from the embargo up to a certain amount; in addition, much of these transfers pass through informal channels and are not registered by U.S. authorities.

17. See also the calculations in Carranza (1992: 153). Though González Gutiérrez (1997) speaks of a 'submerged economy,' generally the term 'informal economy' has become accepted in Cuba, since Carranza first wrote about "the necessity of recognizing the existence of an informal sector" (Carranza 1992: 153). When the government set up official state exchange bureaus in the fall of 1995, even the PCC organ *Granma* announced that they "offer the same rate which rules in the informal foreign exchange market" (*Granma International,* November 1, 1995)—a remarkable term for what before was referred to as criminal speculative activities against the socialist economy.

18. The central publication of Cuba's economic reform debate was the work of Carranza et al. (1995) from the Center for the Study of the Americas (CEA) in Havana. In this, the authors pleaded for a gradual but coherent reform of the economic model, for "moving from the classical model of socialism to another model of socialism (. . .) that accords the market an active although not exclusive or dominant role in the allocation of resources and the general functioning of the economy." At the heart of their proposal was a comprehensive monetary reform to overcome the dollar-peso dualism and to re-establish a national currency of value. On Cuba's monetary dualism, see also Ritter (1995) and Marquetti (1995). For a broader presentation of the academic reform debate and its major publications, see Hoffmann (1996).

19. This was particularly the case during the *Llamamiento*—the call to articulate issues felt as pressing within the Party structures—before the IV Party Congress in 1991; internationally, the case of the Writers' Union President Lisandro Otero, who argued for the re-opening of international agricultural markets, gained some prominence (Otero 1993).

20. As a consequence, the Center for the Study of the Americas (CEA) was put under close Party control, its Director was fired and all its leading academics—among them Carranza, Monreal, Dilla and R. Hernández—were transferred to other institutions.

21. Ibid; this phrase became the headline under which the Party newspaper published the announcement. Whereas before the opening of agricultural markets the government's discourse had maintained that the state supplies distributed via the rationing system were entirely sufficient, now the population could read in the official press that the new markets will help "solve the nutritional problems of the families, since, as is known, the basic rations of the *bodegas* [the distribution centers for rationed products] suffice only for the first half of the month" (Bohemia, October 14, 1994).

22. So named for the Democratic Representative Robert Torricelli who sponsored the bill; official name: Cuban Democracy Act (U.S. Congress 1992).

23. So named for its initiators, the Republican Senator Jesse Helms and the Democratic Representative Dan Burton; official name: The Cuban Liberty and Democratic Solidarity (LIBERTAD) Act of 1996; cf. U.S. Congress 1996. The passing of the Helms-Burton law was preceded by a manifest crisis in U.S.-Cuban relations, when the Cuban Air Force shot down two Cessna planes of the exile organization *Hermanos al Rescate* on February 24, 1996. The four pilots died. While the *Hermanos* had pursued an active strategy of violating Cuban air-space, it is an object of debate whether they were in Cuban airspace at the moment of being shot down. In any case, the deliberate downing of the planes led to outrage in the U.S. public far beyond the Cuban community. Whereas before it had been very much an open question if the Helms-Burton Law would appeal to a majority in Congress, it took no more than ten days after the incident to pass it with broad majorities in both houses of Congress (see Whitehead 1996, Pérez-Stable 1996; see also the interview with Fidel Castro in *Time,* March 11, 1996: 22).

24. The author's personal observations were not only formed while participating in a number of local and municipal acts of this type, but also from attending one of the most prominent mass demonstrations of recent years: the commemoration of the 40th anniversary of the proclamation of the socialist character of the Cuban Revolution, on 16 April 2001, at Havana's central 23rd street and with Fidel Castro as the main speaker. Large numbers of workers and students had gotten off early from school or work and arrived punctually in groups. However, by the time Fidel spoke, most of them had already left. The two street blocks closest to the speakers' stage were largely filled with members of the military or militia, which remained until the end, cheering and raising their rifles (producing the camera shots for the national and international media). Beyond these military spectators, among those that remained more were chatting with their neighbors than listening to the words of the *Comandante*.

25. See Dilla (1999 and 2000). The publication of this thesis was a factor that led to Dilla's dismissal from his academic employment in Cuba. He left the island and is now working for FLACSO in the Dominican Republic.

26. This line of conflict is apparent in a quote from a Miami Herald (March 17, 1990) editorial: "This linkage between a Federal agency and the politically ambitious head of a politically active foundation should be severed. It's unacceptable in principle, and it's damaging to Radio Martí's credibility" (cited in Press 1996a: 18).

27. Money transfers remained subject to specific embargo restrictions. Thus, family remittances from Cuban-Americans, though allowed up to a certain amount, may not be sent through internet services but have to be transferred through more traditional (and often, more costly) ways.

28. This follows the accounts of relatively high-ranking managerial personnel in the Cuban NICT sector, transmitted to the author by personal communication under condition of anonymity. As to the early withdrawal of *Domos,* different explanations circulate: a) the Mexican currency crisis put the

company under unexpected financial pressure so that it was unable to comply with its investment plans; b) *Domos* never planned to comply with its investment plans, but from the very beginning counted on selling its shares at a profitable price; and c) much of the capital behind *Domos* was U.S. capital that tried to circumvent the restrictions that prevented U.S. companies from investing in the Cuban telecommunications sector, and that withdrew due to political pressure from Washington

29. A popular trick was to introduce a prepaid card, dial an international number and after the connection was established, 'tap' the cable of the public telephone and connect it to a regular residential telephone the person carried in a bag; the prepaid-card was withdrawn but the line was maintained and the person kept on talking as long as he wanted without paying.

30. Information from the *Gran Kaiman Teleco* representatives at the *Informática 2000* fair in Havana from May 22–26, 2000.

NOTES TO CHAPTER SIX

1. Cited from Palaima's presentation of LANIC at the XXII Latin American Studies Association (LASA) Congress in Miami, Florida (March 16–18, 2000). The e-mail server was later discontinued, due to the decreasing demand and the fact that the server presented a security problem for LANIC since it was particularly vulnerable to hacker attacks (ibid).

2. After this initial study a number of other RAND publications took up and expanded the argument of increased communications as a means "to open up Cuba's closed system"; see González/Ronfeldt (1994), Arquilla (1994) and González (1996).

3. Cisler had been invited to a librarians' conference in Havana to make a presentation on the possibilities offered by the Internet for this sector. Since he could not access the Internet from Cuba for his presentation, he showed a video showing a computer monitor during an Internet session in which sites and materials relevant for the conference participants were accessed (Cisler 1994).

4. Personal information from a Cuban networking expert in a leading position, Havana, May 2000.

5. See e.g. Castro 1997. Even in the profound crisis of the early 1990s, although Castro called for substituting tractors with animals in agriculture, this was no turn to a luddite ideology: this emergency measure went parallel to continued investments in an ambitious high-tech sector in biotechnology and pharmaceuticals. Also in 1991, in the midst of the crisis, in Havana a central *Palacio de la Computación* was inaugurated as a learning facility for Cuban youths. On signing the guest book, Fidel Castro, directing himself to Cuba's young generation, wrote "I envy you!—Fidel Castro, 7 March 1991," a dedication that since then has been exhibited at the institution's main entrance hall (author's on-site visit on May 25, 2000).

6. At a meeting of the Forum of Latin American and Caribbean Networks (*Foro de Redes Latinoamericanas y del Caribe*) in April 1996, CENIAI Director Jesús Martínez for the first time announced that concrete plans

were elaborated (and had the government's support) for establishing an IP connection to North America (Press 1996: 50)

7. Official title: *Decreto No. 209: Del acceso desde la República de Cuba a redes informáticas de alcance global*; published in: Gazeta Oficial No. 27, 13 September 1996. Cf. also Valdés/Rivera (1999:147f.), Press (1998), as well as www.cubanic.cu/documentos. The following paragraphs are essentially based on Valdés (1997: 7) and on official Cuban documents such as MIC (2000) and Ronda (2000), as well as personal information from Cuban personnel working in NICT related institutions.

8. The text of the government note announcing the creation of the MIC is cited in Gil Morell (2000: 5); for a presentation of the Ministry, its structures and missions, see Rodríguez Gavilán (2000) and Linares (2001) and, of course, the Ministry's website: www.mic.gov.cu.

9. In 1999, CENIAI (now under the name of *CENIAInternet*), Cuba-NIC and the CD-Rom producing company CEDISAC became integrated parts of the new company *Citmatel;* this, however, did not affect their Ministerial affiliation, since *Citmatel* is—as the name already indicates—assigned to the Ministry of Science, Technology and the Environment (CITMA).

10. The number of direct employees of the Ministry is officially as low as 185 (that is less than were previously employed in the MINCOM alone); the number of employees and workers of all institutions and companies assigned to the Ministry is given as 39,072 (Linares 2001). According to the official data, the Ministry's hard currency revenues—presumably to a large extent resulting from ETECSA's earnings from international telephone calls—amount to US$500 million annually (ibid.).

11. A number of projects to provide online Government-to-Citizens (G2C) services are planned (cf. del Puerto 2000); however, given the low rate of NICT access in the population, only a very limited impact of these initiatives should be expected. Chances seem to be higher for e-government initiatives directed at the state and joint-venture enterprises.

12. Another example is given by the declarations of Sergio Pérez, the director of the state Internet service provider 'Teledatos,' in *Granma,* 7 February 2001: "Cuba is a poor and blockaded country—if in Cuba food is rationed and there are shortages in medicine, how shall the citizens' access to the Internet not be limited, too?" (cited in: Notimex 2001); cf. also Grogg (2000).

13. With capital letters in the original; not to be confused with the CubaNet website, which is established in Miami to host articles from dissident journalists on the island (see chapter 4.4.9)

14. Originally these were two separate ISPs, with *Teledatos* belonging to the *Grupo de la Electrónica del Turismo* (GET) and being almost exclusively dedicated to the country's tourism industry.

15. Most of the information in this section is based on numerous conversations with officials and employees working in the sector during the author's field visits to Cuba in May 2000 and April 2001, with anonymity being an indispensable condition for obtaining this type of information.

16. Conversation with this student included access to her diploma thesis in Havana, May 2000.

17. Websites chosen for this test were the above mentioned sites of the dissident journalists (www.cubanet.org) and of the Miami daily *El Nuevo Herald* (www.elherald.com), as well as the homepages of the Cuban-American National Foundation (www.canfnet.org) and the oppositional daily Internet newspaper edited in Madrid, *Encuentro en la Red* (www.cubaencuentro.com).
18. Observations and data from author's field visits.
19. Author's on-site visit and interview, April 14, 2001.
20. Note that 'Security' is written with a capital letter in the original. In the following, the regulations state in particular that any encryption of messages is forbidden and that "the text of the messages may not contain obscene words or phrases which are contrary to the socialist order and morality" (Correos de Cuba 2002). Similarly, in the case of full Internet access, "it is forbidden to access sites and web-pages of obscene character or contrary to the socialist order and morality" (ibid), leaving open to interpretation what precisely is to be understood by these terms.
21. A particular role is played by the *Infomed* network in the 'internationalist missions' in which 1,751 medical doctors and other medical staff have been sent to work in other Third World countries (*Granma*, August 1, 2000). In these missions, which have become an important element in Cuba's foreign policy in the second half of the 1990s, the *Infomed* computer network allows a maximum of connection to the structures of the Cuban health system.
22. Author's conversations with medical staff in Havana, April 2001.
23. For instance, in 2001, *InfoMed* was among the finalists of the Association for Progressive Communications' (APC) 'Premio Betinho,' which is awarded to initiatives making exemplary social use of the NICT (cf.www.apc.org/english/betinho/2001/). In 2002, *Infomed* won the similarly defined 'Stockholm Challenge' prize in the category of health (www.challenge.stockholm.se).
24. Cf. Comisión Cubana de Comercio Electrónico (2000), Herrera/García (2000:28f.) and Lage (2000:3). An overview of judicial problems presented by the NICT is given by the President of the Cuban Association for Law and Informatics, Amoroso Fernández (1999).
25. The reach of the embargo also extends to all other kind of Internet business involving U.S. companies and Cuban-produced goods. For instance, the Internet auctioning site E-bay is not allowed to auction any item produced in Cuba (cf. http://news.bbc.co.uk/1/hi/talking_point/644928.stm).
26. Informal conversations at the *Informática 2000* Fair in Havana, May 2000.
27. Stephane d'Amours, co-founder of the Montreal-based company named 'Silicon Island,' which has established a business partnership with the Cuban state-run software company CenterSoft (González, A. 2001; see also Ashby 2001).
28. Conversation with the author, 23 May 2000. In addition, the information on *CubaWeb* draws on Press (1998) and Valencia Almeida (1997).
29. The 'Asociación de Periodistas Independientes de Cuba' was founded by Yndamiro Restano Díaz; re-christened 'Agencia de Prensa Independiente de Cuba' (APIC), it still exists today. Restano Díaz himself left the island in 1995.

30. E-mail-communication from April 4, 2000, following an interview with the author in Miami on March 21, 2000.
31. The early study of Hill/Hughes (1993: 47–108) on "Citizen Activism in the Age of the Internet" includes an empirical analysis of the Usenet discussion groups, asking if there is evidence of any democratizing effect. However, at the time of their study, not only was such Internet use in Third World countries limited to a handful of activists, but the authors themselves also noted a limitation of their approach's focus on the Usenet newsgroup on Cuba (soc.culture.cuba). Here, the very high level of 'pro-democracy' content of critical and oppositional writing stemmed from the simple fact that it was almost exclusively used by the exile community (ibid: 90).
32. Interview with CubaNet Vice President Rosa Berre in Miami, March 20, 2000.
33. According to CubaNet Vice President Rosa Berre, the funding from NED was around US$20,000 in 1995; around US$40,000 in 1996; around US$75,000 in 1997 as well as in 1998; and around US$65,000 in 1999. From US-AID, CubaNet received around US$98,000 in 1999 and around US$245,000 in 2000. While US-AID funds cannot be transferred to Cuba, NED funds can be transferred to Cubans on the island and, according to information provided by the NED website, in 2000 CubaNet received US$64,000, and in 2001 US$35,000 to "supply humanitarian aid and material assistance to Cuban independent journalists in Cuba" (www.ned.org/grants/grants.html).
34. Interview in Miami, March 20, 2000.
35. Words of Fernando Rojas, spokesman of the Cuban delegation to the International Book Fair in Guadalajara in 2002, cited in Encuentro en la Red (2002). 'Encuentro' actually has provided a list of its financial supporters, which includes the National Endowment for Democracy, the Ford Foundation, the Open Society Institute (all from the United States), the Agency for International Cooperation of the Spanish Foreign Ministry, the Olof Palme International Center (Sweden), and others (ibid). According to its annual report, the National Endowment for Democracy supported the *Asociación Encuentro de la Cultura Cubana* with US$83,000 in 2000 (www.ned.org/grants/grants.html). Far more important financing for the Internet newspaper project, amounting to more than US$800,000, has come from the Ford Foundation (see its grant database under www.fordfound.org). However, for the Cuban government, it may not have been as opportune to take the Ford Foundation's financing as proof for its accusations of subversion, because a number of official Cuban institutions on the island also receive support from the Ford Foundation through their international cooperation projects, e.g. the National Small Peasants Association, ANAP, through a cooperation with the British NGO Oxfam etc. (cf. also for this the Ford Foundation's grant database).

Bibliography

Acosta, Dalia. 2001. *Kuba: Ein Panda-TV für jedes Wohnzimmer. China liefert der Zuckerinsel eine Million Fernsehgeräte.* Inter press service, October 26, 2001.

Aglietta, Michel. 1979. *A Theory of Capitalist Regulation. The U.S. Experience.* London: Verso.

Alexander, Robert. 1990. Import Substitution in Latin America in Retrospect. In *Progress Toward Development in Latin America. From Prebisch to Technological Autonomy,* edited by J.L. Dietz and D. Dilmus, 15–28. Boulder/London: Lynne Rienner.

Alfonso, Pablo. 2000. Cuba tiende la cortina de hierro por la Internet, in: *El Nuevo Herald,* Miami, October 9, 2000.

Alonso Becerra, Beatriz. 2000. Integración de tecnologías. *Ciencia, Innovación y Desarrollo* (5)2: 23–25.

Alonso, Aurelio. 1992. La economía cubana—los desafíos de un ajuste sin desocialización. *Cuadernos de Nuestra América,* (IX)19: 159–174. La Habana: CEA.

Altenburg, Tilman, Philipp Botzenhardt, Andreas Stamm, and Gundula Weitz. 2002. *E-Business und KMU. Entwicklungstrends und Förderansätze.* DIE Berichte und Gutachten 1/2002. Bonn: Deutsches Institut für Entwicklungspolitik.

Altenburg, Tilman, Wolfgang Hein, and Jürgen Weller. 1990. *El desafío económico de Costa Rica. Desarrollo agroindustrial autocentrado como alternativa.* San José: DIE.

Altenburg, Tilman, and Jörg Meyer-Stamer. 1999. How to Promote Clusters. Policy Experiences from Latin America. In *World Development,* Vol. 27, No. 9, 1693–1713.

Altet Casas, Vilma Isabel. 2001. *Las tecnologías de la información y las comunicaciones como factores de éxito en la nueva economía—América Latina en el contexto global.* Ponencia presentada en el XXIII Congreso Internacional de la Asociación de Estudios Latinoamericanos, LASA 2001, Washington D.C., September 6–8, 2001.

Altvater, Elmar. 1987. *Sachzwang Weltmarkt. Verschuldungskrise, blockierte Industrialisierung und ökologische Gefährdung. Der Fall Brasilien.* Hamburg: VSA.

Altvater, Elmar. 1991. *Die Zukunft des Marktes. Ein Essay über die Regulation von Geld und Natur nach dem Scheitern des "real existierenden Sozialismus."* Münster: Westfälisches Dampfboot.

Altvater, Elmar, and Birgit Mahnkopf. 1996. *Grenzen der Globalisierung. Ökonomie, Ökologie und Politik in der Weltgesellschaft.* Münster: Westfälisches Dampfboot. (In Spanish published as: *Las limitaciones de la Globalización—Economia, ecologia y política de la globalización.* México, D.F./Buenos Aires: Siglo XXI, 2002.)

Altvater, Elmar, Kurt Hübner, Jochen Lorentzen, and Raúl Rojas, eds. 1988. *Die Armut der Nationen. Handbuch zur Schuldenkrise von Argentinien bis Zaire.* 2nd ed. Berlin: Rotbuch.

Alvarez, Elena. 1997. La biotecnología en Cuba. Un potencial para nuevas ventajas comparativas. In *Economía y reforma económica en Cuba,* edited by D. Dirmoser and J. Estay, 269–289. Caracas: Nueva Sociedad. (First published in: *Economía Cubana.* Boletín Informativo No 8 del Centro de Investigaciones de la Economía Mundial, La Habana, August 1992.)

Amoroso Fernández, Yarina. 1999. Nuevas tecnologías de la información y la comunicación, valores humanos y derecho. In *Simposio Latinoamericano y del Caribe: Las tecnologías de información en la sociedad,* edited by Instituto Nacional de Estadística, Geografía e Informática. Aguascalientes (México): Instituto Nacional de Estadística, Geografía e Informática.

Amsden, Alice. 1989. *Asia's Next Giant. South Korea and Late Industrialization.* New York: Oxford University Press.

Anfossi Gómez, Andrea, and Clotilde Fonseca Quesada. 1999. *Hacia la construcción de un nuevo modelo de formación docente: el caso de la información educativa en Costa Rica.* San José: Fundación Omar Dengo.

Angeles Torres, María de los. 1988. From Exiles to Minorities. The Politics of Cuban-Americans. In *Latinos and the Political System,* edited by C.F. García, 81–98. Notre Dame: University of Notre Dame Press.

APC [Association for Progressive Communication]. 2001. *APC Internet Rights Charter.* Fourth Draft. http://www.apc.org/english/rights/charter.shtml.

Arias, Oscar. 1987. *Frieden für Zentralamerika.* Frankfurt a. M.: Vervuert.

Arquilla, John. 1994. *A Decision-Modelling Perspective on U.S.-Cuba Relations.* Santa Monica: RAND.

Asamblea Legislativa. 2000. Ley para el mejoramiento de los servicios públicos de electricidad y telecomunicaciones y de la participación del Estado. Informe de la Asamblea Legislativa de Costa Rica. *La Nación,* March 26, 2000.

Ashby, Timothy. 2001. Silicon Island: Cuba's Digital Revolution. *Harvard International Review* 23 (3): 14–18.

Azcuy Henríquez, Hugo. 1995. Estado y sociedad civil en Cuba. *Temas—Cultura, Ideología, Sociedad* No. 4: 105–110. La Habana.

Banco Central de Costa Rica. 2001. *Comentario Balanza de Pagos del 2000.* San José: Banco Central de Costa Rica.

Banco Central de Cuba. 2001. *Informe Económico 2001.* La Habana: Banco Central de Cuba.

Banco Nacional de Cuba. 1995. *Economic Report 1994.* La Habana, August.

Barnola, Luis, and Daniel Pimienta. 2001. MISTICA: A Collective Endeavor In Search of the Social Impact of ICTs in Latin America and the Caribbean. *Techknowlogia: International Journal of Technologies for the Advancement of*

Knowledge and Learning, Vol. 3, N° 4, July-August 2001. http://www.tech-knowlogia.org.

Bastos, Maria Ines. 1992. State Policies and Private Interests. The Struggle over Information Technology Policy in Brazil. In *Hi-Tech for Industrial Development. Lessons from the Brazilian Experience in Electronics and Automation,* edited by H. Schmitz and J. Cassiolato, 239–272. London/New York: Routledge.

Bate, Peter. 1999. *Costa Rica y su nuevo "café" para el siglo XXI. Invierte en su naciente industria informática.* Washington, D.C.: IDB. http://www.iadb.org/id-bamerica/Archive/stoies/1999/ esp/c1099f1.htm.

Becker, Jörg. 1985. *Massenmedien im Nord-Süd-Konflikt.* Frankfurt a. M.: Campus.

Bell, John Patrick. 1971. *Crisis in Costa Rica: The 1948 Revolution.* Austin: University of Texas Press.

Berlage, Michael, and Thomas Schnöring. *1992. Telekommunikation in Osteuropa. Bestandsaufnahme und Entwicklungstrends.* WIK Diskussionsbeiträge Nr. 93. Bad Honnef: Wissenschaftliches Institut für Kommunikationsdienste.

Bermellon, Yizzet, and Dodanys García. 2000. Cuba vende en la red. *El Economista de Cuba Online* 38. La Habana. http://www.eleconomista.cubaweb.cu/2000/nro38/30–114.html.

Bettelheim, Charles, Fidel Castro, and Ernesto Guevara. 1969. *Wertgesetz, Planung und Bewusstsein. Die Planungsdebatte in Cuba.* Frankfurt a. M.: Verlag Neue Kritik.

Bimber, Bruce. 2000. The Gender Gap on the Internet. *Social Science Quarterly,* vol. 81, N° 3: 868–876.

Blasier, Cole. 1987. *The Giant's Rival. The USSR and Latin America.* 2nd rev. ed. Pittsburgh: University of Pittsburgh Press.

Blau, Andrew. 1997. A High Wire Act in a Highly Wired World. Universal Service and the Telecommunications Act of 1996. In *The Social Shaping of Information Superhighways. European and American Roads to the Information Society,* edited by H. Kubicek et al., 247–263. Frankfurt a. M./New York: Campus/St. Martin's Press.

Boas, Taylor C. 1998. www.cubalibre.cu? The Internet and its Prospects for a Democratic Society in Cuba. *Stanford Journal of International Relations,* (1)1, Summer-Fall. http://www. stanford.edu/group/sjir/issues/1.1/www.cubalibre. cu/body.html.

Boas, Taylor C. 2000. The Dictator's Dilemma? The Internet and U.S. Policy toward Cuba. *The Washington Quarterly* (23)3, Summer: 57–67.

Bomkamp, Dana, and Maria Soler. 2000. *Information Technology in Cuba.* Washington, D.C.: American University. http://www.american.edu/carmel/ms4917a/HW percent20Mftg.htm.

Booth, John A. 1998. *Costa Rica. Quest for Democracy.* Boulder: Westview.

Borja, Arturo. 1995. *El estado y el desarrollo industrial. La política méxicana de cómputo en una perspectiva comparada.* México D.F.: Centro de Investigación y Docencia Económicas.

Bortagaray, Isabel, and Scott Tiffin. 2000. *Innovation Clusters in Latin America.* Paper presented at the 4th International Conference on Technology Policy and

Innovation, Curitiba, Brazil, August 28–31, 2000. http://in3.dem.ist.utl.pt/downloads/cur2000/papers/S11P01.PDF.

Bourne, Peter. 1988. *Fidel Castro.* Düsseldorf: Econ. (Orig. in English as: Castro. A biography of Fidel Castro. London: Macmillan, 1987)

Boyer, Robert. 1990. *The Regulation School. A Critical Introduction.* New York: Columbia University Press.

Brundenius, Claes, and Pedro Monreal. 1998. *The Future of the Cuban Economic Model: The Longer View.* Paper presented to the workshop on "Globalization, Changing Paradigms and Development Options in the Third World: Cuba and Vietnam," Copenhagen: Centre for Development Research, June 11–13, 1998.

Bulmer-Thomas, Victor. 1987. *The Political Economy of Central America since 1920.* Cambridge (U.K.): Cambridge University Press.

Bulmer-Thomas, Victor. 1998. The Latin American Economies, 1929–1939. In *Latin America. Economy and Society since 1930,* edited by L. Bethell, 65–114. Cambridge (U.K.): Cambridge University Press.

Burchardt, Hans-Jürgen, ed. 2000. *La última reforma agraria del siglo. La agricultura cubana entre el cambio y el estancamiento.* Caracas: Nueva Sociedad.

Camacho, Kemly. 2001. *The Internet: A Tool for Social Change? Elements of a Necessary Discussion.* San José: Fundación Acceso. http://www.acceso.or.cr/publica/socialchange.shtml.

Cancio, Wilfredo. 1996. El periodismo en Cuba—otra vuelta de tuerca. Prácticas comunicativas y desafíos profesionales bajo el modelo de prensa socialista. *Encuentro de la Cultura Cubana* 2, Madrid: 31–38.

Caprosoft. 2001. *La industria del software in Costa Rica.* Online ppt-document. http://www.prompex.gob.pe/prompex/Inf_Sectorial/Apesoft/Caprosoft.pdf.

Cardoso, Fernando Henrique, and Enzo Faletto. 1988. *Dependencia y Desarrollo en América Latina. Ensayo de interpretación sociológica.* 23rd ed. (First edition 1969). México: Siglo XXI.

Carranza Valdés, Julio. 1992. Los retos de la economía cubana. *Cuadernos de Nuestra América,* (IX)19, La Habana: CEA: 131–158.

Carranza Valdés, Julio, Pedro Monreal, and Luis Urdaneta Gutiérrez. 1995. *La restructuración de la economía cubana. Una propuesta para el debate,* La Habana: Editorial Ciencias Sociales. (Also under the same title by Nueva Sociedad, Caracas, 1997. In English as: Cuba: Restructuring the Economy—A Contribution to the Debate. London: Institute of Latin American Studies, 1998.)

Carvajal Pérez, Marvin. 2002. *La protección de los datos personales en Costa Rica.* Paper presented at "Jornadas Iberoamericanas de Protección de Datos," San Lorenzo de El Escorial (Spain), May 20, 2002. http://www.democraciadigital.org/ derechos/arts/0207datos.html.

Casacó, Luis. 2002. *Cuba's Software Industry.* Manuscript for a forthcoming book on the Cuban economy edited by Archibald Ritter, 284–297.

Castells, Manuel. 2000. *The Rise of the Network Society. The Information Age: Economy, Society, Culture, Vol. I.,* 2nd rev. ed. (First edition 1996). Oxford/U.K.: Blackwell.

Castells, Manuel. 1997. *The Power of Identity. The Information Age: Economy, Society, Culture, Vol. II.* Oxford/U.K.: Blackwell.

Castells, Manuel. 1998. *End of Millennium. The Information Age: Economy, Society, Culture, Vol. III.* Oxford/U.K.: Blackwell.

Castro, Fidel. 1972. *La Revolución Cubana.* La Habana: Ediciones Era.

Castro, Fidel. 1991. *Presente y futuro de Cuba. Entrevista concedida a la revista "Siempre!."* La Habana: Oficina de Publicaciones del Consejo de Estado.

Castro, Fidel. 1993a. Discurso en la clausura del 40 aniversario del asalto al Cuartel Moncada. *Granma,* July 28, 1993: 3–7.

Castro, Fidel. 1993b. *La Historia me absolverá.* Redacted and annotated version of the speech held in 1953. La Habana: Oficina de Publicaciones del Consejo de Estado.

Castro, Fidel. 1997. *La ciencia en Cuba: Impetuoso desarrollo.* Discurso en el acto por el Día de la Ciencia Cubana. La Habana: Editora Política.

Castro, Fidel. 2001. *Discurso pronunciado por Fidel Castro, Presidente de la República de Cuba, en el acto en conmemoración del aniversario 40 de la proclamación del carácter socialista de la Revolución, efectuado en la esquina de 12 y 23, Vedado, Plaza de la Revolución, el 16 de abril del 2001.* La Habana: MINREX. http://www.cubaminrex.cu/informacion/discfidel_12y23.htm.

Castro, Raúl. 1996. Informe del Buró Político en el V. Pleno del Comité Central del Partido, March 23, 1996. *Granma Internacional,* April 10, 1996: 4–8.

Castro, Vanessa. 2000. Entrevista a la Lic. Vanessa Castro (Transcript of Interview by Fundación Acceso, December 14, 1999. In *Regulación y uso de las nuevas tecnologías de información y comunicación (NTIC) en el cambio de los procesos de transformaciones políticas y económicas en América Latina: El caso de Costa Rica* (mimeo), edited by Fundación Acceso, 71–74. San José: Fundación Acceso.

CEDISAC/Prensa Latina. 1997. *Todo de Cuba* (CD-Rom). La Habana: CEDISAC.

CENAT [Centro Nacional de Alta Tecnología]. 1999. *Costa Rica en el siglo XXI: Tecnología de Información para el Desarrollo* (mimeo of conference contributions). San José: CENAT.

CEPAL [Comisión Económica para América Latina y el Caribe]. 1969. *El pensamiento de la CEPAL.* Santiago de Chile: Editorial Universitaria.

CEPAL. 1997. *La Economía Cubana: Reformas estructurales y desempeño en los noventa.* México D. F.: Fondo de Cultura Económica.

CEPAL. 1998. *CEPAL cincuenta años* (número extraordinario de la Revista de la CEPAL). Santiago de Chile: CEPAL.

CEPAL. 2002. *Estudio Económico de América Latina y el Caribe 2001–2002.* Santiago de Chile: CEPAL.

Chang, Laurence, and Peter Kornbluh. 1992. *The Cuban Missile Crisis 1962.* A National Security Archive Documents Reader. New York: The New Press.

Chéneau-Luquay, Annie. 2002. Afrika am Netz. Kabel, Satellit und Telezentren. *Le Monde Diplomatique,* Januar 2002: 11.

Cisler, Steve. 1994. *Our LAN in Havana: Networking with People and Computers in Cuba.* http://www.geo.unipr.it/~davide/cuba/computer/LAN-trip.html.

Cleaver, Harry. 1995. *The Zapatistas and the Electronic Fabric of Struggle.* http://www.eco.utexas.edu/faculty/Cleaver/zaps.html. (A shorter version was published in: John Holloway and Eloína Peláez, eds. Zapatista! Reinventing Revolution in Mexico. London: Pluto Press, 1998.)

Clinton, Bill. 2000. *America Has a Profound Stake in what Happens in China* (Transcript: President Clinton on U.S.-China Trade Relations). The White House, Office of the Press Secretary, March 8, 2000. http://usinfo.state.gov/regional/ ea/uschina/clint308.htm.

Clippinger, John. 1976. *Who gains by Communication Development?* Studies of Information Technologies in Developing Countries. Working Paper 76–1. Cambridge: Harvard University Press.

Comisión Cubana de Comercio Electrónico. 2000. *Nuevo Plan de Trabajo de la Comisión Cubana de Comercio Electrónico.* Unpublished working document.

Comité Estatal de Estadística. 1989. *Anuario Estadístico 1989.* La Habana: Comité Estatal de Estadística.

Córdova, Armando. 1973. *Strukturelle Heterogenität und wirtschaftliches Wachstum.* Frankfurt a. M.: Suhrkamp.

Correos de Cuba. 2002. *Reglamento interno de las oficinas de correos para el servicio de navegacion y correo electronico.* http://www.correosdecuba.cu/ term&cond.asp.

CPSR [Computer Professionals for Social Responsibility]. 2002. *Women and Computing.* http://www.cpsr.org/program/gender/index.html.

del Puerto, Roberto. 2000. *Estrategia de informatización del país.* Ponencia en la Primera Reunión Nacional de Informatización de la Sociedad, September 29–30, 2000. La Habana: Ministerio de la Informática y las Comunicaciones. Unpublished working document.

del Puerto, Roberto. 2004. *Intervención en el programa de televisión 'Mesa Redonda Informativa: Internet: Mitos y realidades. Cuba en la Red'* (Cubavisión), January 22, 2004. Accessed via World Wide Web: http://www. mesaredonda.cu.

Delgado Q., Félix. 2000. La economía de Costa Rica en 1999. In *Los retos políticos de la reforma económica en Costa Rica,* edited by R. Jiménez, 3–44. San José: Academia de Centroamérica.

Dilla, Haroldo. 1999. Genossen und Investoren. Der Übergang Kubas. *Prokla— Zeitschrift für kritische Sozialwissenschaft* 117(29)4: 627–646.

Dilla, Haroldo. 2000. The Cuban Experiment. *Latin American Perspectives* 110, Vol. 27, No.1: 33–44.

Dilla, Haroldo, ed. 1995. *La democracia en Cuba y el diferendo con los Estados Unidos.* La Habana: CEA.

Dilla, Haroldo, ed. 1996. *La participación en Cuba y los retos del futuro.* La Habana: CEA.

Dirmoser, Dietmar, and Jaime Estay, eds. 1997. *Economía y reforma económica en Cuba.* Caracas: Nueva Sociedad.

Drake, William. 1997. Public Interest Groups and the Telecommunications Act of 1996. In *The Social Shaping of Information Superhighways. European and American Roads to the Information Society,* edited by H. Kubicek et al., 173–198. Frankfurt a. M./New York: Campus/St. Martin's Press.

Drake, William, Shanthi Kalathil, and Taylor C. Boas. 2000. *Dictatorships in the Digital Age: Some Considerations on the Internet in China and Cuba.* Washington D.C.: The Carnegie Endowment for International Peace. http://www.cisp.org/imp/october_2000/10_00drake.htm.

Durán, Alfredo. 1995. Testimony to the U.S. Senate, June 14, 1995. *Cuban Affairs/Asuntos Cubanos,* (II)1–2, Spring/Summer: 23.

Eckstein, Susan Eva. 1994. *Back from the Future. Cuba under Castro.* Princeton: Princeton University Press.

EFE. 2004. *Cuba afirma que tiene 98.000 usuarios de internet.* EFE news, January 16, 2004.

EIU [The Economist Intelligence Unit]. 2002. *Country Report Cuba.* London: The Economist.

EKA. 1999. Mitos de la reforma en telecomunicaciones. *EKA—Portfolio de Inversiones,* N° 141, 6/99: 12–13.

El Nuevo Herald. 1999. *El registro de propiedades será gratuito y en la internet.* September 22, 1999.

El Nuevo Herald. 2001a. *Los disidentes abren una página en la internet.* December 8, 2001.

El Nuevo Herald. 2001b. *Cuba le está sacando provecho a la internet.* September 2, 2001.

El Semanario. 2001. *El Combo—un año después.* Suplemento. April 2001.

Encuentro en la Red. 2002. *Editorial: La Asociación Encuentro rebate acusaciones de La Habana.* December 6, 2002. http://www.cubaencuentro.com.

Encuentro en la Red. 2004. *Editorial: ¡Vade retro, Internet!* January 23, 2004. http://www.cubaencuentro.com.

Enzensberger, Hans Magnus. 1969. Bildnis einer Partei. Vorgeschichte, Struktur und Ideologie der PCC. *Kursbuch* 18: 192–216.

Ernst, Manfred, and Sönke Schmidt. 1986. Die costaricanische Linke in historischer Perspektive—zwischen revolutionärer Utopie und Pragmatismus. In *Demokratie in Costa Rica—ein zentralamerikanischer Anachronismus?,* edited by M. Ernst and S. Schmidt , 108–121. Berlin: FDCL.

Eßer, Klaus, Wolfgang Hillebrand, Dirk Messner, and Jörg Meyer-Stamer. 1996. *Systemic Competitiveness. New Governance Patterns for Industrial Development.* London: Frank Cass.

Evans, Peter. 1979. *Dependent Development. The Alliance of Multinational, State and Local Capital in Brazil.* Princeton: Princeton University Press.

Everard, Jerry. 2000. *Virtual States. The Internet and the Boundaries of the Nation-State.* New York/London: Routledge.

Fabian, Horst. 1981. *Der kubanische Entwicklungsweg. Ein Beitrag zum Konzept autozentrierter Entwicklung.* Opladen: Westdeutscher Verlag.

Fajnzylber, Fernando. 1983. *La industrialización trunca de América Latina.* México D.F.: Nueva Imagen.

Farah Calderón, Walter. 2001. *El caso de Costa Rica.* Documento para el Taller "Garantizando la Participación Ciudadana en la Era Digital: Políticas Públicas y Derechos Internet en América Latina y el Caribe," November 19–23, 2001. http://www.apc.org/apps/img_upload/5ba65079e0c45cd29dfdb3e618dda73/taller_costa_rica.doc.

FCC [Federal Communications Commission]. 2002. *The FCC's Universal Service Support Mechanisms.* http://www.fcc.gov/cgb/consumerfacts/universalservice.html.

FECON [Federación Costarricense para la Conservación del Ambiente]. 2000. *Posición de la Federación Costarricense para la Conservación del Ambiente frente al Proyecto de Ley para el Mejoramiento de Ser vicios Públicos de Electricidad y Telecomunicaciones y de la Participación del Estado.* San José: FECON.

Feinsilver, Julie M. 1992. Will Cuba's Wonder Drugs lead to Political and Economic Wonders? Capitalizing on Biotechnology and Medical Exports. *Cuban Studies* Vol. 22: 79–111.

Feinsilver, Julie M. 1995. Cuban Biotechnology. The strategic success and commercial limits of a First World approach to development. In *Biotechnology in Latin America: Politics, Impacts and Risks,* edited by N. Patrick Peritore and Ana Karina Galve-Peritore, 97–125. Wilmington: Scholarly Resources Books.

Ferdinand, Peter, ed. 2000. *The Internet, Democracy and Democratization.* London: Frank Cass. (Book version of the special issue of *Democratization,* Vol. 7.1).

Fernández Martínez, José Antonio. 2003. *Las telecomunicaciones en Cuba: Presente y retos del futuro.* Presentación en la VII Cumbre de Presidentes AHCIET (Asociación Hispanoamericana de Centros de Investigación y Empresas de Telecomunicaciones). La Habana, November 2003. http://www.ahciet.net/agenda/ponencias/2003/CumbrePresidentes/ETECSA-CumbrePresidentesAHCIETNoviembre2003JoseA.Fernandez.ppt.

Figueres, José María. 2000. Keynote Address. In *The Power of Ideas. Building Tomorrow's Global Knowledge Economies,* edited by Deutsche Stiftung für internationale Entwicklung, 18–28. Berlin: DSE.

Fischbach, Rainer. 2001. In den Fängen einer vernetzten Zukunft. *Freitag* (Berlin), August 17, 2001.

Fischbach, Rainer. 2002. Frühes Ende einer Epoche. *Freitag* (Berlin), July 26, 2002.

Forteux, Michel. 1999. La communauté cubaine des États-Unis: d'éxile á immigré, une nouvelle identité. *Cahiers des Amériques Latines,* 2ème série, N° 31/32: 197–210.

Fox, Elizabeth, ed. 1988. *Media and Politics in Latin America.* London: Sage.

Fox, James / Monge, Ricardo. 1999. *United StatesID en el desarrollo de Costa Rica. Impacto de 50 años de actividades.* San José: Academia de Centroamérica.

Freeman, Christopher. 1996. The Two-Edged Nature of Technological Change: Employment and Unemployment. In *Information and Communication Technologies. Visions and Realities,* edited by William H. Dutton, 19–36. Oxford/New York: Oxford University Press.

Frieden, Jeffry. 1981. Third world indebted industrialization. International finance and state capitalism in Mexico, Brazil, Algeria and South Korea. *International Organization* 35: 3, 407–432.

Fritz, Barbara. 2002. *Entwicklung durch wechselkurs-basierte Stabilisierung? Der Fall Brasilien.* Marburg: Metropolis.

Fundación Acceso. 2000. *Regulación y uso de las nuevas tecnologías de información y comunicación (NTIC) en el cambio de los procesos de transformaciones políticas y económicas en América Latina: El caso de Costa Rica.* San José: Fundación Acceso (mimeo; in part published as Fundación Acceso's "Documento de Trabajo No. 4").

Fürst, Edgar. 1990. Politik der Strukturanpassung in Costa Rica, 1982–1988. In *Entwicklungsprobleme Costa Ricas,* edited by L. Ellenberg and A. Bergemann, 171–194. Saarbrücken/Ft. Lauderdale: Breitenbach.

G 8 [Group of the Eight Leading Industrialized Countries]. 2000. *Okinawa Charter on Global Information Society.* http://www.dotforce.org/reports/it1.html.

García Luis, Julio. 1990. En una cuerda fina y tensa (entrevista). *Juventud Rebelde,* October 21, 1990: 8–9.

Gazeta Oficial. 1996. Decreto No. 209: Del acceso desde la República de Cuba a redes informática de alcance global. *Gazeta Oficial* No. 27, September 13, 1996.

German, Christiano. 1999. *Der Weg Brasiliens in das Informationszeitalter.* Sankt Augustin: Konrad-Adenauer-Stiftung.

Gil Morell, Melchor. 2000. Nueva realidad informática cubana. *Metánica—Revista de la Industria Cubana Siderúrgica y Mecánica,* La Habana, (6)1, January–April: 5–7.

Gómez, Ricardo. 2000. The Hall of Mirrors. The Internet in Latin America. *Current History,* Vol. 99, No. 634 (February), 72–77.

Gómez, Ricardo, and Benjamín Casadiego. 2002. *Letter to Aunt Ofelia. Seven Proposals for Human Development Using New Information and Communication.* http:// www.idrc.ca/pan/ricardo/publications/ofelia_eng.htm.

Gómez, Ricardo, Patrik Hunt, and Emanuelle Lamoureux. 1999. Telecentros en la mira: ¿Cómo pueden contribuir al desarrollo social? *Revista Latinoamericana de Comunicacion CHASQUI,* No. 66, June 1999. http://www.idrc.ca/pan/chasquiSP.htm (in English: Wondering about Telecentres: can they contribute to sustainable development in Latin America? http://www.idrc.ca/pan/chasquiSP_e.htm).

Gómez, Ricardo, and Juliana Martínez. 2001. *The Internet . . . why? And what for? Thoughts on Information and Communication Technologies for Development in Latin America.* San José: Fundación Acceso. http://www.acceso.or.cr/PPPP/.

González Gutiérrez, Alfredo. 1997. La economía sumergida en Cuba. In *Economía y refroma económica en Cuba,* edited by D. Dirmoser and J. Estay, 239–256. Caracas: Nueva Sociedad.

González Planas, Ignacio. 2000. La conectividad es clave (entrevista). *Giga—La revista cubana de computación* 3: 4–7.

González, Ángel. 2001. Silicon Island: A Cuban Fantasy? *Wired News* (online), June 6, 2001. http://www.wired.com/news/infostructure/0,1377,44279,00.html.

González, Edward. 1996. *Cuba—Clearing Perilous Waters?* Santa Monica: RAND.

González, Edward, and David Ronfeldt. 1992. *Cuba Adrift in a Postcommunist World.* Santa Monica: RAND.

González, Edward, and David Ronfeldt. 1994. *Storm Warnings for Cuba.* Santa Monica: RAND.

Gore, Al. 1994. *Global Information Infrastructure. An Agenda for Cooperation.* Speech at the International Telecommunication Union Meeting, March 21, 1994. http:// www.ifla.org/documents/infopol/us/goregii.txt.

Gregorio, Carlos G. 2002. Privacidad en América Latina. *Boletín Políticas y Derechos en Internet,* Agosto de 2002. Asociación para el Progreso de las Comunicaciones (APC). http://lac.derechos.apc.org/boletin.shtml?-1-Privacidad.

Grieco, Joseph. 1984. *Between dependency and autonomy. India's experience with the international computer industry.* Berkeley: University of California Press.

Grogg, Patricia. 2000. *Kuba: Internetzugang für alle—Regierung fördert kollektive Nutzung.* Inter press service, October 20, 2000.

Guehenno, Jean-Marie. 1995. *The End of the Nation State.* Minneapolis: University of Minnesota Press.

Guevara, Ernesto 'Che.' 1982. *Bolivianisches Tagebuch.* 11th ed. München: Trikont

Günther, Wilfried, and Heinz Uhlig. 1992. *Telekommunikation in der DDR. Die Entwicklung von 1945 bis 1989.* Bad Honnef: Wissenschaftliches Institut für Kommunikationsdienste.

Gutiérrez Araya, Shirley. 2002. *Lluvia de cuestionamientos por triángulo de solidaridad.* San José: Tiquicia.com. http://www.tiquicia.com/articulos/nacionales/actualidad/samb220502.asp.

Habermas, Jürgen. 1998. *Die postnationale Konstellation. Politische Essays.* Frankfurt a. M.: Suhrkamp. (English translation as: The Post-National Constellation. Cambridge: Polity 2001.)

Haldenwang, Christian von. 2002. *Electronic Goverment und Entwicklung. Ansätze zur Modernisierung der öffentlichen Politik und Verwaltung.* Bonn: Deutsches Institut für Entwicklung.

Hamelink, Cees. 1995. *World Communication: Disempowerment and Self-empowerment.* London: ZED.

Hamelink, Cees. 1997. *New Information and Communication Technologies, Social Development and Cultural Change.* Geneva: UNRISD.

Hamelink, Cees. 1999. *ICTs and Social Development: The Global Policy Context.* Geneva: UNRISD.

Hamelink, Cees. 2002a. *Some observations from the Plenary Keynote on Monday July 1.* Programa de la Sociedad Civil—PrepCom 1, Geneva, July 1–5 2002. http:// www.geneva2003.org/home/events/documents/hamelink.doc.

Hamelink, Cees. 2002b. Interview without title. *Connexion—The WSIS PrepCom 1 daily newsletter,* edited by the Conference of NGOs, C.O.N.G.O. http://www.congo.org.

Hanson, Gordon. 2001. *Should Countries Promote Foreign Direct Investment?* UNCTAD G-24 Discussion Paper Series No. 9, February 2001, Geneva: UNCTAD.

Hargittai, Eszter. 2002. Second-Level Digital Divide: Differences in People's Online Skills. *First Monday. Peer-Reviewed Journal on the Internet* Vol. 7, No 4; April 1, 2002. http://www.firstmonday.org/issues/issue7_4/hargittai/index.html.

Hernández Castillo, José Alberto. 1999. Comments on 'The Political Economy of the Internet in Cuba' by Valdés and Rivera. In *Cuba in Transition Vol. 9,* edited by the Association for the Study of the Cuban Economy (ASCE), 155–156. http:// www.lanic.utexas.edu/la/cb/cuba/asce/cuba9/castillo.pdf.

Hernández, Rafael. 1994. La sociedad civil y sus alrededores. *La Gaceta de Cuba,* No. 1, La Habana: UNEAC.

Herr, Hansjörg. 2000. Das chinesische Akkumulationsmodell und die Hilflosigkeit der traditionellen Entwicklungstheorien. *Prokla—Zeitschrift für kritische Sozialwissenschaft* 119 (30) June 2000: 181–210.

Herrera, Juan Carlos, and Dodanys García. 2000. Los caravanas viajan vía módem. *Ciencia, Innovación y Desarrollo* (5)2: 26–29.

Hershberg, Eric, Jorge Monge, and Juan Pablo Pérez Sainz, eds. 2003. *From Coffee to Semi-conductors: Costa Rica's Strategy for Industrial Upgrading and Equity.* San José: FLACSO-Costa Rica.

Herzog, Roman. 2002. *Politik und Ökonomie des Internet in Argentinien und Peru.* Doctoral thesis, Free University of Berlin. http://www.diss.fu-berlin.de/2002/190.

Herzog, Roman, Bert Hoffmann, and Markus Schulz. 2002. *Internet und Politik in Lateinamerika. Regulierung und Nutzung der Neuen Informations- und Kommunikationstechnologien im Kontext der politischen und wirtschaftlichen Transformationen.* Schriftenreihe des Instituts für Iberoamerika-Kunde, Hamburg, 55. Frankfurt a. M.: Vervuert.

Hess Araya, Christian. 2002a. *Derecho a la intimidad y autodeterminación* (Online-Publication of "Democracia Digital"). http://www.democraciadigital.org/derechos/arts/0201intimidad.html.

Hewitt de Alcántara, Cynthia. 2001. *The Development Divide in a Digital Age.* UNRISD Technology, Business and Society Programme Paper No. 4, August 2001. Geneva: UNRISD.

Heyman, Timothy, and Ricardo Setti. 2001. The Economic and Business Dimension. In *The Future of the Information Revolution in Latin America: Proceedings of an International Conference,* edited by G.F. Treverton and L. Mizell, 15–22. Santa Monica: RAND. http://www.rand.org/publications/CF/CF166.1/ CF166.1.ch3.pdf.

Hilbert, Martin R. 2001. *Latin America on its Path into the Digital Age: Where are We?* Serie Desarrollo Productivo No. 104. Santiago de Chile: CEPAL.

Hill, Kevin A., and John E. Hughes. 1996. *Cyberpolitics. Citizen Activism in the Age of the Internet.* Lanham: Rowman & Littlefield.

Hirschkop, Ken. 1996. Democracy and the New Technologies. *Monthly Review* (48)3, July/August: 86–98.

Hirschman, Albert O. 1958. *The Strategy of Economic Development.* New Haven: Yale University Press.

Hirschman, Albert O. 1970. *Exit, Voice, and Loyalty: Responses to Decline in Firms, Organizations, and States.* Cambridge: Harvard University Press.

Hirschman, Albert O. 1971. The Political Economy of Import-Substituting Industrialization in Latin America. In *A Bias for Hope. Essays on Development and Latin America, by Albert O. Hirschman,* 85–123. New Haven: Yale University Press.

Hirschman, Albert O. 1981. A Generalized Linkage Approach to Development, with Special Reference to Staples. In *Essays in Trespassing: Economics and Politics and Beyond,* by Albert O. Hirschman, 59–97. Cambridge: Cambridge University Press.

Hirschman, Albert O. 1992. Abwanderung, Widerspruch und das Schicksal der Deutschen Demokratischen Republik. *Leviathan* 3: 330–358. (English version as: Exit, Voice, and the Fate of the German Democratic Republic: An Essay in Conceptual History. *World Politics* 45: 173–202, 1993.)

Hobday, Michael. 1990. *Telecommunications in Developing Countries: The Challenge from Brazil.* New York/London: Routledge.

Hobsbawm, Eric. 1995. *Age of Extremes. The Short Twentieth Century 1914–1991.* London: Michael Joseph.

Hoffmann, Bert, ed. 1995. *Cuba. Apertura y reforma económica. Perfil de un debate.* Caracas: Nueva Sociedad. (Orig. in German as: Wirtschaftsreformen in Kuba. Konturen einer Debatte, Frankfurt a. M.: Vervuert, 1994.)

Hoffmann, Bert. 1996. Cuba: La reforma desde adentro que no fue. *Notas* No 9. Frankfurt: Vervuert 1996, 48–65. (Reprinted in *Encuentro de la Cultura Cubana* 10, Madrid 1998, 71–84.)

Hoffmann, Bert. 1997. Helms-Burton und kein Ende? Auswirkungen und Perspektiven für Kuba, die USA und Europa. *Lateinamerika. Analysen-Daten-Dokumentation* 33: 35–50. Hamburg: Institut für Iberoamerika-Kunde.

Hoffmann, Bert. 1999. Las ONG en Cuba: la sociedad civil en el socialismo y sus límites. In *Sociedad civil en América Latina: representación de intereses y gobernabilidad,* edited by P. Hengstenberg, K. Kohut and G. Maihold, 338–351. Caracas: Nueva Sociedad.

Hoffmann, Bert. 2000. *Kuba.* München: C. H. Beck.

Hoffmann, Bert. 2001. Transformation and Continuity in Cuba. *Review of Radical Political Economics,* Vol. 33, No. 1, 1–20.

Hoffmann, Bert. 2002. El cambio imposible. Cuba como 'asunto interméstico' en la política de EE.UU.: consecuencias y perspectivas. In *El triángulo atlántico: América Latina, Europa y Estados Unidos en un sistema internacional cambiante,* edited by K. Bodemer, W. Grabendorff and W. Jung. Sankt Augustin: Konrad-Adenauer-Stiftung.

Hofmann Jeanette. 2000. Und wer regiert das Internet? Regimewechsel im Cyberspace. In *Global@home. Informations- und Dienstleistungsstrukturen der Zukunft. Jahrbuch Telekommunikation und Gesellschaft 2000,* edited by H. Kubicek et al., 67–78. Heidelberg: Hüthig.

Huffschmid, Anne. 2001. Tomar la palabra y no el poder. El discurso zapatista y la opinión pública. In *Violencia y regulación de conflictos en América Latina,* edited by K. Bodemer et al., 123–136. Caracas: Nueva Sociedad.

Huntington, Samuel P. 1991. *The Third Wave. Democratization in the late Twentieth Century.* Oklahoma: University of Oklahoma Press.

Hunya, Gábor. 1995. *Transformation of the transport and telecommunications infrastructure in East-Central Europe.* Forschungsberichte 214. Wien: WIIW.

Hurtienne, Thomas. 1984. *Theoriegeschichtliche Grundlagen des sozialökonomischen Entwicklungsdenkens, Band II: Paradigmen sozialökonomischer Entwicklung im 19. und 20. Jahrhundert.* Saarbrücken/Ft. Lauderdale: Breitenbach.

Hurtienne, Thomas. 1986. Fordismus, Entwicklungstheorie und Dritte Welt. *Peripherie. Zeitschrift für Politik und Ökologie der Dritten Welt,* No 22/23: 60–110.

Hutter, Michael. 2001. Der 'Digital Divide'—ein vorübergehender Zustand. In *Internet@future: Technik, Anwendungen und Dienste der Zukunft. Jahrbuch Telekommunikation und Gesellschaft 2001,* edited by H. Kubicek et al., 362–370. Heidelberg: Hüthig.

ICE [Instituto Costarricense de Electricidad]. 1997a. *El ICE y la Electrificación en Costa Rica 1949–1996.*San José: ICE.

ICE. 1997b. *El ICE y las telecomunicaciones en Costa Rica 1963–1995.* San José: ICE.

ICE. 1997c. *Estaciones Terrenas.* San José: ICE.

IDC [International Data Corporation]. 1999. *Advanced Web Users around the World (Project Atlas).* htp://www1.worldcom.com/global/resources/cerfs_up/presentations/1999/project_atlas/1.

IDICT [Instituto de Información Científica y Tecnológica]. 2001. *La Política Nacional de Información en Cuba.* http://www.idict.cu/asuntos/politica.htm.

Iglesias, Carlos. 2000. *Cuba: Virtualidad real.* La Habana: World Data Service, WDS-0359, October 16, 2000.

IIJusticia. 2002. *El Impacto de Las Tecnologias de Informacion y Comunicacion (TIC) sobre la Protección de los Datos Personales en las Américas.* http://www.iijusticia.edu.ar/privacidad/Paises.htm.

ILO [International Labour Office]. 2001. *World Employment Report 2001. Life at Work in the Information Economy* (+ CD-Rom with Background Papers). Geneva: ILO.

IMF [International Monetary Fund]. 2001. *The Information Technology Revolution. World Economic Outlook 2001.* Washington D.C.: IMF.

InfoDev. 2000. *The Networking Revolution. Opportunities and Challenges for Developing Countries* (InfoDev Working Paper). Washington D.C.: The World Bank Group, June 2000. http://www.infodev.org/library/WorkingPapers/NetworkingRevolution.pdf.

Instituto Alvin Toffler. 1999. *Telecomunicaciones—una llave para el desarrollo.* San José: CINDE/SIECA.

IRELA [Instituto de Relaciones Europeo-Latinoamericanas], ed. 1996. *Documentación del Seminario 'El refuerzo del embargo de EEUU contra Cuba—Implicaciones para el comercio y las inversiones,' Sitges, July 8–10, 1996.* Madrid: IRELA.

ISOC [Internet Society]. 2000. *A Brief History of the Internet (v3.31).* http://www.isoc.org/internet/history/brief.shtml.

ITU [International Telecommunication Union]. 1984. *The Missing Link. Report of the Independent Commission for World Wide Telecommunications Development.* Geneva: ITU.

ITU. 1989. *The Changing Telecommunications Environment: Policy Considerations for the Members of the ITU. Report of the Advisory Group on Telecommunications Policy.* Geneva: ITU.

ITU. 1991. *Restructuring of Telecommunications in Developing Countries. An Empirical Investigation with ITU's Role in Perspective.* Geneva: ITU.

ITU. 1999. *Challenges to the Network—Internet for Development. Executive Summary.* Geneva: ITU.

ITU. 2000. *Americas Telecommunications Indicators 2000.* Geneva: ITU.

ITU. 2001. *ITU Internet Reports—IP Telephony.* Geneva: ITU.

ITU. 2002a. *World Telecommunication Development Report 2002: Reinventing Telecoms.* Geneva: ITU.

ITU. 2002b. *World Telecommunication Development Report 2002: Reinventing Telecoms. Executive Summary.* Geneva: ITU. http://www.itu.int/ITU-D/ict/publications/wtdr_02/index.html.

ITU. 2002c. *3G Mobile Policy. The Cases of Chile and Venezuela* (Authors: Ben A. Petrazzini and Martin Hilbert). Geneva: ITU. http://www.itu.int/osg/spu/ni/3G/casestudies/chile-venezuela/Chile-Venezuela.PDF.

ITU. 2002d. *Trends in Telecommunication Reform 2002. Effective Regulation.* Geneva: ITU.

ITU. 2003. *World Telecommunication Development Report 2003: Access Indicators for the Information Society.* Geneva: ITU.

Jiménez, Ronulfo, ed. 2000. *Los retos políticos de la reforma económica en Costa Rica.* San José: Academia de Centroamérica.

Kahin, Brian, ed. 1992. *Building Information Infrastructure.* New York: McGraw-Hill.

Kahin, Brian, and James Keller, eds. 1995. *Public Access to the Internet.* Cambridge: The MIT Press.

Kahin, Brian, and Charles Nesson, eds. 1997. *Borders in Cyberspace. Information Policy and the Global Information Infrastructure.* Cambridge: The MIT Press.

Kahin, Brian, and Ernest J. Wilson III., eds. 1996. *National Information Infrastructure Initiatives. Visions and Policy Design.* Cambridge: The MIT Press.

Kalathil, Shanthi, and Taylor C. Boas. 2001. *The Internet and State Control in Authoritarian Regimes: China, Cuba, and the Counterrevolution.* Washington D.C.: Carnegie Endowment for International Peace. http://www.ceip.org/files/pdf/ 21KalathilBoas.pdf.

Kalathil, Shanthi, and Taylor C. Boas. 2003. *Open Networks, Closed Regimes. The Impact of the Internet on Authoritarian Rule.* Washington D.C.: Carnegie Endowment for International Peace.

Kedzie, Christopher. 1997. *Communication and Democracy. Coincident Revolutions and the Emergent Dictator's Dilemma.* Santa Monica: RAND.

Kisch, Egon Erwin. 1935. *Changing Asia.* New York: Alfred A Knopf. (Orig. 1932 in German as: Asien gründlich verändert.)

Kleinsteuber, Hans J., ed. 1996. *Der ‚Information Superhighway.' Amerikanische Visionen und Erfahrungen.* Opladen: Westdeutscher Verlag.

Kleinwächter, Wolfgang. 2001. Global Governance in the Information Age: ICANN as a Policy Pilot Project for Co-Regulation and a New Trilateralism? In *Global Democracy and the ICANN Elections*, edited by H. Klein. Atlanta: Georgia Institute of Technology.

Kraemer, Kenneth L., and Jason Dedrick. 1996. IT and Economic Development. International Competitiveness. In *Information and Communication Technologies. Visions and Realities*, edited by W. H. Dutton, 319–334. Oxford/New York: Oxford University Press.

Kranzberg, Melvin. 1985. The Information Age: Evolution or Revolution? In *Information, Technologies and Social Transformation*, edited by B.R. Guile. Washington/D.C.: National Academy of Engineering.

Krennerich, Michael. 2000. Internet in Costa Rica. Am Vorabend der Öffnung des Telekommunikationsmarktes. *Lateinamerika. Analysen-Daten-Dokumentation* 43: 46–57. Hamburg: Institut für Iberoamerika-Kunde.

Krueger, Anne O. 1974. The Political Economy of the Rent-Seeking Society. *American Economic Review*, vol. 64, No 3: 291–303.

Kubicek, Herbert. 2001. Gibt es eine digitale Spaltung? Kann und soll man etwas dagegen tun? In *Internet@future: Technik, Anwendungen und Dienste der Zukunft. Jahrbuch Telekommunikation und Gesellschaft 2001,* edited by H. Kubicek et al., 371–377. Heidelberg: Hüthig.

Kummels, Ingrid. 1995. La cotidianidad difícil. Consideraciones de una etnóloga sobre la crisis y la cultura popular. In Cuba: Apertura y reforma económica. Perfil de un debate, edited by B. Hoffmann, 131–144. Caracas: Nueva Sociedad.

Lage, Carlos. 1996. Mientras mayores sean las dificultades mayor será el estímulo a nuestra inteligencia y a nuestro trabajo (Intervención de Carlos Lage en el V. Pleno del Comité Central del Partido Comunista de Cuba, el día 23 de marzo de 1996). *Granma Internacional,* April 10, 1996: 9–12.

Lage, Carlos. 2000. *Discurso de Carlos Lage en Informática 2000.* La Habana: Informática 2000.

Lagos, Marta. 2001. Between Stability and Crisis in Latin America. How People View Democracy. *Journal of Democracy* Vol. 12, No 1, January 2001: 137–145.

Lamprecht, Rudi. 2002. Setting Standards. *Connect-World Trends Annual Issue 2002.* http://www.connect-world.com/past_issues/trends/2002/trends/r_lam-precht_SIEMENS_2002.asp.

Larraín B., Felipe, Luis F. López-Calva, and Andrés Rodríguez-Clare. 2000. *Intel. A Case Study of Foreign Direct Investment in Central America.* Working Paper No. 58. Cambridge: Harvard University, Center for International Development. http://www2.cid.harvard.edu/cidwp/058.pdf.

Linares, Ramón. 2001. *Objetivos, funciones y estructura del MIC.* Ponencia en la Primera Reunión Nacional de Informatización de la Sociedad, 29–30 Sept 2000, Ministerio de la Informática y las Comunicaciones. Unpublished working document.

Lindegaard, Klaus, and Leiner, Vargas. 2002. *New Economies and Innovation for Developing Countries. The Case of Intel in Costa Rica.* Paper presented at the DRUID Summer Conference on "Industrial Dynamics of the New and Old Economy—Who is embracing whom?," Copenhagen/Elsinore June 6–8, 2002. http://www.druid.dk/conferences/summer2002/Papers/vargas_lindegaard.pdf.

Linz, Juan, and Alfred Stepan. 1996. *Problems of Democratic Consolidation. Southern Europe, South America, and Post-Communist Europe.* Baltimore/London: The Johns Hopkins University Press.

Lipietz, Alain. 1985a. Akkumulation, Krisen und Auswege aus der Krise. *Prokla. Zeitschrift für Politische Ökonomie und sozialistische Politik* 58: 109–137.

Lipietz, Alain. 1985b. *Mirages et Miracles. Problèmes de l'industrialisation dans le Tiers-Monde.* Paris: La Découverte.

List, Friedrich. 1982. *Das nationale System der Politischen Ökonomie.* Berlin: Akademie-Verlag. (First published in Stuttgart/Tübingen 1841.)

Lizano, Eduardo. 1987. *Desde el Banco Central.* San José: Academia de Centroamérica.

Lizano, Eduardo. 1997. *Deuda Interna. Documentos, Notas y Comentarios.* San José: Academia de Centroamérica.

Lizano, Eduardo. 2000. Política Económica y Desarrollo Nacional. In *Los retos políticos de la reforma económica en Costa Rica,* edited by R. Jiménez, 179–220. San José: Academia de Centroamérica.

Lizano, Eduardo, and Norberto Zuñiga. 1999. *Evolución de la economía de Costa Rica durante el período 1983–1998: Ni tan bien, ni tan mal.* San José: Academia de Centroamérica.

López, Félix. 2002. Llamada pública. *Granma,* October 23, 2002.

Lulei, Wilfried. 1997. Vietnam. Wirtschaftsreformen und außenpolitische Öffnung. In *Jahrbuch Dritte Welt 1997,* edited by J. Betz and S. Brüne. München: C.H. Beck.

Lüthje, Boy. 2001. Silicon Valley: Vernetzte Produktion, Industriearbeit und soziale Bewegungen im Detroit der "New Economy." *Prokla. Zeitschrift für kritische Sozialwissenschaft* No. 122, vol. 31/1: 79–102.

Maislinger, Andreas, ed. 1986. Costa Rica. Politik, Gesellschaft und Kultur eines Staates mit ständiger aktiver und unbewaffneter Neutralität. Innsbruck: Inn-Verlag.

Mansell, Robin and Uta Wehn, eds. 1998. *Editors Knowledge Societies. Information Technology for Sustainable Development.* Oxford: Oxford University Press. http:// www.sussex.ac.uk/spru/ink/knowledge.html.

Marcuse, Peter. 2002. Entpolitisierte Globalisierungsdiskussion. Information-szeitalter und Netzwerkgesellschaft bei Manuel Castells. *Prokla—Zeitschrift für kritische Sozialwissenschaft* 127, Vol. 32, No. 2, June 2002: 321–344.

Marquetti, Hiram. 1995. *La liberalización de la circulación de divisa en Cuba: Resultados y problemas.* La Habana (mimeo).

Márquez González, Norma. 1999. Joven Club: Un apoyo a la cultura informática. *Giga—La revista cubana de computación,* La Habana, 3: 4–5.

Marrero, Rafael. 2000. *Estado actual y perspectivas de las telecomunicaciones.* Ponencia en la Primera Reunión Nacional de Informatización de la Sociedad, 29–30 Sept 2000, Ministerio de la Informática y las Comunicaciones. Unpublished working document.

Martínez Trujillo, Carlos. 2000. *Perspectivas del correo como soporte de nuevos servicios para el ciudadano.* Ponencia en la Primera Reunión Nacional de Informatización de la Sociedad, 29–30 Sept 2000, Ministerio de la Informática y las Comunicaciones. Unpublished working document.

Martínez, Jesús. 1996. [e-mail communication without title]. Reprinted in *Lateinamerika. Analysen-Daten-Dokumentation* 43: 135. Hamburg: Institut für Iberoamerika-Kunde.

Martínez, Jesús. 1999. The Net in Cuba. *Matrix News,* (9)1, Jan. http://www.matrix.net/publications/mn/mn0901.html.

Martínez, Juliana. 2001. *Investigación sobre el impacto de internet en América Central. Análisis de entornos nacionales.* San José: Fundación Acceso.

Martinussen, John. 1997. *Society, State & Market. A Guide to Competing Theories of Development.* London: ZED.

Mesa-Lago, Carmelo. 1993. Cuba. Un caso único de reforma anti-mercado. Retrospectiva y perspectivas. *Pensamiento Iberoamericano* 22–23, Tomo II: 65–100.

Mesa-Lago, Carmelo. 1994. *Historia Económica de la Cuba Socialista*. Madrid: Alianza.

Mesa-Lago, Carmelo. 2000. *Market, Socialist, and Mixed Economies, Comparative Policy and Performance: Chile, Cuba, and Costa Rica*. Baltimore: John Hopkins University.

Messner, Dirk. 1997. *The Network Society. Economic Development and International Competitiveness as Problems of Social Governance*. London: Frank Cass.

Meyer-Stamer, Jörg. 1996. *Technologie und industrielle Wettbewerbsfähigkeit. Allgemeine Überlegung und Erfahrungen aus Brasilien*. Schriften des Deutschen Instituts für Entwicklungspolitik 109, Köln: Weltforum. (Published in English as: Technology, Competitiveness and Radical Policy Change. The Case of Brazil. London: Frank Cass 1997.)

MIC [Ministerio de la Informática y las Comunicaciones]. 2000. *Primera Reunión Nacional de Informatización de la Sociedad, 29–30 Sept 2000*. Unpublished working document.

MIC. 2001a. Objetivos estratégicos del Ministerio de la Informatica y las Comunicaciones para el período 2001–2003. http://www.cubagob.cu/des_eco/mic/mic_objetivos/objetivos_2001_2003.htm.

MIC. 2001b. *Tecnologías de la información y las comunicaciones—estrategia cubana*. Unpublished working document. (Presentation in a Research Center in Havanna in March 2001.)

MIC. 2004. *Resolución 180/2003*. La Habana: MIC.

MICIT [Ministerio de Ciencia y Tecnología]. 2001. *Inauguración del Plan Piloto para el Establecimiento de la Red Internet Avanzada y Firma del Decreto para la Creación de la Red Nacional de Investigación Avanzada, April 18, 2001*. San José: MICIT.

MIDEPLAN [Ministerio de Planificación y Política Económica]. 2000. *Telecomunicaciones, Rindiendo Cuentas* No. 10, March.

Miles, Ian. 2001. Information technology: A Continuing Revolution. In *World Employment Report 2001. Life at Work in the Information Economy*, edited by International Labour Organization (CD-Rom version, Background Papers). Geneva: ILO.

Minkner, Mechthild. 1990. Zur Problematik der Verschuldung und der internationalen Finanzbeziehungen. In *Entwicklungsprobleme Costa Ricas*, edited by L. Ellenberg and A. Bergemann, 283–294. Saarbrücken/Ft. Lauderdale: Breitenbach.

Minkner, Mechthild. 1999. Costa Rica: Internationalisierung der Wirtschaft mit Bananen und Mikroprozessoren. *Brennpunkt Lateinamerika* 19/99. Hamburg: Institut für Iberoamerika-Kunde.

MISTICA. 2002. *Trabajando la Internet con una visión social* [trabajo colectivo de a Comunidad Virtual Mistica]. http://funredes.org/mistica/castellano/ciberoteca/tema-tica/esp_doc_olist1.html.

Monge, Ricardo. 2000. La economía política de un intento fallido de reforma en telecomunicaciones. In *Los retos políticos de la reforma económica en Costa Rica*, edited by R. Jiménez, 273–318. San José: Academia de Centroamérica.

Monge, Ricardo and Federico Chacón. 2002. *Bridging the Digital Divide in Costa Rica. Access to and Use of Information and Communication Technologies.* Serie Costa Rica Digital 01. San José: CAATEC.

Monreal, Pedro. 1999. Migration und Überweisungen—Anmerkungen zum Fall Kuba. In *Migrationen—Lateinamerika Analysen und Berichte Band 23,* edited by K. Gabbert et al., 73–96. Bad Honnef: Horlemann.

Monreal, Pedro. 2000. *Los dilemas de las trayectorias económicas de Cuba. Apuntes sobre una polémica* (mimeo). La Habana: Centro de Investigaciones de Economía Internacional. (An abridged version was published in English as: Development as Unifinished Affair: Cuba After the "Great Adjustment" of the 1990s. *Latin American Perspectives* 124. Vol 29, No. 3, May 2002: 75–90.)

Mueller, Milton, and Zixiang Tan. 1997. *China in the Information Age. Telecommunications and the Dilemmas of Reform.* Westport: Praeger.

Müller-Plantenberg, Urs. 1972. Technologie und Abhängigkeit. In *Imperialismus und strukturelle Gewalt. Analysen über abhängige Reproduktion,* edited by D. Senghaas, 335–355. Frankfurt a. M.: Suhrkamp.

Müller-Plantenberg, Urs. 1990. Was heute Sozialismus sein könnte. *Lateinamerika Nachrichten* 198, Berlin.

Muse, Robert. 1996. Legal and Practical Implications of Title III of the Helms-Burton Law. In *Documentación del Seminario 'El refuerzo del embargo de EEUU contra Cuba—Implicaciones para el comercio y las inversiones,' Sitges, July 8–10, 1996,* edited by IRELA. Madrid: IRELA.

Negroponte, Nicholas. 1995. *Being Digital.* London: Hodder and Stoughton.

Nelson, Roy. 1999. *Intel's Site Selection Decision in Latin America,* Thunderbird Case Series A03–99–0016, edited by The American Graduate School of International Management. http://www.t-bird.edu/pdf/about_us/case_series/a03990016.pdf.

Nordenstreng, Kaarle, Enrique Gonzales Manet, and Wolfgang Kleinwächter. 1986. *New International Information and Communication Order. A Sourcebook.* Prague: International Organization of Journalists.

Norris, Pippa. 2001. *Digital Divide? Civic Engagement, Information Poverty and the Internet in Democratic Societies.* Cambridge: Cambridge University Press.

Notimex. 2001. *Asegura Cuba que limita acceso a Internet por causas económicas.* February 7, 2001.

NTIA [National Telecommunications and Information Administration, U.S. Department of Commerce]. 1995. *Falling Through the Net. A Survey of the 'Have Nots' in Rural and Urban America.* http://www.ntia.doc.gov/ntia-home/fallingthru.html.

NTIA. 2000. *Falling Through the Net. Toward Digital Inclusion. A Report on Americans' Access to Technology Tools.* http://search.ntia.doc.gov/pdf/fttn00.pdf.

Nuhn, Helmut. 2001. Biotechnologie als Entwicklungsstrategie in Kuba. Der Aufbau eines medizinisch-pharmazeutischen Produktionskomplexes und seine außenwirtschaftliche Bedeutung. In *Kubas Weg aus der Krise. Neuorganisation der Produktion von Gütern und Dienstleistungen für den Export,* edited by G. Mertins and H. Nuhn, 145–169. Marburg/Lahn: Marburger Geographische Gesellschaft.

Ó Siochrú, Seán, and Bruce Girard (with Amy Mahan). 2002. *Global Media Governance*. Lanham: Rowman & Littlefield.

O'Donnell, Guillermo. 1973. *Modernization and Bureaucratic-Authoritarianism: Studies of South American Politics*. Berkeley: Institute of International Studies.

O'Donnell, Guillermo. 1979a. State and Alliances in Argentina, 1956–1976. *Journal of Development Studies* 15, no. 1, 1979. Cited from the reprint in: *Counterpoints. Selected Essays on Authoritarianism and Democratization*, by Guillermo O'Donnell, 3–33. Notre Dame: University of Notre Dame Press, 1999.

O'Donnell, Guillermo. 1979b. Tensions in the Bureaucratic-Authoritarian State and the Question of Democracy. In *The New Authoritarianism in Latin America*, edited by D. Collier, 285–318. Princeton: Princeton University Press.

O'Donnell, Guillermo. 1998. Poverty and Inequality in Latin America: Some Political Reflections. In *Poverty and Inequality in Latin America. Prospects and Challenges*, edited by V. Tokman and G. O'Donnell, 49–71. Notre Dame: Notre Dame University Press.

O'Donnell, Guillermo. 1999a. Polyarchies and the (Un)Rule of Law in Latin America. In *The Rule of Law and the Underprivileged in Latin America*, edited by J. Méndez, G. O'Donnell, and P. S. Pinheiro, 303–338. Notre Dame: University of Notre Dame Press.

O'Donnell, Guillermo. 1999b. *Democratic Theory and Comparative Politics*. Working Paper P 99–004. Berlin: Wissenschaftszentrum Berlin für Sozialforschung.

O'Donnell, Guillermo. 2000. *Democracy, Law and Comparative Politics*. Working Paper 274. Notre Dame: University of Notre Dame, Helen Kellog Institute for International Studies.

O'Donnell, Guillermo, Philippe C. Schmitter, and Laurence Whitehead, eds. 1986. *Transitions from Authoritarian Rule: Prospects for Democracy*. Baltimore: Johns Hopkins University Press (4 volumes).

OECD [Organisation for Economic Cooperation and Development]. 2001. *Understanding the Digital Divide*. Paris: OECD.

Ohmae, Kenichi. 1995. *The End of the Nation State: The Rise of Regional Economies*. New York: Free Press.

Otero, Lisandro. 1993. *La utopía cubana desde adentro*. México, D.F.: Siglo XXI.

Panos Institute. 1998. *The Internet and Poverty*. Panos Briefing No. 28 (written by Duncan Pruett with James Deane). London: Panos Institute.

PCC [Partido Comunista de Cuba]. 1998. Resolución Económica del V Congreso del Partido Comunista de Cuba. *Granma Internacional*, February 22, 1998: 5–12.

Pérez Jr., Louis A. 1995. *Cuba. Between Reform and Revolution*. 2nd ed. New York/Oxford: Oxford University Press.

Pérez Rojas, Niurka, Ernel González Mastrapa, and Miriam García Aguiar, eds. 1996. *UBPC. Desarrollo rural y participación*. La Habana: Universidad de La Habana.

Pérez, Lisandro. 1999. De Nueva York a Miami. El desarrollo demográfico de las comunidades cubanas en Estados Unidos. *Encuentro de la Cultura Cubana* 15 (Madrid): 13–23.

Pérez-López, Jorge F. 1995. *Cuba's Second Economy. From Behind the Scenes to Center Stage.* New Brunswick: Transaction Publishers.

Pérez-Stable, Marifeli. 1993. *The Cuban Revolution. Origins, Course and Legacy.* New York/Oxford: Oxford University Press.

Pérez-Stable, Marifeli. 1996. Misión Cumplida: De cómo el gobierno cubano liquidó la amenaza del diálogo. *Encuentro de la Cultura Cubana* No 1: 25–31. Madrid.

Pimienta, Daniel. 2000a. *La 'Mistica' del Trabajo Social Colaborativo en la Internet.* http://funredes.org/mistica/castellano/panlacpdf.ZIP.

Pimienta, Daniel. 2000b. *[No title]* contribution to the MISTICA mailing list on October 13, 2000 (http://funredes.org/mistica/castellano/emec/produccion/ memoria5/ 0195.html).

Pimienta, Daniel. 2002. *Alocución de Daniel Pimienta para la introducción del PrepCom en Ginebra,* July 1, 2002. http://www.geneva2003.org/home/ events/documents/ gen_pimienta_es.htm.

Pinzler, Petra, and Christian Tenbrock. 2001. Stars im Sturz. Die New Economy wird zum Ballast für die amerikanische Wirtschaft. *Die Zeit* 12/2001.

Pohjola, Matti. 1998. *Information Technology and Economic Development. An Introduction to the Research Issues.* Working Papers No 153. Helsinki: United Nations University / World Institute for Development Economics Research.

Porter, Michael E. 1990. *The Competitive Advantage of Nations.* New York: Free Press.

Porter, Michael E. 2000. Clusters and the new economics of competition. In *World View: Global Strategies for the New Economy,* edited by J.E. Garten, 201–226. Cambridge: Harvard Business School Press.

Portes, Alejandro, and Alex Stepick. 1993. *City on the Edge. The Transformation of Miami.* Berkeley: University of California Press.

Porto, José Rubens Dória, and Marcio Pochmann. 2001. Information and Telecommunications Systems in Brazil. In *World Employment Report 2001. Life at Work in the Information Economy,* edited by International Labour Organization (CD-Rom version). Geneva: ILO.

Pratt, Christine. 2000. C.R. Cheers Cheap, Fast Internet. *The Tico Times,* December 11, 2000. http://www.infocostarica.com/news/11-december-00.html.

Prebisch, Raúl. 1949. *El desarrollo económico de la América Latina y algunos de sus principales problemas.* Reprinted in: Mauro Marini, Ruy: La Teoría Social Latinoamericana. Textos escogidos. México D.F.: UNAM 1994.

Prebisch, Raúl. 1963. *Towards a Dynamic Development Policy for Latin America.* New York: United Nations.

Presidencia de la República. 2001. *Agenda Digital. Un impulso hacia la sociedad del conocimiento.* San José: Presidencia de la República.

Press, Larry. 1996a. *Cuban Telecommunication Infrastructure and Investment.* Paper presented at the Conference of the Association for the Study of the Cuban Economy, Miami, August 1996. http://som.csudh.edu/fac/lpress/dev-nat/nations/cuba/asce.htm.

Press, Larry. 1996b. *Cuban Telecommunications, Computer Networking, and U.S. Policy Implications.* Santa Monica: RAND.

Press, Larry. 1998. The Internet in Cuba. In *The Global Diffusion of the Internet Project. An Initial Inductive Study,* edited by the MOSAIC Group. http://som.csudh.edu/ cis/lpress/devnat/nations/cuba/cubasy.htm.

Press, Larry. 2001. Surveying the Latin American Infrastructure. In *The Future of the Information Revolution in Latin America,* edited by G.F. Treverton and L. Mizell, 5–14. Santa Monica: RAND. http://www.rand.org/publications/CF/ CF166.1/ CF166.1.ch2.pdf.

Press, Larry, William A. Foster, and Seymour E. Goodman. 1999. *The Internet in China and India,* INET '99 Proceedings, Internet Society, 1999. http://www.isoc.org/inet99/proceedings/3a/3a_3.htm.

Preston, William, Edward S. Herman, and Herbert Schiller. 1989. *Hope and Folly: the United States and UNESCO, 1945–1985.* Minneapolis: University of Minnesota Press.

Proenza, Francisco J., Roberto Bastidas-Buch, and Guillermo Montero. 2001. *Telecentros para el desarrollo socioeconómico y rural en América Latina y el Caribe.* Washington D.C.: FAO/UIT/BID. http://www.iadb.org/regions/tele-centros/index.htm.

Prößdorf , Henrik. 1997. Options and Reforms in a Political Economic Perspective. In *Telecommunications Take-Off in Transition Countries,* edited by K.-E. Schenk. Aldershot: Avebury.

Proyecto de Ley. 2000. *Ley de Derecho de Acceso a Internet, Expediente No. 14.029.* http://www.asamblea.go.cr/boletines/Proyectos/comisiones/economi-cos/14029.doc.

Proyecto Estado de la Nación. 2000. *Estado de la Nación en desarrollo humano sostenible.* San José: Proyecto Estado de la Nación.

Przeworski, Adam. 1991. *Democracy and the Market. Political and Economic Reforms in Eastern Europe and Latin America.* Cambridge: Cambridge University Press.

Pye, Lucian. 1963. *Communications and Political Development.* Princeton: Princeton University Press.

Quah, Danny. 1999. *The Weightless Economy in Economic Development.* Working Papers No 155. Helsinki: United Nations University / World Institute for Development Economics Research.

RACSA [Radiografía Costarricense S.A.]. 2002. *Demanda de Internet del país cubierta en 100%* (July 2002). http://www.racsa.co.cr.

RACSA. 2003. *Disminuye la Brecha Digital.* Press Release, December 2003. http://196.40.31.19/menu/racsanoticias/brecha_digital.htm.

Reporters sans Frontières. 2001a. *The enemies of the Internet.* http://www.rsf.fr.

Reporters sans Frontières. 2001b. *Acoso, exilio, encarcelamiento. Un centenar de periodistas independientes frente al Estado.* http://www.rsf.fr.

Reportèrs sans Frontiéres. 2004. *Cuba—el gobierno va a acorralar a los internautas "no autorizados."* Press release, January 14, 2004.

República de Cuba. 1976. *Constitución de la República de Cuba [as approved in 1976].* http://www.georgetown.edu/pdba/Constitutions/Cuba/cuba1976.html.

República de Cuba. 1992. *Constitución de la República de Cuba [including the Constitutional Reform of July 1992].* http://www.cuba.cu/gobierno/consti.htm.

Reuben Soto, Sergio. 1988. *Ajuste estructural en Costa Rica. Estudio socio-económico de una década.* San José: Editorial Porvenir.

Rieff, David. 1995. From Exiles to Immigrants. *Foreign Affairs* (74)4, July/August: 76–89.

Ritter, Archibald. 1995. The Dual Currency Bifurcation of Cuba's Economy in the 1990's. Causes, Consequences and Cures. *CEPAL Review,* No. 57: 113–131.

Rodríguez Armas, Gisela. 2003. *La biotecnología importa a la economía cubana.* World Data Service, January 1, 2003.

Rodríguez-Clare, Andrés. 1996. Multinationals, Linkages and Economic Development. *American Economic Review* 86: 852–873.

Rodríguez Gavilán, Agnerys. 2000. Informática y comunicaciones. Garantía de la excelencia en los servicios. *Juventud Rebelde,* May 21, 2000.

Rodríguez, Francisco, and Ernest J. Wilson III. 2000. *Are Poor Countries Losing the Information Revolution?* InfoDev Working Paper. http://www.infodev.org/library/ WorkingPapers/wilsonrodriguez.doc.

Rogers, Everett M., and Judith K Larsen. 1984. *Silicon Valley Fever: Growth of High-Tech Culture,* New York, NY: Basic Books.

Rohn, Walter. 2002. *Regelung versus Nichtregelung internationaler Kommunikationsbeziehungen. Das Beispiel der UNESCO-Kommunikationspolitik.* ISR-Forschungsbericht 26. Wien: Verlag der Österreichischen Akademie der Wissenschaften.

Rohter, Larry. 2002. Brazil Sets an Example in Computerizing Its National Elections. *New York Times,* October 30, 2002.

Rojas, Rafael. 1997. Políticas invisibles. *Encuentro de la Cultura Cubana* 6/7 (Madrid): 24–35.

Ronda, Gral. Alejandro. 2000. *Seguridad de las redes informáticas.* Ponencia en la Primera Reunión Nacional de Informatización de la Sociedad, 29–30 Sept 2000, Ministerio de la Informática y las Comunicaciones. Unpublished working document.

Ronfeldt, David F., John Arquilla, Graham E. Fuller, and Melissa Fuller. 1998. *The Zapatista 'Social Netwar' in Mexico.* Santa Monica: RAND.

Rovira Mas, Jorge. 1988. *Estado y política económica en Costa Rica 1948–1970.* 3rd ed. San José: Editorial Porvenir.

Rovira Mas, Jorge. 1989. *Costa Rica en los años '80.* 3rd ed. San José: Editorial Porvenir.

Ruggiero, Roberto. 2002. ICANN: Can I? Sociedad Civil y gobierno del Internet. *Boletín Electrónico de Derechos en Internet,* No 1, Julio 2002, edited by Asociación para el Progreso de las Comunicaciones. http://www.apc.org/espanol/rights/lac/bolarchivo.shtml.

Rushton, Mark. 2000. *Information & Communication Technologies and Cuba: Building for the Future.* Project Report; Cida Awards Programme for Canadians. http:// www.nscuba.org/JCCE/report.html.

Sáenz, María, and Nora Galeano. 1996. *Estado de las telecomunicaciones en Centroamérica. Impresiones sobre la situación actual.* San José: CRIES and Fundación Acceso.

Sala Constitucional de la Corte Suprema de Justicia. 2000. *Sentencia de Sala Cuarta que declaró inconstitucional el primer debate del 'Combo ICE' (Exp.:*

00–002411-CO-E). http://www.nacion.co.cr/In_ee/ESPE-CIALES/Leyes/tele-com.sala.html.

Sánchez Villaverde, Ricardo. 1995. *La informatización de la sociedad: un arma de guerra del carril II*. La Habana: CID FAR.

Sánchez Villaverde, Ricardo. 1996. Las nuevas tecnologías de la información—un análisis político. *Cuba Socialista* 4/96: 39–51.

Sassen, Saskia. 1998. The State and the Global City. Notes toward a Conception of Place-Centered Governance. In *Globalization and its Discontents. Essays on the New Mobility of People and Money*, by Saskia Sassen, 195–218. New York: The New Press.

Saunders, Robert J., Jeremy J. Warford, and Björn Wellenius. 1994. *Telecommunications and Economic Development*. 2nd ed. Baltimore: The Johns Hopkins University Press.

Saxenian, Anna-Lee. 1994. *Regional Advantage. Culture and Competition in Silicon Valley and Route 128*. Cambridge: Harvard University Press.

Scheeres, Julia. 2001. Cuba Not so Libre with the Net. *Wired News*, February 23, 2001. http://www.wired.com/news/politics/0,1283,41940,00.html.

Schiller, Dan. 2000. *Digital Capitalism*. Cambridge: MIT Press.

Schmidt, Sönke. 1986. Der Internationale Währungsfonds in Costa Rica: Die Strangulierung der nationalen Souveränität. In *Demokratie in Costa Rica—ein zentralamerikanischer Anachronismus?*, edited by E. Fürst and S. Schmidt, 59–85. Berlin: FDCL.

Schmitz, Hubert, and José Cassiolato, eds. 1992. *Hi-Tech for Industrial Development. Lessons from the Brazilian Experience in Electronics and Automation*. London/New York: Routledge.

Schramm, Wilbur. 1964. *Mass Media and National Development. The Role of Information in Developing Countries*. Stanford: Stanford University Press.

Schulz, Markus S. 1998. Collective Action across Borders. Opportunity Structures, Network Capacities, and Communicative Praxis in the Age of Advanced Globalization. *Sociological Perspectives*, Vol. 41, No 3: 587–616.

Schvarzer, Jorge. 1998. *La implantación de un modelo económico: La experiencia argentina entre 1975 y el 2000*. Buenos Aires: A-Z editora.

Schvarzer, Jorge. 2000. *External Dependency and Internal Transformation: Argentina Confronts the Long Debt Crisis*. UNRISD Social Policy and Development Programme Paper 1. Geneva: UNRISD.

Schwartzman, Simon (with Eduardo Krieger, Fernando Galembeck, Eduardo Augusto Guimarães, Carlos Osmar Bertero). 1995. Science and Technology in Brazil: A New Policy for a Global World. In *Science and Technology in Brazil: A New Policy for a Global World*, Vol. I., edited by S. Schwartzman, 1–56. Brasilia: Fundação Getulio Vargas.

Seligson, Mitchell A. 1984. *El campesino y el capitalismo agrario de Costa Rica*. 2nd ed. San José: Editorial de Costa Rica. (Orig. in English as: Peasants of Costa Rica and the Development of Agrarian Capitalism. Madison: University of Wisconsin Press, 1980.)

Seligson, Mitchell A. 2002. Trouble in Paradise? The erosion of system support in Costa Rica, 1978–1999. *Latin American Research Review* Vol. 37, No. 1: 160–185.

SIME [Ministerio de la Industria Sideromecánica y Electrónica]. 1997. *Líneamientos estratégicos para la informatización de la sociedad cubana.* La Habana: SIME.

Singer, Paul. 1996. São Paulo—Industrielle Krise und Deindustrialisierung. In *Offene Rechnungen. Lateinamerika Analysen und Berichte 20,* edited by K. Gabbert et al., 117–139. Bad Honnef: Horlemann.

Skidmore, Thomas E. 1993. *Television, Politics, and the Transition to Democracy in Latin America.* Baltimore: Johns Hopkins University Press.

Smith, Anthony. 1980. *The Geopolitics of Information: How Western Culture Dominates the World.* New York/Oxford: Oxford University Press.

Smith, Peter H. 1996. *Talons of the Eagle. Dynamics of U.S.-Latin American Relations.* New York/Oxford: Oxford University Press.

Smith, Wayne. 1987. *Closest of Enemies: A Personal and Diplomatic Account of the Castro Years.* New York City: W.W. Norton.

Snow, Anita. 2001. *Government Blames Cubans' Limited Web Access on Poor Networks.* AP news, March 4, 2001.

Sojo, Ana. 1984. *Estado empresario y lucha política en Costa Rica.* San José: Editorial Universitaria Centroamericana.

Solís, Ottón. 2001. Soy proteccionista (interview). *Semanario Universidad,* April 20, 2001: 3–5.

Solís, Ottón. 2002. Red Internet avanzada. Debemos vencer todas las presiones contra este gran proyecto. *La Nación,* April 18, 2002.

Spar, Debora. 1998. *Attracting High Technology Investment: Intel's Costa Rican Plant.* Foreign Investment Advisory Service Occasional Paper 11. Washington D. C.: World Bank.

Stahl, Karin. 1987. Kuba. *Eine neue Klassengesellschaft?* Heidelberg: VSA.

Stallman, Richard. 2002. *UNESCO y el Software Libre.* http://www.unesco.org.uy/informatica/consorcio/stallman.html.

Stamm, Andreas. 2002. Technologie und Innovation. Verpasst Lateinamerika den Anschluss an die Wissensgesellschaft? In *Jahrbuch Lateinamerika 2002,* edited by Institut für Iberoamerika-Kunde, Hamburg, 67–88. Frankfurt a. M.: Vervuert.

Tablada Pérez, Carlos. 1987. *El Pensamiento Económico del Che,* La Habana: Casa de las Américas.

Tacsan, Rodolfo. 2001. The Potentials of Leapfrogging in Costa Rica. In *World Employment Report 2001. Life at Work in the Information Economy,* edited by International Labour Organization (background paper on CD-Rom version). Geneva: ILO.

Talero, Eduardo, and Philip Gaudette. 1996. *Harnessing Information for Development. A Proposal for a World Bank Group Strategy.* Washington D.C.: The World Bank. http://www.worldbank.org/html/fpd/harnessing/ hid1.html.

Téramond, Guy de. 1995. *Interconexión de Costa Rica a las Grandes Redes de Investigación Bitnet e Internet.* http://www.crnet.cr/documentos/interco.html.

Téramond, Guy de. 2000. Transcript of Interview, January 6, 2000. In *Regulación y uso de las nuevas tecnologías de información y comunicación (NTIC) en el cambio de los procesos de transformaciones políticas y económicas en América Latina: El caso de Costa Rica,* edited by Fundación Acceso, 84–86. San José: Fundación Acceso.

Thorp, Rosemary. 1984. *Latin America in the 1930s.* Oxford: Oxford University Press.

Tigre, Paulo Bastos, and Antonio Jose Junqueira Botelho. 2001. Brazil Meets to Global Challenge. IT Policy in a Post-Liberalization Environment. *The Information Society,* Vol. 17, No. 2.

Trejos, Marta. 1986. Zur aktuellen politischen Lage. In *Demokratie in Costa Rica— ein zentralamerikanischer Anachronismus?* edited by M. Ernst and S. Schmidt, 26–39. Berlin: FDCL.

UN Department of Humanitarian Affairs. 1993. *Cuba—Neuromyelopathy Epidemic June 1993* UN DHA Situation Reports 1–5. Published June 1, 1993, Geneva.

UNCTAD [United Nations Conference on Trade and Development]. 2001. *E-Commerce and Development Report 2001.* Geneva: UNCTAD.

UNCTAD. 2002. *E-Commerce and Development Report 2002.* Geneva: UNCTAD.

UNDP [United Nations Development Programme]. 2000. *Human Development Report 2000: Human Rights and Human Development.* Oxford/New York: Oxford University Press. http://www.undp.org/hdr2000/english/HDR2000. html.

UNDP. 2001. *Human Development Report 2001: Making New Technologies Work for Human Development.* Oxford/New York: Oxford University Press.

UNDP. 2002. *Human Development Report 2002. Deepening Democracy in a Fragmented World.* Oxford/New York: Oxford University Press.

UNDP. 2003. *Human Development Report 2003. Millennium Development Goals: A Compact among Nations to End Human Poverty.* Oxford/New York: Oxford University Press.

UNDPEPA / ASPA [United Nations Division for Public Economics and Public Administration / American Society for Public Administration]. 2002. *Benchmarking E-government: A Global Perspective. Assessing the Progress of the UN Member States.* New York: UNDPEPA / ASPA. http://unpan1.un.org/ intradoc/groups/public/documents/un/unpan003984.pdf.

UNESCO [United Nations Educational, Scientific and Cultural Organization]. 1984. *Many voices, one world: Communication and society, today and tomorrow: The Mac Bride Report.* Paris: UNESCO.

UNESCO. 1988. *A Documentary History of a New World Information and Communication Order seen as an Evolving and Continuous Process 1975–1986.* Communication and Society No. 19. Paris: UNESCO.

UNESCO. 1989. *Communication in the Service of Humanity. Resolution 104 adopted by the General Conference at its twenty-fifth session.* Paris: UN-ESCO. http://www.unesco.org/ webworld/com_media/bastxt/en/human.htm.

UNESCO. 1999. *World Communication and Information Report 1999–2000.* Paris: UNESCO. http://www.unesco.org/webworld/wcir/en/report.html.

UNIDO [United Nations Industrial Development Organization]. 1998. *National Information Infrastructure in International Perspective* (authored by G. Harindranath and J. Liebenau). Vienna: UNIDO.

Unimer-La Nación. 2000a. *Transformación del ICE* [Encuesta sobre el *Combo* del ICE] http://www.nacion.co.cr/ln_ee/encuestas/unimer/6–2000/Parte5.htm. Summary in: La Nación, June 25, 2000.

Unimer-La Nación. 2000b. *Manifestaciones Sociales* [Encuesta sobre el *Combo* del ICE]. http://www.nacion.co.cr/ln_ee/encuestas/unimer/6–2000/parte7.htm. Summary in: La Nación, June 25, 2000.

Urra González, Pedro, and Baudilio Jardines Méndez. 2000. Salud para todos: ¿Reto Internet? *Metánica—Revista de la Industria Cubana Siderúrgica y Mecánica,* La Habana (6)1, Jan.-Apr.: 37–41.

U.S. Congress. 1983. *Radio Broadcasting to Cuba Act.* Public Law 98–111, enacted October 4, 1983.

U.S. Congress. 1992. *Cuban Democracy Act of 1992.* Public Law 102–484, enacted October 23, 1992. http://www.state.gov/www/regions/wha/cuba/democ_act_1992.html.

U.S. Congress. 1996. U.S. Congress H.R. 927: Cuban Liberty and Democratic Solidarity (LIBERTAD) Act of 1996. ftp://ftp.loc.gov/pub/thomas/c104/h927.enr.txt.

U.S. Department of Commerce. 1993. *The National Information Infrastructure: Agenda for Action.* September 15, 1993. http://www.ifla.org/documents/infopol/us/nii.txt.

U.S. Presidency. 2002. Fact Sheet U.S.-Central America Free Trade Agreement. http://www.whitehouse.gov/news/releases/2002/01/20020116–11.html.

Valdés Menéndez, Ramiro. 1999. Efectos en la sociedad de la integración de las telecomunicaciones y la telemática. Conferencia dictada por el Comandante de la Revolución Ramiro Valdés Menéndez en el marco del II Taller Internacional 'Electrónica 99' durante el evento 'Metánica 99.' Palacio de las Convenciones, Julio 15 de 1999. *Giga—La revista cubana de computación* 5/99: 4–9.

Valdés Vivó, Raúl. 1997. Se trata de Pirañas. *Granma Internacional,* December 21, 1997.

Valdés, Nelson P. 1997. *Cuba, the Internet, and U.S. Policy.* Cuba Briefing Paper Series No 13. Washington D.C.: Georgetown University, Center for Latin American Studies, The Caribbean Project.

Valdés, Nelson P. 2002. Cuba y la tecnología de la información. *Temas—Cultura Ideología y Sociedad* (Havanna) No 31, October-December: 57–71.

Valdés, Nelson P., and Mario A. Rivera. 1999. The Political Economy of the Internet in Cuba. *Cuba in Transition Vol. 9,* edited by Association for the Study of the Cuban Economy (ASCE), 141–154. http://www.lanic.utexas.edu/la/cb/cuba/asce/cuba9/valdes.pdf.

Valencia Almeida, Marelys. 1997. Internet. Como la imprenta para el medioevo. *Granma Internacional,* December 21, 1997: 8–9.

Villasuso, Juan Manuel, ed. 1992. *El nuevo rostro de Costa Rica. Un análisis de los principales cambios culturales, sociales, económicos y políticos de Costa Rica durante los últimos años.* San José: CEDAL/Friedrich-Ebert-Stiftung.

Villasuso, Juan Manuel. 1998. Política económica y social en tiempos de la transformación estructural. In *Política social y descentralización en Costa Rica,* edited by UNICEF, 171195. San José: UNICEF.

Wachs, Friedhelm. 1988. Polen: Mit Gewalt und Reform aus der Krise. In *Die Armut der Nationen. Handbuch zur Schuldenkrise von Argentinien bis Zaire,* edited by E. Altvater et al., 242–252. 2nd ed. Berlin: Rotbuch.

Wacker, Gudrun. 2000. *Hinter der virtuellen Mauer. Die VR China und das Internet.* Berichte des Bundesinstituts für Ostwissenschaftliche Studien, 6–2000. Köln: BIOST.

Warschauer, Mark. 2002. Reconceptualizing the Digital Divide. *First Monday. Peer-Reviewed Journal on the Internet.* Vol. 7, No 7.: http://www. firstmonday.org/ issues/issue7_7/ warschauer/index.html.

Wellenius, Björn. 1992. *Telecommunications. World Bank Experience and Strategy.* World Bank Discussion Papers No. 192. Washington D.C.: World Bank.

Wellenius, Björn, and Peter Stern, eds. 1991. *Implementing Reforms in the Telecommunications Sector. Lessons from Recent Experience.* Washington D.C.: World Bank.

Wellenius, Björn, Peter Stern, Timothy Nulty, and Richard Stern, eds. 1989. *Restructuring and Managing the Telecommunications Sector. A World Bank Symposium.* Washington D.C.: World Bank.

Whitehead, Laurence. 1996. What Europe can do in Response to Helms-Burton. In *Documentación del Seminario 'El refuerzo del embargo de EEUU contra Cuba—Implicaciones para el comercio y las inversiones,' Sitges, July 8–10, 1996,* edited by IRELA. Madrid: IRELA.

Williams, Robin, and David Edge. 1996. The Social Shaping of Technology. In *Information and Communication Technologies. Visions and Realities,* edited by W. H. Dutton, 53–68. Oxford/New York: Oxford University Press.

Williamson, John. 1990. What Washington Means by Policy Reform. In *Latin American Adjustment. How Much has Happened?* by John Williamson, 7–20. Washington D.C.: Institute for International Economics.

Wilson, Bruce M. 1998. *Costa Rica. Politics, Economics, and Democracy.* Boulder: Lynne Rienner.

Winzerling, Werner. 2002. Linux und Freie Software. Eine Entmystifizierung. *Prokla—Zeitschrift für kritische Sozialwissenschaft* 126: 37–56.

World Bank. 1996. *Costa Rica: A Strategy for Foreign Investment in Costa Rica's Electronics Industry.* Washington D.C.: World Bank.

World Bank. 1998. *Knowledge for Development. World Development Report 1998/99.* New York/Oxford: Oxford University Press.

Wriston, Walter. 1997. Bits, Bytes and Diplomacy. *Foreign Affairs* 76/5 (September/ October):172–182.

WTO [World Trade Organization]. 1996. *Reference Paper [on the Agreement on Basic Telecommunication].* http://www.wto.org/english/news_e/pres97_e/ref-pap-e.htm.

Zakon, Robert. 2002. *Hobbes' Internet Timeline (v5.6).* http://www.zakon.org/ robert/internet/timeline/.

Index